Lean and Cleaner Production

Peng Wu • Sui Pheng Low

Lean and Cleaner Production

Applications in Prefabrication to Reduce Carbon Emissions

Springer

Peng Wu
School of Engineering and Technology
Central Queensland University
Rockhampon, QLD, Australia

Sui Pheng Low
Department of Building
School of Design and Environment
National University of Singapore
Singapore, Singapore

ISBN 978-3-642-42061-0 ISBN 978-3-642-42062-7 (eBook)
DOI 10.1007/978-3-642-42062-7
Springer Heidelberg New York Dordrecht London

Library of Congress Control Number: 2013957380

Printed on acid-free paper

Springer is part of Springer Science+Business Media (www.springer.com)

Abstract

Climate change has emerged as one of the most pressing environmental issues in recent years. The construction industry contributes to the increase in the level of carbon dioxide (CO_2) in many aspects. For example, the cement sector alone accounts for 5 % of global man-made CO_2 emissions. Manufacturing of raw materials (e.g. cement and steel) and chemicals have considerable impact on CO_2 emissions.

In tandem with the cleaner production concept, the lean production philosophy has proven to be effective in increasing environmental benefits by eliminating waste, preventing pollution and maximizing value to owners. However, an in-depth investigation of the lean concept's role in reducing carbon emissions should be conducted before any recommendations can be made. Prefabrication systems are believed to have the potential for better environmental performance and have been adopted by the construction industry to meet the challenges posed by sustainable development. However, there remains many areas in the prefabrication systems that can be improved in order to achieve sustainability, such as site layout, work flow and inventory control. This study therefore seeks to identify the non-value adding activities in precast concrete production and installation to reduce carbon emissions. The non-value adding activities identified in this study can be used to help guide the precasters' and contractors' decision-making process to meet the challenges of global climate change.

Four stages in the precast concrete production cycle are investigated, which are site layout management, supply chain management, production management and stock management. In addition, four stages in the precast concrete erection cycle are investigated, which are site layout management, transportation management, stock management and erection management. The importance of the non-value adding activities identified in this study is rated by a weighted factor model using both the non-parametric tests (for precasters) and the parametric tests (for contractors). The results suggest that many lean principles can be applied in precast concrete factories and in the construction sites to reduce carbon emissions, e.g. the pull system, total quality control and benchmarking.

In addition to the data collected from the survey work, four case studies (one precaster and three contractors) are presented in this study. Various theoretical and practical implications and conclusions of this study are provided for precasters, contractors and regulatory authorities. It is argued that the lean production philosophy can be used to achieve low-carbon production and low-carbon installation in terms of eliminating non-value adding activities from waste of raw materials, waste of finished products as well as inappropriate production/erection arrangements. The lean improvements will enable precasters and contractors to perform better in many sustainability-related rating systems, such as the Singapore Green Labelling Scheme, and the Building and Construction Authority (BCA) Green Mark Scheme provided for under the Building Control Act. It also suggests that the practitioners should pay special attention to the "continuous improvement" characteristics of the lean concept to focus on long-term improvement.

Keywords: Sustainability, Prefabrication, Climate change, Lean, Carbon emissions

Contents

Abbreviations

AIA	American Institute of Architects
ASCE	American Society of Civil Engineers
ASTM	American Society of Testing and Materials
BCA	Building and Construction Authority
BRE	Building Research Establishment
BREEAM	Building Research Establishment Environmental Assessment Method
CFC	Chlorofluorocarbons
CSR	Corporate Social Responsibility
DETR	Department of the Environment, Transport and the Regions
EFDB	Emission Factor Database
EMA	Energy Market Authority
EPA	Environmental Protection Agency
GBI	Green Building Initiative
GDP	Gross Domestic Product
GHGs	Greenhouse Gases
GNP	Gross National Product
HDB	Housing and Development Board
HKSAR	Hong Kong Special Administrative Region
ICE	Inventory of Carbon and Energy
IPCC	Intergovernmental Panel on Climate Change
IUCN	International Union for Conservation of Nature
JIT	Just-In-Time
LCA	Life Cycle Analysis
LCC	Life Cycle Costing
LEED	Leader in Energy and Environmental Design
MCDM	Multi-Criteria Decision-Making
MtF	Make-to-Forecast
NCCC	National Climate Change Committee
NEA	National Environment Agency
OECD	Organisation for Economic Co-operation and Development
QC	Quality Control

QFD	Quality Function Deployment
SDC	Sustainable Development Charity
SEC	Singapore Environment Council
SETAC	Society for Environmental Toxicology and Chemistry
SGLS	Singapore Green Labelling Scheme
TPM	Total Productive Maintenance
TPS	Toyota Production System
TQC	Total Quality Control
TQM	Total Quality Management
UNEP	United Nations Environment Programme
UNFCCC	United Nations Framework Convention on Climate Change
USGBC	US Green Building Council
USGS	US Geological Survey
WBCSD	World Business Council for Sustainable Development
WCED	World Commission on Environment and Development
WRI	World Resources Institute
WSA	World Steel Association

List of Figures

List of Tables

Chapter 1
Introduction

1.1 Introduction

Climate change is said to be one of the biggest threats to future development. According to the Intergovernmental Panel on Climate Change (IPCC 2007), 11 of the last 12 years (1995–2006) ranked among the 12 warmest years in the instrumental record of global surface temperature since 1850. At the same time, global average sea level has risen since 1960 at an average rate of 1.8 mm/year and since 1993 at 3.1 mm/year, which has considerable impact on future development, especially on countries like Singapore which is surrounded by sea from all sides (IPCC 2007). Billions of people are exposed to natural disaster risks, including weather-related disasters that take lives, damage infrastructure and natural resources, and disrupt economic activities (Pelling et al. 2004). There is broad consensus that global climate change has been caused by an increase in greenhouse gas (GHG) emissions from both natural and man-made sources (Environment Agency 2005). However, human activity is believed to be the most significant source of emissions, which is mainly caused by fossil fuel consumption such as petrol, gas, oil and diesel.

The building sector is the largest source of greenhouse gas emissions in the US, as shown in Fig. 1.1. According to the American Institute of Architects (2007), it is estimated that nearly 50 % of all the greenhouse gas emissions are generated by buildings and their construction in terms of the energy used in the production of materials, transportation of materials from production factories to construction sites, as well as energy consumed in the operation stage. According to the United Nations Environment Programme (UNEP 2007), the life cycle of energy consumption in buildings can be divided into five phases, from production to demolition. The first phase is related to the production of construction materials, which is referred to as "embodied energy" (also known as "Cradle-to-Gate"). Building is a complex combination of different materials, contributing to embodied energy of the whole building. In the construction industry, concrete, steel and aluminum are considered as materials with high embodied energy due to the complexity of the materials and large amount of processes required for their manufacture. The second and third phases correspond to

P. Wu and S.P. Low, *Lean and Cleaner Production: Applications in Prefabrication to Reduce Carbon Emissions*, DOI 10.1007/978-3-642-42062-7_1,

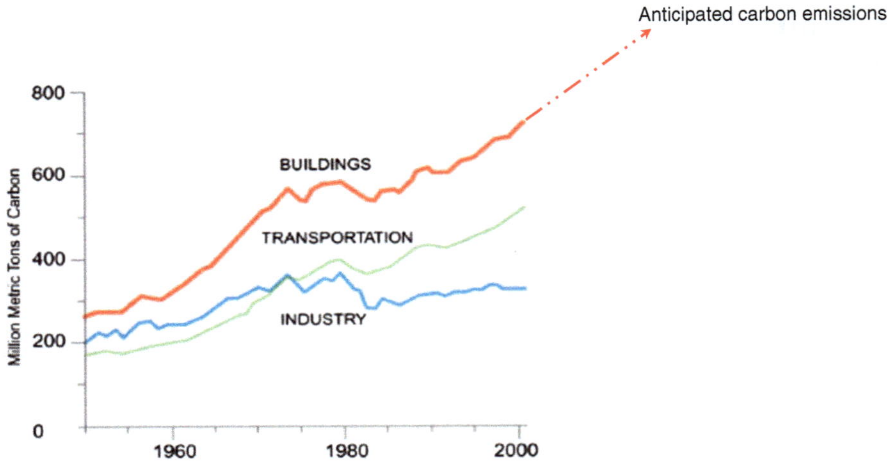

Fig. 1.1 U.S. CO_2 emissions by sectors (Adapted from: AIA 2007)

the energy used to transport construction materials from production factories to construction sites, as well as the building construction, which are referred to as grey energy and induced energy respectively. The fourth consumption phase is the operation stage of the building, which corresponds to the energy consumption in the running of the building and is often referred to as operation energy. Finally, energy is consumed in the demolition stage as well as in recycling of the parts. The building sector is responsible for almost 50 % of the greenhouse gas emissions but there is considerable potential to control and cut down the emissions (AIA 2007).

Most studies relating to carbon emissions reduction are concerned about technical innovations, many of which are highly costly and may take several decades before promising breakthroughs can actually happen. Little has been done about how management improvements can help to eliminate carbon emissions. This can be seen from one phenomenon that all major green building rating systems, including LEED, Green Globes and BCA Green Mark, rely heavily on innovative design, technologies and materials. Only limited credits are allocated to management practices (Wu and Low 2010). However, not all production and construction activities are always efficient. There remains many non-value adding activities which consume energy, leading to carbon emissions. The role of project management, which is represented through the cutting down of non-value adding activities (also known as the lean philosophy), in reducing carbon emissions in the precast concrete industry, should be identified.

1.2 Problem Statement

Global actions have been made to reduce GHGs emissions to achieve long term sustainable development. According to Stern (2007), most actions that have been taken to reduce carbon emissions are focused on technical issues, including:

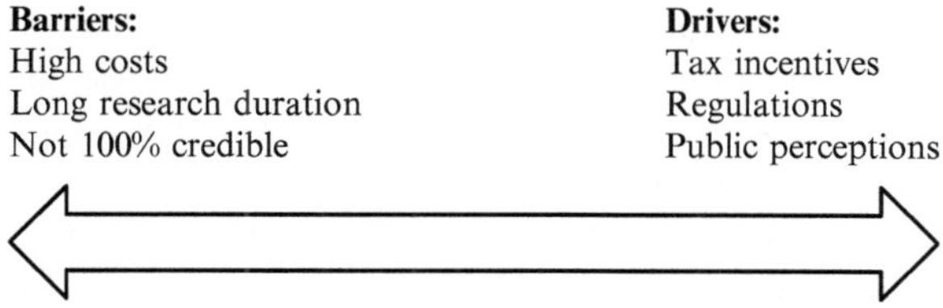

Fig. 1.2 The need for research into other viable options

1. Increased energy efficiency;
2. Changes in demand for energy intensive technologies;
3. Adoption of clean power, heat and transport technologies;
4. Carbon pricing and budgeting through tax, trading and regulations;
5. Supporting innovation and deployment of low carbon technologies; and
6. Remove barriers to energy efficiency: inform, educate and persuade individuals to change their behaviour.

Kruse (2004) stated that general strategies that are currently adopted in the construction industry to address climate change include:

1. Government taxation and regulations, which include rewards for energy efficiency, raising energy efficiency standards for construction, as well as calling for increased transparency in energy consumption (e.g. Singapore National Climate Change Strategy 2008).
2. Voluntary targets. The construction industry should set reporting metrics, while individual companies should set targets (Rehan and Nehdi 2005).
3. Process and technology innovation (Spence and Mulligan 1995).
4. Adopting low carbon fuels (Hendriks et al. 1999).
5. Identifying alternative low carbon raw materials (Ellis 2004).
6. CO_2 capture and sequestration (Herzog 2001).
7. Emissions trading (Szabo et al. 2006).

However, these actions are not always feasible in the construction industry, especially when global recognition to reduce carbon emissions is still in its infancy. It can take years before the costs of adopting clean power and energy efficient materials and resources are affordable to construction companies. In addition, the development of innovation and deployment of low carbon technologies cannot be done once and for all. It is a long term improvement and may take several decades before promising breakthroughs can actually happen, which seems to be contradictory with the current situation that reducing carbon emissions is imperative. Carbon pricing is not sufficient to the industry on the scale and pace required as future pricing policies of governments and international regulatory bodies cannot be 100 % credible (Stern 2007). Thus, investigation of other viable, affordable and beneficial options for the construction industry is important. This is the reason why management improvements are investigated in this study to provide a more cost effective solution to the current urgency in reducing carbon emissions, as illustrated in Fig. 1.2.

In addition, the literature and information relating to the carbon inventory in the Singapore construction industry are very limited. Although world-wide average data can be applied in Singapore to obtain first-level calculations of the carbon emissions from construction activities, these calculations are not accurate enough, because one needs to consider that Singapore is a small country which relies heavily on the import of resources. The Singapore-specific emissions factors are necessary for the country to develop its own carbon inventory. Methodologies and results presented in the study will offer useful information for Singapore to develop its own carbon inventory.

The conventional economic analysis of the environmental impacts, especially in carbon related studies, are based on cost-benefits analysis (e.g. Peck and Tiesberg 1992; Manne et al. 1995). However, this cost-benefit analysis may lead to several problems, which include:

- Conventional economic analysis of the environmental impacts gives less importance to flows that take place in the future (Broome 1992; Price 1993, 1996). As Padilla (2004, p. 527) stated: "the application of conventional discounting devalues and practically removes from the analysis the impacts that occur in the distant future in such a way that for these models the maintenance of the necessary conditions for life far in the future is of negligible present value". There are certain rights of future generations that should be respected and be taken account in the analysis (Padilla 2002).
- Monetary compensation is not always appropriate when evaluating the environmental impacts. The intrinsic values of the ecosystems are not well understood and analysts tend to use the monetary values to determine the preferred policies (Lave and Dowlatabadi 1993). Many authors argued that extreme care should be paid when conducting value judgements in environmental studies (Padilla 2002, 2004; Broome 1992).
- Conventional economic analysis of the environmental impacts assumes that the Earth and all its resources, including the climatic system, belong to the present, and that it has the right to do with whatever the analysis shows as acceptable – including the right to destroy them (Padilla 2002). According to Padilla (2002), the analysts tend to undervalue the losses and overvalue the economic gains, hence leading to the recommendation that either say emissions control should be mild, or that there should be no control, at least in the short term.

It seems that the evaluation method of the environmental impacts, especially in the carbon-related studies, should evolve to overcome the problems stated above. A new measurement should be introduced to the evaluation system.

1.3 Research Aim and Objectives

Originated from the Toyota Production System, the lean production philosophy was developed as a new way of thinking which advocates reducing or eliminating non-value adding activities, as well as improving the efficiency of value adding ones at the same time. The lean philosophy can be considered as a new way to design and make things differentiated from mass and craft forms of production by the objectives and techniques applied on the shop floor, in design and along supply chains (Howell 1999).

By applying the lean principles in the construction industry, non-value adding activities, which consume energy and generate carbon emissions in the production, delivery and construction processes can be identified and eliminated. Unlike other carbon reducing techniques, such as introducing high performance building components to reduce energy consumption, which often incorporate high investment costs, the lean principles are more amenable at a managerial level. It seeks to build up a sustainable managerial environment which promotes an aggressive search for non-value adding activities, which are often referred to as wasteful, inefficient and ineffective activities. For this reason, the lean principles correspond well with the cleaner production concept. According to Yacooub and Fresner (2006), cleaner production is a company-specific, preventive environment protection initiative that is intended to minimize waste and emissions as well as to maximize production outputs. By analyzing material flows and energy consumption in an organization, management identifies avenues for emissions and waste minimization using source reduction strategies in the production processes. Source reduction strategies can include organizational improvement and technological enhancement (Yacooub and Fresner 2006).

Based on these discussions, the aim of this study is to:

> Apply the lean principles to reduce carbon emissions in the Singapore precast concrete industry and provide a management tool-kit for both precasters and contractors in Singapore to reduce carbon emissions without incurring high investment costs.

In order to achieve the aim stated above, the following are the specific research objectives:

1. **To identify the non-value adding activities which are closely related to unnecessary carbon emissions in the precast concrete industry.**

   Previous studies (Alwi et al. 2002; Lee et al. 1999) have been carried out on non-value adding activities in the construction industry, which is closely related to this study, but not at the same abstraction level. This study seeks to identify the non-value adding activities in the precast concrete industry. These non-value adding activities are identified by applying the lean principles to the precast concrete production and installation cycle. In addition, only the non-value adding activities that may increase the carbon emissions level are identified and assessed in this study. Based on the research aim, there is a need to identify those non-value adding activities which consume energy and generate carbon emissions. In addition, the ranked order of the importance of each waste would need to be investigated.
2. **To quantify the recommended improvements by examining how much carbon emissions can be reduced.**

   The construction industry is an industry which lacks change (Elhag 2006). Innovative carbon reducing technologies may take several years before promising breakthroughs can actually happen. Although construction companies are facing pressure from regulatory bodies to reduce carbon emissions, it cannot change the fact that they would remain cautious when choosing the techniques or management improvements to be used, especially those that may bring about significant changes to the organizational structure and daily routine operations. In this case, quantitative measurements are able to provide a persuasive set of data to convince the precasters and contractors to change their behaviors to reduce carbon emissions. There are many lean techniques, e.g. Just-in-Time, Total Quality Control, which can be applied to achieve the objective.

3. **To provide a management tool-kit for both precasters and contractors.**
   Applying a single lean principle might not be totally new in the construction industry. Companies and organizations are applying one or several lean construction principles in their daily operations, though they may not always realize it. However, it could be a real challenge to implement the lean principles as a holistic package. This study seeks to take the first step to apply the lean principles in the context of reducing carbon emissions to the value chain of the precast concrete industry. A comprehensive checklist to help both precasters and contractors improve will be provided at the end of the study.
4. **To add to the Singapore-specific carbon inventory by identifying the embodied carbon of precast concrete components.**
   Carbon inventory is the amount of CO_2 caused by a process or an organization at a point of time. It is of critical importance for the construction industry to achieve sustainable development. Identifying the embodied carbon of precast concrete components is related to the objective of quantifying carbon improvements through lean principles, as explained in objective 2. As calculation of the carbon emissions in the precast concrete products is provided in this study, both the procedures and results will contribute to the knowledge of the Singapore-specific carbon inventory. Precast concrete components are typically adopted in precast concrete projects to reduce construction duration. The Singapore Construction 21 Committee (1999) proposed the use of prefabrication in the construction industry to achieve high productivity. In addition, the production processes of precast concrete components are quite similar, which means that the application of the lean principles on one of the products can be adopted in the production process of another precast concrete product with minimum modification.
5. **To develop a relative measurement of the environmental impacts, especially in carbon-related evaluation.**
   Conventional economic analysis of the environmental impacts may have a few problems, as discussed in Sect. 1.2. A relative measurement of the carbon emissions using the lean concept may be appropriate to address the problem. The applicability of the relative measurement to construction materials other than precast concrete products is also provided. It should be noted that carbon emissions are the focus of this study. Not all environmental impacts are considered.

1.4 Scope of the Study

This study is driven by the rising recognition of global climate change and the need of the construction industry to be sustainable. Although it seeks to serve as a bridge to link sustainable development with the lean principles, it is necessary to define several specific boundaries. Sustainable development is a large concept involving many subdivisions, from economic, social to environmental sustainability. Previous studies (e.g. Ekins 1992; SDC 2002) have added several new values to the concept of sustainable development, including the four capital models and the five capital models. In addition, while the lean principles are being recognized by more and more professionals in the construction industry, it should be acknowledged that the lean concept has its origin in the automobile industry. In order to achieve the aim and objectives of this study, several specific boundaries are identified below:

1. **Research focus.** This study focuses on applying the lean principles in the precast concrete industry for two reasons. Firstly, precast concrete components have benefits compared to traditional construction materials. These are easy and quick to install, which can significantly reduce construction time. In addition, other advantages including fire resistance, elimination of formwork and propping on-site also enables the precast concrete industry to expand quickly. According to Elhag (2006), the precast concrete market has grown by 8 % a year since 2000 in the UK market. The demand for precast concrete components grows significantly as "system building" types of construction grew in popularity (Glass 2000). This is also the case in Singapore where there is a large demand for public housing projects which are highly reliant on "system building". Second, the lean principles originated from the automobile industry and had been applied into the production and manufacturing system for several decades. Construction processes which use precast concrete components have many similarities with the manufacturing processes, which means that lean principles can be applied with minimum modifications to test their applicability in reducing carbon emissions.
2. **Measurable factors.** This study intends to link the lean principles with one aspect of sustainable development, which is environmental sustainability. Sustainable development consists of economic, social and environmental sustainability. However, this study seeks to limit the boundaries to carbon emissions, due to rising recognition of global climate change and the fact that the construction industry may face increasing regulatory pressure to cut down on carbon emissions.
3. **Precasters and contractors as targeted audience.** After this study, a management practice tool-kit (checklist) is provided for precasters and contractors. The tool-kit includes the most important non-value adding activities in precast concrete factories and in construction sites. However, this does not mean that the findings are not applicable to other stakeholders. For example, regulatory bodies can also benefit from this study by incorporating management practices into their green building rating systems.

The boundaries also have a direct impact on the methodology in this study, including data collection, the case studies chosen and interpretation of results which will be discussed in the following chapters.

1.5 Significance and Contribution of the Study

Firstly, the application of the lean principles in the precast concrete industry is considered as the major contribution of this study. As stated previously, the application of the whole lean production concept in the precast concrete is limited. Most precasters and contractors seem to have adopted a single lean principle in their daily production and construction activities. For example, the application of the total quality control concept is very common in the production process of precast concrete products. Precasters are aware of the lean concept and the benefits that come along with the application of the lean concept. However, the overall application of the lean production concept is limited. According to Koskela (1992), the lean production concept has 11 principles. Womack and Jones (1996) have concluded 6 principles of the lean production concept. Applying the lean thinking to deal with a specific environmental impact is the major significance of this study.

Secondly, extending the application of the lean production concept to reduce carbon emissions is also of significance. As has been discussed earlier, unlike other carbon reducing technologies, which rely heavily on technologies, applying the lean concept to reduce carbon emissions seems to be a better solution to cut down on carbon emissions at lower investment costs. In addition, the lean production concept advocates a "continuous improvement" principle that may benefit the company for long-term development. According to Howell and Ballard (1998), implementing lean thinking is a developmental process because applying lean thinking will lead to change in every aspect of project and company management and this change in the mental model cannot be a one-off effort.

Thirdly, the model developed in this study (research objective 2) to calculate carbon emissions in the precast concrete factories and in construction sites is useful to calculate carbon emissions for the construction industry in Singapore. It is especially helpful to calculate the embodied emissions of precast concrete components. For example, the model to calculate the embodied emissions of precast concrete columns is provided in this study. The model, including the processes and data sources, will be useful to help precasters and regulatory bodies to calculate the embodied emissions of other precast concrete components.

Fourthly, the study may provide practical implications. Although the application of the lean production concept is designed to address the problem of carbon emissions in this study, the framework proposed may have practical implications for applying the lean production concept to address the other problems in the precast concrete industry. For example, the lean production concept can also be applied to reduce solid waste. Lean thinking is able to help improve the management process to reduce delivery time of finished products.

In addition, by introducing the lean concept to environmental labeling programmes, this study proposes a new presentation of the eco-label information associated with the traditional LCA approach. This new presentation will help overcome a few problems brought about by using the LCA approach, such as no consideration of future activities in current evaluation.

Lastly, there are a few other points that will be of significance in this study. For example, the relationship between the lean production concept and the objective of being "green" is investigated in this study. The embodied carbon emissions of the precast concrete columns will contribute to build the carbon inventory of construction materials in Singapore.

1.6 Description of Chapters

This book is organized into 12 chapters in the following sequence, as shown in Fig. 1.3.

Chapter 1 is an introductory chapter which explains the aim and objectives of this study. This chapter also offers the scope of the study, which can influence the research methodology, data collection, data analysis, and more importantly, the operationalized measureables.

Chapter 2 presents a review of the literature on the concept of sustainable development and its impact on the construction industry, focusing on the environmental aspects. The calculation process for carbon emission levels is explained in the chapter.

Chapter 3 provides detailed explanation of the lean production philosophy and its application in the construction industry. In addition, the similarities and differences between Just-In-Time (JIT) and the lean philosophy are identified in this chapter. This is based on the fact that both have originated from the Toyota Production System, but have since evolved into different philosophies.

Chapter 4 describes the background of the precast concrete industry that this study focuses on, including the process of production, transportation and erection. More importantly, the applicability of the lean production philosophy in the precast concrete industry to reduce carbon emissions is identified in this chapter. Two pilot studies are also provided in this chapter.

Chapter 5 investigates the theoretical background of the study. Economics theories and management theories that are related to the study are explained in this chapter. In addition, the conceptual framework is provided in this chapter.

Chapter 6 provides the research methodology and operationalized measureables. Followed by the conceptual framework which has been provided in Chap. 5, the research methodology of this study is described in this chapter, including data collection and data analysis methods.

Chapter 7 presents the empirical study regarding the application of the lean philosophy in the precast concrete factories to reduce carbon emissions. Many

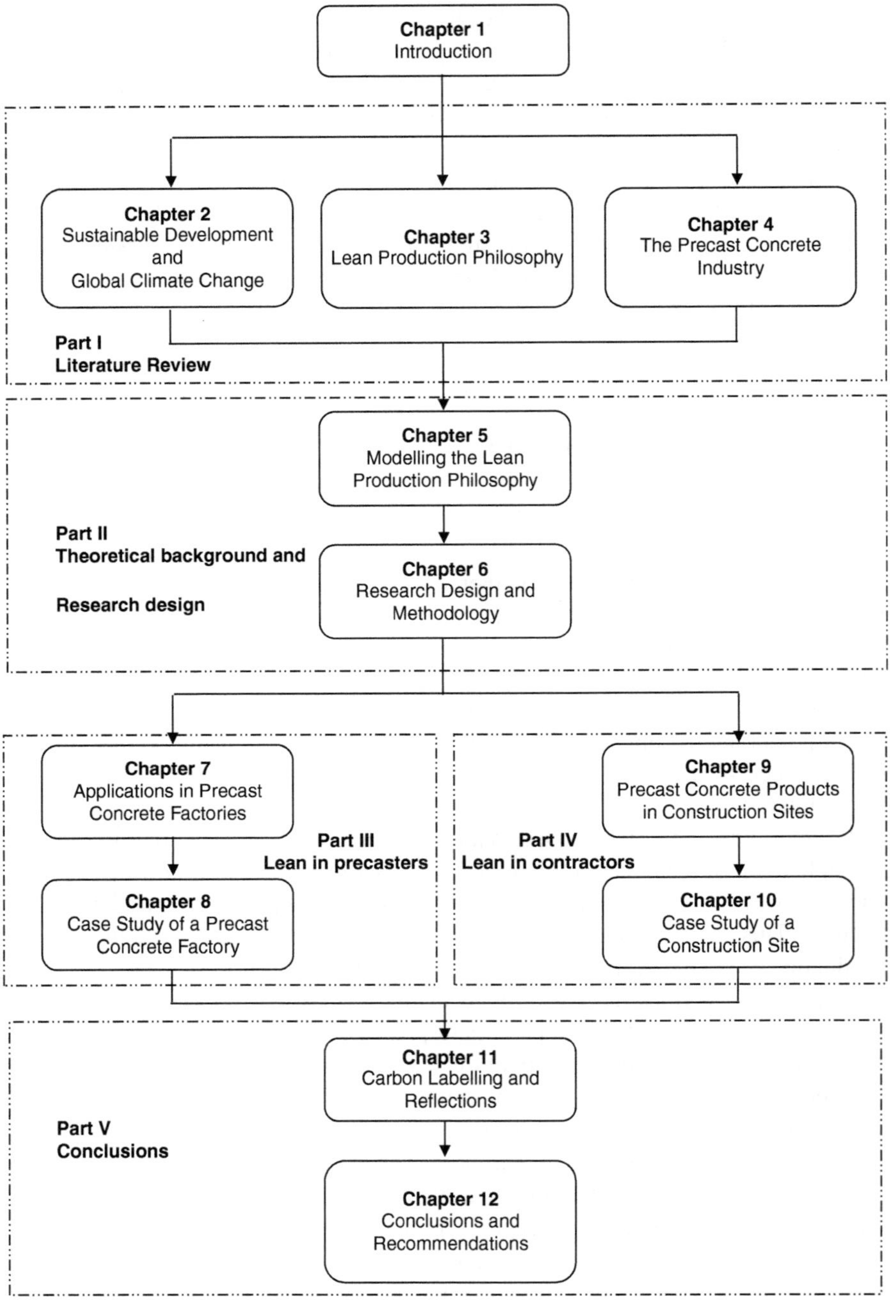

Fig. 1.3 Structure of the book

non-value adding activities are identified and ranked by the precasters. The data obtained from questionnaires and semi-structured interviews are used in Chap. 7.

Chapter 8 provides a case study relating to how precasters can adopt the lean concept to reduce carbon emissions in the precast concrete factories. The carbon emissions from different non-value adding activities are both quantitatively assessed and qualitatively described.

Chapter 9 presents the empirical study regarding the application of the lean philosophy in the construction sites to reduce carbon emissions. Many non-value adding activities are identified and ranked by the contractors. The data obtained from questionnaires and semi-structured interviews were used in Chap. 9.

Chapter 10 provides a case study relating to how contractors can adopt the lean concept to reduce carbon emissions in the precast concrete factories. The carbon emissions from different non-value adding activities are both quantitatively assessed and qualitatively described.

Chapter 11 presents the discussion and implications of this study. The contribution of this study to a few knowledge areas, such as the value concept and the evaluation of environmental impacts is provided in this chapter. The implications of this study to the precasters, contractors and regulatory authorities are also provided.

Chapter 12 covers the conclusions of this study, including summaries, theoretical and practical contributions to low-carbon production and construction. Limitations of this study and suggestions for future research are also provided in this chapter.

References

Alwi S, Hampson K, Mohamed S (2002) Waste in the Indonesian construction project. In: Proceedings of the 1st international conference of CIB W107 – creating a sustainable construction industry in developing countries, Spier, South Africa

American Institute of Architects (2007) Architects and climate change. [Online] Available at: http://www.aia.org. Accessed 20 Feb 2008

Broome J (1992) Counting the costs of global warming. White Horse Press, Cambridge

Ekins P (1992) A four-capital model of wealth creation. In: Ekins P, Max-Bneef M (eds) Real-life economics: understanding wealth creation. Routledege, London/New York, pp 147–155

Elhag G (2006) Business improvement through a structured approach to sustainability in the precast concrete flooring industry. Unpublished PhD thesis, Loughborough University

Ellis G (2004) Industrially interesting approaches to low-CO_2 cements. Cem Concr Res 34 (9):1489–1498

Environment Agency (2005) The climate is changing – time to get ready. Online available at: http://www.environment-agency.gov.uk/?lang=_e. Accessed 9 Mar 2008

Glass J (2000) The future for precast concrete in low-rise housing. British Precast Concrete Federation, Leicester

Hendriks CA et al (1999) Emission reduction of greenhouse gases from the cement industry. International Energy Agency Greenhouse Gas R&D Programme, Cheltenham

Herzog HJ (2001) What future for carbon capture and sequestration. Environ Sci Technol 35 (7):148A–153A

Howell G (1999) What is lean construction. In: Tommelein ID (ed) Proceedings of IGLC-7, U.C. Berkeley, CA
Howell G, Ballard G (1998) Implementing lean construction: understanding and action. In: Formoso CT (ed) Proceedings of IGLC-6, Guaruja, Brazil
IPCC (2007) Climate change 2007: synthesis report. [Online] Available at: http://www.ipcc.ch/ipccreports/ar4-syr.htm. Accessed 9 Mar 2008
Koskela L (1992) Application of the new production philosophy to construction. CIFE technical report#72. Stanford University, Stanford
Kruse C (2004) IIGCC briefing note: climate change and the construction sector. [Online] Available at: http://www.iigcc.org/docs/PDF/Public/ConstructionSector_2004.pdf. Accessed 17 June 2008
Lave LB, Dowlatabadi H (1993) Climate change policy: the effects of personal beliefs and scientific uncertainty. Environ Sci Technol 27(10):1962–1972
Lee SH, Diekmann JE, Songer AD, Brown H (1999) Identifying waste application of construction process analysis. In: Tommelein ID (ed) Proceedings of IGLC-7, U.C. Berkeley, CA
Manne A, Mendelsohn R, Richels R (1995) MERGE: a model for evaluating regional and global effects of GHG reduction policies. Energy Policy 23(1):17–34
National Climate Change Secretariat (2008) National climate change strategy. Available from: http://app.nccs.gov.sg/page.aspx?pageid=123&secid=57. Cited 28 Nov 2013
Padilla E (2002) Intergenerational equity and sustainability. Ecol Econ 41(1):69–83
Padilla E (2004) Climate change, economic analysis and sustainable development. Environ Value 13(4):523–544
Peck SC, Tiesberg TJ (1992) CETA: a model for carbon emissions trajectory assessment. Energy J 13(1):55–77
Pelling M, Maskrey A, Ruiz P, Hall L (2004) Reducing disaster risk: a challenge for development. United Nations Development Bank, Bureau for Crisis Prevention and Recovery, New York
Price C (1993) Time, discounting and value. Blackwell, Oxford
Price C (1996) Discounting and project appraisal: from the bizarre to the ridiculous. In: Kirkpatrick C, Weiss J (eds) Cost-benefit analysis and project appraisal in developing countries. Edward Elgar, Cheltenham, pp 90–101
Rehan R, Nehdi M (2005) Carbon dioxide emissions and climate change: policy implications for the cement industry. Environ Sci Policy 8(2):105–114
SDC (2002) Models of sustainable development. [Online] Available at: http://www.forumforthefuture.org.uk/aboutus/sdtools_page398.aspx#FCM. Accessed 25 Aug 2007
Singapore Construction 21 Committee (1999) Construction 21: reinventing construction. Ministry of Manpower and Ministry of National Development, Singapore
Spence R, Mulligan H (1995) Sustainable development and the construction industry. Habitat Int 19(3):279–299
Stern N (2007) The economics of climate change: the Stern review. Cambridge University Press, New York
Szabo L, Hidalgo I, Ciscar JC, Soria A (2006) CO_2 Emission trading within the European Union and Annex B countries: the cement industry case. Energy Policy 34(1):72–87
UNEP (2007) Buildings and climate change: status, challenges and opportunities. [Online] Available at: http://www.uneptie.org/pc/sbc/documents/Buildings_and_climate_change.pdf. Accessed 20 Mar 2008
Womack JP, Jones DT (1996) Lean thinking. Simon & Schuster, New York
Wu P, Low SP (2010) Project management and green buildings: lessons from the rating systems. J Prof Issues Eng Educ Pract 136(2):64–70
Yacooub A, Fresner J (2006) Half is enough. An introduction to cleaner production. LCPC Press, Beirut

Chapter 2
Sustainable Development and Global Climate Change

2.1 Introduction

The key factors governing sustainable development are poverty, population, pollution, participation, policy and market failures, and prevention and management of disasters, which can be regarded as the major pillars on which sustainable development rests (Rogers et al. 2008). The United Nations Environment Programme's GEO-4 (UNEP 2007) report identifies the following stark figures:

1. Around 1 billion people, 15 % of the global population, are affected by land degradation, as a result of human activities;
2. Around half of the world's rivers are seriously polluted;
3. About 23 % of mammals and 12 % of bird species are currently regarded as globally threatened;
4. Depletion of the ozone layer has now reached record levels. In September 2006, the ozone hole over Antarctica covered more than 29 million km^2; and
5. Concentrations of carbon dioxide stand at 380 ppm, much higher than the pre-industrial (eighteenth century) level of 280 ppm.

The role of construction in achieving sustainable development involves a dilemma (Carpenter 2001). The construction process is regarded as activities which harness nature, consume energy and resources to service the human beings. In the current process, more materials and resources are consumed than nature can supply. On the other hand, construction activities are essential to satisfy the demands from increasing populations and developing economies. The dilemma leads to a reconsideration of the relationships between environment, construction and sustainable development.

This chapter aims to investigate the concept and requirements of sustainable development, and its implications for the construction industry. More importantly, according to the aim of this study, the relationship between the construction industry and global climate change will be addressed in this chapter. In addition,

P. Wu and S.P. Low, *Lean and Cleaner Production: Applications in Prefabrication to Reduce Carbon Emissions*, DOI 10.1007/978-3-642-42062-7_2,

the role of project management in achieving sustainable construction is investigated. The literature review in this chapter aims to answer the following questions:

1. What is sustainable development and what is the connection between sustainable development and the construction industry?
2. How is global recognition of climate change, especially recognition for cutting down carbon emissions affecting the construction industry and the construction companies?
3. What is sustainable construction and how would management improvements help to achieve sustainable construction?

2.2 The Concept of Sustainable Development

Sustainable development has been evolving since 1972, when the international community firstly explored the connection between quality of life and environmental quality at the United Nations Conference on the Human Environment in Stockholm (Rogers et al. 2008). The concept of sustainable development was first defined in the Brundtland Report as (WCED 1987, p. 43):

> development that meets the needs of the present without compromising the ability of future generations to meet their own needs.

The Brundtland Report highlighted the importance of meeting human's economic and social needs with full consideration of the natural environment. However, this definition only drew out the guidelines rather than principles. There were nearly 200 interpretations of the concept of sustainable development, and over the past few years, many professionals have come to acknowledge the one developed by the UK government in 1999 as a milestone. In *A Better Quality of Life* (DETR 1999), the UK government stated that sustainable development needed to meet four objectives:

1. social progress that recognizes the needs of everyone;
2. effective protection of the environment;
3. prudent use of natural resources; and
4. maintenance of high and stable levels of economic growth and employment.

It was widely believed that there were three sub-dimensions in the concept of sustainable development, which were environmental, economic and social dimensions, also known as the triple bottom line, and were often applied to evaluate the success of a project or a programme. According to Schuber and Stormer (2007), the starting point of the triple bottom line was the conflicting relation between economy and ecology, implying a trade-off between economic and environmental objectives defining social welfare. In addition, the social dimension was added to the concept as a third pillar, which was considered as long-term development policies and responsibilities.

2.2.1 Economic Sustainability

Indefinite growth was impossible, that was unsustainable, if it relied on the depletion of global resources and was inequitable if it involved gains for some at the expense of others (Carpenter 2001). The growth rate was restrained by the capacity of other resources, including but not limited to natural resources. In order to fully understand the concept of economic sustainability, several definitions are examined.

Definition 1:

> The core idea of sustainability is that current decisions should not impair the prospects for maintaining or improving future living standards. This implies that our economic systems should be managed so that we can live off the dividends of our resources. (Repetto 1986, p. 10)

Repetto (1986) clearly identified one core area of economic sustainability, which implied that the society or economies should be developed at a certain rate that was decided by the capacity of the natural environment, or the capacity of the man-made environment, plus the managed capacity for expansion (Rogers et al. 2008). This statement also implies that it was not always beneficial for the economies to develop at a faster pace, which had similar implications on construction companies. Most firms nowadays put too much attention on performance and ignore economic sustainability which was reflected by their capacity to deal with such performance relating to organizational structure, partnering, accounting system, etc. Good performance in a single year cannot guarantee long term development in the following years. The indicators for evaluation should not be centred only on performance, but also the capacity of companies to deal with such performance.

Definition 2:

1. Definition 2a:
   Sustainable economic growth means that real GNP per capita is increasing over time and the increase is not threatened by "feedback" from either biophysical impacts (pollution, resource degradation) or from social impacts. (Pearce et al. 1989, p. 33)
2. Definition 2b:
   Sustainable development argues for: (1) development subject to a set of constraints which set resource harvest rates at levels not higher than managed natural regeneration rate, and (2) use of the environment as a "waste sink" on the basis that waste disposal rates should exceed rates of managed or natural assimilative capacity of the ecosystem. (Pearce 1988, p. 58)

Pearce's (1988) definition described the capacities which restrained economic development described by Pearce et al. (1989), including social and environmental considerations. Generally speaking, economies should be managed under such capacity to minimize both environmental and social impacts.

Definition 3:

> Sustainable development means basing developmental and environmental policies on a comparison of costs and benefits and on careful economic analysis that will strengthen environmental protection and lead to rising and sustainable levels of welfare. (Pezzey 1992, p. 8)

Although the capacity mentioned in Definition 2 was determined by social and environmental factors, it was not a simple process to analyze current social and environmental conditions to find out the exact capacity. The evaluation process should be carefully carried out, where comparisons and scenario simulations might be helpful. In this definition, "the sustainable levels of welfare" aimed at referring to social sustainability. However, social sustainability remained more of a concept rather than a mature application, which will be discussed in the following sections.

2.2.2 Environmental Sustainability

> Sustainable development is about maintenance of essential ecological processes and life support systems, the preservation of genetic diversity, and the sustainable utilization of species and ecosystems. (WRI, IUCN and UNEP 1992, p. 4)

Environmental sustainability was firstly recognized by some professionals in 1972 when the oil crisis happened. Environmental impacts played a very important role on global survival due to its high risks, large affected areas and deep effect on human survival. The major environmental impacts included:

1. **Climate Change:** Climate change was defined as "change of climate which is attributed directly or indirectly to human activity that alters the composition of the global atmosphere and which is in addition to natural climate variability observed over comparable time periods" (IPCC 2001, p. 2). Industry and human activities, represented by using fossil fuels, produced greenhouse gases which affected the composition of the atmosphere. The most significant source was carbon emissions, which caused considerable threat to human development, including sea level rises, death of humans and loss of biodiversity (IPCC 2001).
2. **Ozone Depletion:** Excessive use and production of Chlorofluorocarbons (CFC) or other similar items had a significant impact on the ozone layer (UNEP 2003). The chemical shorthand for these reactions is listed as follow:

$$\mathrm{Cl} + \mathrm{O_3} \rightarrow \mathrm{ClO} + \mathrm{O_2}$$
$$\mathrm{ClO} + \mathrm{O} \rightarrow \mathrm{Cl} + \mathrm{O_2}$$
$$\text{In sum } \mathrm{O_3} + \mathrm{O} \rightarrow \mathrm{O_2} + \mathrm{O_2}$$

Where:

O = Oxygen atom
O_2 = Oxygen
O_3 = Trioxygen

The result was that the recombination keeps decreasing the amount of O_3, which was the dominant chemical in the ozone layer. A decrease in the ozone layer would increase the level of UVB radiation (one kind of Ultra Violet radiation), which could cause skin cancer. In addition, a number of economically important species of plants, such as rice, depended on cyanobacteria residing in their roots for the retention of nitrogen. Cyanobacteria were very sensitive to UV light and would be affected by the increase of UV radiation (UNEP 2003).

3. **Mineral extraction:** Every year, tonnes of aggregates, cement, concrete, plaster and other cementitious materials were consumed in the global market. All of these were non-renewable. Waste generated from mineral extraction would cause contamination. Strong economic growth, which was driven by globalization and technical advances, had often taken a toll on the environment and natural resources (OECD 2001). In most cases, these resources were not efficiently used and could lead to more consumption. Sustainable development aimed at balancing the conflict between strong economic growth and environmental degradation.
4. **Waste:** Tonnes of construction and demolition wastes, including clay and subsoil, were produce annually. With an increasing rate of consumption, the amount of waste would rise unless better waste management strategies were carried out in construction sites. In addition, construction and demolition wastes were harmful to the soil if no actions were taken. Soil contamination could be a result of waste if there was no recycling and reuse of demolition and construction wastes.

Without appropriate consideration, pressures on the environment and natural resources were expected to increase over the next 20 years due to economic and demographic trends (OECD 2001). Although technology innovations and efficiency improvements could help support the high demand, which came along with strong economic growth, there were always resource limitations, or capacities which could not be exceeded.

2.2.3 Social Sustainability

Social development referred to the improvements in both individual wellbeing and the overall welfare of society, which resulted from increases in social capital – typically, the accumulation of capacity for individuals and groups of people to work together to achieve shared objectives (Munasinghe and Swart 2005). Accordingly, social capital, which was determined by the quantity and quality of social interactions, including the level of mutual trust and extent of shared social norms, tended to grow with greater use, unlike economic and environmental capitals which were depreciated by usage. Social sustainability was drawn from environmental sustainability by enlarging the concept of habitats to a man-made environment such as communities, villages and cities. Factors that contributed to social sustainability included (Rogers et al. 2008):

1. Reducing vulnerability and maintaining the health of social and culture system, and their ability to withstand shocks (Bohle et al. 1994);
2. Enhancing social capital and value by advocating education, strengthening the level of trust and extending the shared social norms; and
3. Weakening the social factors that will harm the development of social capital, such as irresponsible behaviors to the community, environment, or criminal acts in some cases.

One important concept which is drawn from social sustainability was "corporate social responsibility (CSR)" (WBCSD 1999), which had a significant role in enhancing social capital and weakening harmful social factors. CSR was highlighted based on its importance of raising public awareness of reducing carbon emissions. As stated previously, the construction industry was an industry which lacked the momentum to change. Profit-oriented companies were unwilling to change their organizational structure and to invest in carbon reducing technologies, unless they were forced to do so by regulations. This was unlike socially responsible companies who were willing to do so pro-actively. Sustainability was a "nice to have" rather than a "must have" item in business activities (Tang and Yeoh 2007). One importance factor, which restrained the adoption of sustainability in daily business activities, was the economic consideration that by doing so would cost too much and would actually undermine the company's commercial position (Tang and Yeoh 2007). In other words, the commercial position was the bottom line of current business activities. If the bottom line was harmed by implementing sustainability, few companies would adopt sustainable operations unless they were forced to do so.

According to Wood (1991), the basic idea of CSR was that business and society were interwoven rather than distinct entities. Decisions made according to the business environment, activities carried out by staff, customers and suppliers had a significant impact on society which could not be neglected. Like economic sustainability, the essence of CSR could be identified by several definitions:

Definition 1:

> CSR is the continuing commitment by business to behave ethically and contribute to the economic development while improving the quality of life of the workforce and their families as well as of the local community and society at large. (Moir 2001, p. 18)

In this definition, the interaction between business activity and its impact on society was highlighted. As stated previously, business and society were intertwined with each other rather than served as distinct entities. Economic performance could not be the only indicator when business decisions were being made. Although environmental considerations were embedded in improving the quality of life of the workforce, their families as well as the local community and society at large, this definition had its limitations. First of all, the scopes of CSR varied with the background of the companies. For examples, when making decisions, construction companies would act differently with companies from other business sectors, such as investment, financing and servicing. The latter types of organizations tended to focus more on improving social welfare. Although the improvement of social welfare was not neglected by construction companies in decision-making, they were usually more concerned with environmental impacts due to the uniqueness of the construction industry. In addition, environmental factors today had global rather than local impacts. A chemical leakage in Russia might affect the business of downstream countries, such as Germany and France. Global warming would threaten the development, or even the survival of coastal cities by rising sea levels. Broadly speaking, CSR imposed a responsibility of global environmental concerns to the construction companies.

Definition 2:

> We all need to assess the impact our business makes on society and ensure that we balance the economic, environmental and social aspects of everything we do. (Moody-Stuart 1999, p. 2)

CSR aimed at systematically considering the needs of all stakeholders, including economic, social and environmental considerations, which are elaborated as follows (Business Impact 2000):

1. To treat employees fairly and equitably.
2. To operate ethically and with integrity.
3. To respect basic human rights.
4. To sustain the environment for future generations.
5. To be a caring neighbor in their communities.

The construction industry had a significant impact on the world. The approaches adopted by construction companies should be examined by the three fundamental drivers which would affect the development of the local and global society. Environmental considerations should be especially highlighted in their decision-making processes based on the nature of the construction industry which consumed energy and natural resources. By redefining the CSR for the construction industry, the following considerations were essential (Moir 2001):

1. To systematically consider the environmental impacts that are brought with the construction processes to local and global community.
2. To reduce the environmental impacts under such capacity that is determined by the economic situation of the company.
3. To raise the awareness of sustainability from all stakeholders.

It should be noted that CSR should not impose extra responsibility beyond the economic capacity of the company. CSR aimed to balance the investments to reduce environmental impacts in tandem with the company's economic situations. This was especially important for small construction companies which had limited economic revenue. Based on this consideration, this study aimed at developing affordable management improvements which could help construction companies cut down carbon emissions at lower investment costs.

The definitions in the three domains of sustainable development were closely related to construction companies who were trying to reduce their carbon emission levels. Firstly, CSR was the continuing commitment which regulated construction companies to behave ethically so that they could balance the economic, environmental and social aspects of activities. Secondly, when construction companies have made decision to gradually reduce carbon emissions, the strategies should be carefully analyzed based on a comparison of costs and benefits. In addition, the strategies must be based on proper ecological knowledge, such as climate change, air pollution and so on, to ensure that they were effective to maintain essential ecological processes. Therefore, "sustainable development" in this study for construction companies was defined as "continuing committed to reduce carbon emissions based on proper ecological knowledge and a comparison of costs and benefits from economic, environmental and social aspects of activities".

2.3 Sustainable Construction and the Green Building

Sustainable construction has always been used interchangeably with other terms, such as green, lean, high performance and so on. According to the Conseil International du Batiment (Kibert 2008, p. 6), the goal of sustainable construction was defined as "...creating and operating a healthy built environment based on resource efficiency and ecological design". It was a holistic solution to achieve the concept of sustainable development in the project life cycle, including project planning, designing, constructing and operating. The concept of sustainable construction was closely related to the green building, which was often developed under the guidelines of a rating system. This provided guidance on the measurements and could offer recognition and validation of that level of commitment. This section investigates current green building rating systems to identify the sustainable construction practices which could be adopted in the construction industry to achieve sustainable development.

2.3.1 Green Building Rating Systems – History

Green building emerged during the late nineteenth and early twentieth centuries (Cassidy 2003). However, it was until recently that green building has gained widespread development. In the 1980s, under the cover of sustainable development (Rees 1989) and sustainable design (St. John 1992), green building has proved to be successful in contributing towards sustainability. It was also during the time when publications, including the concepts, practices, rating criteria and other green building related ideas started to emerge. Green building referred to tailoring a building and its placement on the site to suit the local climate, site conditions, culture, and community in order to reduce resource consumption, augment resource supply, and enhance the quality and diversity of life (Adler et al. 2006). It was a division under the umbrella of sustainable development, which was characterized by Sara Parkin of the British Environmental Initiative Forum for the Future as "a process that enables all people to realize their potential and improve their quality of life in ways that protect and enhance the Earth's life support systems" (Forum for the Future 2008). Another concept that was widely accepted around the globe was the definition from the Brundtland Report, which referred to sustainable development as development that meets the needs of the present without compromising the ability of future generations to meet their own needs (WCED 1987). In accordance with the three aspects of sustainable development, which were economic, social and environmental, green building could benefits human well-being, community, environmental health and life cycle cost (Adler et al. 2006).

Green building rating systems were designed to assess and evaluate the performance of either the whole building or a specific division of the building from planning, designing, constructing and operations. In so far as the assessment and evaluation criteria were concerned, rating systems, guidelines and standards could be categorized into two groups: those which concentrated on specific building components or areas, and those which identified the building as a whole evaluation entity (Adler et al. 2006). As the concentrations of different rating systems varied, the same building could be green credited by one, while failed to be credited by another at the same time. The rating process seemed to be a more subjective review than it was thought to be.

Among the whole-building rating systems, the Leader in Energy and Environmental Design (LEED), which was developed by the U.S. Green Building Council (USGBC), was the most commonly used whole-building rating system around the world (Smith et al. 2006). This study focused on the LEED-New Construction rating system, which was the premier and most heavily used green rating system in the United States presently (Haselbach 2008). The LEED-NC rating system was applicable to new commercial construction and major renovation projects (USGBC 2008). The LEED included six categories, all of which carried a specific amount of rating credits. Different certification from LEED certified, LEED Silver, LEED Gold to LEED Platinum was issued to the project in proportion to the amount of credits that were allocated to the project.

Due to different market concentration and its relative newness, the Green Globes, on the other hand, did not attract as much attention as the LEED did (Smith et al. 2006). It originated from a British version, which was the BREEAM (Building Research Establishment Environmental Assessment Method), and was adapted and introduced into the Canadian market as the BREEAM Green Leaf rating system. This system was known as the origin of the Green Globes. Similar to the LEED, the Green Globes had seven areas of assessment as well as other supplements. A rating certification of between one and four Green Globes was issued to the project based on the scores achieved.

The BCA Green Mark was a green building rating system developed by the Building and Construction Authority (BCA) and supported by the National Environment Agency in Singapore. It aimed to assess and evaluate a building for its environmental impact and performance based on five key rating criteria. Similar to other rating systems, buildings could be awarded a Platinum, Gold^{PLUS}, Gold or certified rating depending on the points scored (BCA 2008).

There were relatively few publications that directly compare the LEED and the Green Globes. The Athena Sustainable Materials Institute (2002) has completed a study focusing on a comparison of the LEED and the Canadian version of the Green Globes with the conclusion that there was a close relationship between the two systems in terms of contents, goals and weightings of rating criteria. The Natural Resources Defense Council has released a report to focus on the increased stringency and credibility of the LEED rating system over the flexibility and ease of use of the Green Globes system (Bright 2005). In addition, the Carpenters Industrial Council has conducted a research to investigate the comprehensiveness, comparability and credit/point balance between the LEED and the Green Globes (Smith et al. 2006). The BCA Green Mark was released in January, 2005. There was no previous study which focused on comparing the Green Mark with the other two more popular rating systems due to its regional limitation. However, the BCA Green Mark should not be overlooked because of the current eco-city initiative undertaken jointly by Singapore and China in Tianjin. The Tianjin Eco-city project was likely to be the first and most significant building project at a macro level to focus on environment-friendly features that could provide valuable eco-lessons for other countries world-wide (BCA 2008).

2.3.2 Green Building Rating Systems – Overview

This section aims to compare the LEED, the Green Globes and the BCA Green Mark at a general level. This comparison would be useful to highlight the differences between the three rating systems. In order to reflect the more common practices, the most popularly adopted versions were chosen for comparison, which were the LEED for new construction and major renovations version 2.2 (referred to as LEED 2.2), the GBI proposed American National Standard 01-2008P (referred to as the Green Globes) and the BCA Green Mark for non-residential building version 3.0 (referred to as the BCA Green Mark 3.0). At first sight, it seems that all three rating systems shared a lot of common areas in terms of rating categories and allocation of points. As can be seen from Table 2.1, all three categories seem to be closely aligned, in areas relating to Energy, Water and Indoor

Table 2.1 Points allocation of LEED 2.2, Green Globes and BCA Green Mark 3.0

Assessment area \ Green rating system		LEED 2.2	Green Globes	BCA Green Mark 3.0
Project management	T	N/I	100 (10.0 %)	32 (22.9 %)
	M		50	Together with water, indoor environment and other: Max Cap: 50/min: 20
Site	T	14 (20.3 %)	120 (12.0 %)	N/I
	M		24	
Energy	T	17 (24.6 %)	300 (30.0 %)	79 (56.4 %) (exclude bonus points)
	M		Path A:150/ Path B:100	Max Cap: 50/ Min: 30
Water	T	5 (7.3 %)	130 (13.0 %)	14 (10.0 %)
	M		26	See project management
Resources/ materials	T	13 (18.8 %)	145 (14.5 %)	N/I
	M		29	
Emissions and storage of hazadous materials	T	N/I	45 (4.5 %)	N/I
	M		9	
Indoor environment	T	15 (21.7 %)	160 (16.0 %)	8 (5.7 %)
	M		32	See project management
Others	Innovation and design process: 5 points (7.3 %)		N/I	Other green features: 7 points (5 %)

Notes:

Path A – performance design option
Path B – prescriptive design option
T – total possible points
M – minimum points required
N/I – not included

Environment. In addition, LEED 2.2 and Green Globes set Site and Resources/ Materials as common assessment areas. Both Green Globes and BCA Green Mark 3.0 highlighted Project Management as one of the assessment category, although the latter placed Project Management under the heading of Environmental Protection. The allocation of credits/points to different categories was only a few percentages off when these three systems were compared, among which Energy took up the highest weighting. In addition, although LEED 2.2 did not set the minimum credits required to achieve certification, there was a certain amount of prerequisites under each area that must be implemented, albeit without being allocated with credits.

However, when the rating systems were examined more closely, there were a few noteworthy differences. The Green Globes' web-based self-assessment tool could be completed by any team member with a general knowledge of the building's parameters, and was able to provide both the preliminary and final ratings during the assessment (Smith et al. 2006). However, the process to achieve LEED 2.2 and the BCA Green Mark 3.0 certification was more rigid, both of which would require application, registration, preliminary meeting, assessment and final allocation of points/credits. In addition, the categorization of the criteria under each rating system varied. The same entry might be classified into different categories. For example, both LEED 2.2 and the Green Globes put public transport under the heading of Site, while the BCA Green Mark 3.0 addressed the same criteria in the section on Environmental Protection. Previous studies carried out by Smith et al. (2006) reallocated the assessment areas to form a system based on eight generic "sustainability" categories, including (1) Energy Use; (2) Water Use; (3) Pollution; (4) Material/Product Inputs; (5) Indoor Air Quality and Occupant Comfort; (6) Transport; (7) Site Ecology; and (8) Other Sustainable Design. Adopting the same consideration, in this study, rating criteria were reallocated accordingly. However, only those related to project management were reorganized, in accordance with the objectives of this chapter, as lean management might contribute to the knowledge of project management.

2.3.3 Green Building Rating Systems – In-depth Investigation

As stated previously, this section focuses on the project management areas in order to provide the industry with guidance to achieve sustainable construction. At first sight, it would seem that the Green Globes offered a better package which identified project management process as one major rating area. Evaluation points were allocated from the pre-design, construction to post construction stage. In the pre-design stage, the Green Globes suggested that individuals that represented different disciplines, which included the owner's representative, green design and delivery coordinator, architect, contractor, civil engineer and other stakeholders, were to be involved in a planning session to discuss and establish performance goals and measurements. The planning session could be in the form of a meeting, charrette or workshop (GBI 2008). The performance goals and measurements that were decided in the pre-design planning sessions were to be applied in the construction and post-construction stage. In addition, there were 48 points, which represents nearly 50 % of the credits in the project management process, that would be allocated to the commissioning process, which included the structural system, HVAC system, electrical system, etc. The reason was that with the commissioned systems, quality could be assured and environmental protection practice would be applied from the start.

In a similar fashion, the BCA Green Mark 3.0 identified sustainable construction and environmental management practice as important rating criteria when allocating credits. Project management took up to 21 points, which represented 13.2 % of the total credits. However, instead of allocating credits to the processes which stimulated sustainable or green practices, the BCA Green Mark 3.0 focused on allocating credits to these practices themselves, such as using concrete more efficiently and introducing sustainable materials and products in building construction. LEED 2.2 for new construction and major renovations, published in October 2005, seems to have many similarities with the BCA Green Mark 3.0, both of which intended to highlight sustainable practices rather than the process to achieve these practices. In addition, the sustainable construction and environmental management practice adopted in LEED 2.2 and the BCA Green Mark were quite similar. Both encouraged building reuse, if there were existing building structures, as well as construction waste management plans.

When evaluating the LEED 2.2 and the BCA Green Mark 3.0, there might be a problem of "point chasing" from project management credits which had lower investment costs over the other credits that were more difficult to obtain. Project management in LEED 2.2 and the BCA Green Mark 3.0 took around 20 % of the whole project management credits. Other than that, few credits had been allocated to interaction, communication and collaboration during the project life cycle. The problems were even worse in the BCA Green Mark 3.0, even though it had a similar section on mandatory requirements as the prerequisites in LEED 2.2. A minimum of 20 points was allocated to Parts 2, 3, 4 and 5 as a whole, which were Water Efficiency, Environmental Protection (Project management practice), Indoor Environmental Quality and Other Green Features respectively. The project could fulfill the minimum points requirements by simply obtaining the credits from project management practice, which was allocated up to 21 points. Thus, a project could be Green Mark certified even if the areas of Water Efficiency and Indoor Environmental Quality were overlooked. This would obviously harm the development, especially when the BCA Green Mark 3.0 was identified as a whole building rating systems.

On the other hand, the Green Globes set a minimum percentage of points required for each environmental assessment area to address the problem that a project might be certified without considering the whole building performance. In addition, instead of using the all-or-nothing approach which was adopted by LEED 2.2, the Green Globes offered the possibility to gain partial credits for implementing certain technologies which might lower the assessment levels (Smith et al. 2006). However, this did not prevent the newly reviewed Green Globes from being regarded as one of the leading whole building rating systems that took both the processes and results into consideration.

In summary, as shown in Table 2.2 (Wu and Low 2010), the Green Globes allocated 62.7 % of the credits to project management process, where LEED 2.2 and the BCA Green Mark 3.0 only allocated 20–30 % of the credits to project management. It seems that the Green Globes embraced the strategy that it was better to do something instead of nothing at all, which was quite similar to previous findings (Smith et al. 2006). LEED 2.2 and the BCA Green Mark 3.0 seemed to have a many strategies and grey areas (as shown in Table 2.2) in common. Both relied heavily on project management practice (78.6 % for LEED 2.2 and 71.4 % for Green Mark) and yet showed a lack of the processes to achieve sustainable construction. All the credits in the project management process were taken by commissioning and certification activities.

Table 2.2 Comparison of LEED 2.2, Green Globes and BCA Green Mark 3.0 in the area of project management

Bold	=	Major credits	PM	Project Management	IDP	Innovation and Design Process	GDDC	Green Design and Delivery Coordination
Bold italics	=	Subdivisions of credits or strategies employed	EA	Energy and Atmosphere	MR	Materials and Resources	EM	Environmental Management
(grey shading)	=	No equivalent	PR	Prerequisite	SS	Sustainable Site	MRT/LRT	Mass Rapid Transit/Light Rapid Transit

LEED 2.2

Section	Code	Item	Credits	% of PM	% of total
Project Management Process		**Project management process**	3.0	21.4%	4.3%
		EA, IDP and MR	3.0		
	EA PR 1	Fundamental commissioning of the building energy system	0.0		
	EA 3	Enhanced commissioning	1.0		
	IDP 2	LEED accredited professional	1.0		
	MR 7	Certificated woods	1.0		
Project Management Practice		**Project management practice**	11.0	78.6%	15.9%
		Public transport	2.0		
	SS 4.1	Public transport access	1.0		
	SS 4.2	Bicycle storage and changing rooms	1.0		
		Materials and Resources	9.0		
	MR 4.1-4.2	Recycled content	2.0		
	MR 5.1-5.2	Regional materials	2.0		
	MR 1.1-1.3	Building reuse	3.0		
	MR PR 1	Storage and collectin of recyclables	0.0		
	MR 2.1-2.2	Construction waste management	2.0		

Green Globes

Section	Code	Item	Points	% of PM	% of total
Project Management Process		**Project management process**	106.0	62.7%	10.6%
	6.1	***Coordination and benchmarking***	28.0		
		GDDC pre-design green design meetings	4.0		
		GDDC performance goals	10.0		
		GDDC progress meetings for design	6.0		
		GDDC progress meetings for construction	8.0		
		Suggested documentation	0.0		
	6.2	***Environmental management during construction***	16.0		
		Environmental management systems	5.0		
		Building materials and building envelope	2.0		
		Indoor air quality	9.0		
		Suggested documentation	0.0		
	6.3	***Whole building commissioning***	48.0		
		Pre-commissioning	3.0		
		Whole building commissioning	39.0		
	10.3.2	Certification of wood-based products	6.0		
		Suggested documentation	0.0		
	6.4	***Environmental management- post construction***	14.0		
		Operations and maintenance manuals	14.0		
		Suggested documentation	0.0		
Project Management Practice		**Project management practice**	63.0	37.3%	6.3%
		Public transport	8.0		
	7.1.1.2	Located within a 0.25mi of a public transport	5.0		
	7.1.1.3	Bicyle path, bicyle rack, sidewalk access	3.0		
		Materials/Resources	55.0		
	10.1.2/ 10.2.3	Recycled content (assemblies/furnishings, finishes and fit-outs)	21.0		
	10.1.3-4/ 10.2.4-5	Transportation (assemblies/furnishings, finishes and fit-outs)	14.0		
	10.4	Reuse of existing structures	18.0		
	10.5.2	Operations-related recycling programs	2.0		
	10.5.4	Suggested documentations-waste minimization	0.0		
	10.5.4	Suggest documentations-other	0.0		

BCA Green Mark 3.0

Section	Code	Item	Points	% of PM	% of total
Project Management Process		**Project management process**	6.0	28.6%	3.8%
	3-1/3-3	***Sustainable construction/EM practice***	6.0		
	3-3 c	Developer, main building, etc. that are ISO 14000 certificated	1.0		
	3-3 b	Certificated Green Mark Manager or Certificated Green Mark Professional	3.0		
	3-3 d	Building quality assessed under Construction Quality Assessment System	1.0		
	3-1 c(i)	Environmental friendly products certificated under The Singapore Green Labelling Scheme	1.0		
Project Management Practice		**Project management practice**	15.0	71.4%	9.4%
		Public transport accessibility	2.0		
	3-4 a	Good access to MRT/LRT or bus stops	1.0		
	3-4 b	Adequate bicycles parking lots	1.0		
		Environmental protection	13.0		
	3-1 c(ii)	Products with at least 30% recycled content	4.0		
	3-1 a	More efficient concrete usage	4.0		
	3-1 b	Conservation of existing building structure	2.0		
	3-3 f	Provision of facilities for collection and storage of recyclable waste	1.0		
	3-3 a	Implement environmental friendly programmes	1.0		
	3-3 f	Provision of building users' guide	1.0		

2.3.4 *Project Management and Sustainable Construction*

The term "green building" was defined as a building that provided the specified building performance requirements while minimizing disturbance to and improving the functioning of local, regional, and global ecosystems both during and after its construction and specified service life (ASTM 2006). However, much attention has been paid on green design over green construction. Even if green construction was considered in project life cycle, as explained in LEED 2.2 and BCA Green Mark 3.0, it was the results rather than the processes that were being focused upon.

As shown earlier, project management took 20 % of the credits in the rating systems. This route was easy and had a low investment cost to implement during the project life cycle. This section investigates the project management area to achieve sustainable construction, in terms of both process and practice.

2.3.4.1 Scope

There were two strands of concern relating to the management of sustainable construction. One related to the management of the project from a macro perspective to ensure that environmental management practice was effective to fulfill the requirements of green building. The key areas of concern in this section included:

- Project management process in the project life cycle to achieve sustainable construction.
- Delivery of relevant aspirations to different project parties without the sustainability baton, which is referred to as the "green baton" being dropped at key points in the process.
- Construction management practice during the construction period to achieve the green objective.
- Feedback and documentation of the project life cycle for continuous improvement.

2.3.4.2 Project Management Process

Although green building rating systems might not be 100 % credible to evaluate how green the building was, they did provide useful information on good practices and measures to achieve green objectives. In the pre-design stage, the level of understanding of, and commitment to, sustainability varied with different parties, including client, architect, engineers and other stakeholders (Halliday 2008). Individuals that represented the disciplines listed above should be involved in a planning session to discuss green issues and performance goals (GBI 2008). The planning session during the pre-design stage was of critical importance to realize the goal of sustainability because it was the starting point to achieve sustainability. Client had to clearly identify the needs; architects, engineers and project managers had to figure out the project envelope; and environmental engineer and quantity surveyor should research into sustainability issues and life cycle costing. Different parties contributed their knowledge into the process to identify the performance goals, such as site issues, water efficiency, indoor environmental quality, environmentally responsible construction activities, etc.

The construction life cycle could be divided into a few set periods from briefing to handover of the project. Problems existed when the project team transferred the performance goals into the following period. Varied interests always led to the green baton. According to Liddell (2006), green baton meant that at each stage in the project life cycle, where there was a high risk of the sustainability baton being dropped throughout the process, the sustainability agenda was most vulnerable. Hence, as construction moves on, more and more green performance goals are dropped. At the end of the construction period, when the project was handed over to the client, it might be far from what have been expected, in a negative manner. In order to safeguard the green baton during the project life cycle, Halliday (2008) proposed that the project should start with high aspirations, or green performance standards, and involve the widest range of interests in the design process. In addition, a smoother flow or handover of the baton, which could be realized by regular progress meetings during the project life cycle, was of critical importance to the success of green building.

2.3.4.3 Construction

The construction period was often lengthy and easily exposed to all kinds of pollutions, which could influence both the local and national environment, depending on the nature of the project. Three types of inputs were discussed in this section, which were labor, equipment and material. All inputs were of critical importance to improve the overall productivity and reduce waste in the construction period. Basically, green construction was about planning and scheduling. Other than introducing the low-emission vehicles and improving fuel efficiency, green construction aimed at planning and scheduling to fulfill the green project requirements, which was always less technology and design related and was often overlooked by the industry. According to Glavinich (2008), another consideration was the need to minimize site disturbance during construction or to protect materials and equipment from contamination during the construction process. Hence, planning and scheduling was about fulfilling the green project requirements with high efficiency and low interruption (sometimes referred to as "waste").

1. **Labor.** Projects were carried out by people. As a skilled workforce was a must for the success of a project, this workforce must be cared for and preserved just like any other resources used in the construction industry (Glavinich 2008). Training was essential to raise awareness of sustainable construction and reduce waste. According to EPA (2004), if driven carelessly, even a perfectly maintained vehicle would pollute more than necessary. Thus, it was necessary and beneficial for the operators to understand and follow the sustainable operating conventions. In the Green Globes, training and awareness was also

allocated up to five points under the heading of environmental management system with action plan, communication and emergency preparedness and response. Being treated with equity and respect in terms of training, learning and meaningful jobs, collaboration of different parties could be achieved.

2. **Equipment.** The most commonly adopted option to restrict equipment movement to prevent site disturbance was a proper site layout. One target of green construction was to minimize site disturbance either to preserve the natural environment or to cause minimum interruption to local environment around the project, which required restricting equipment movement. In addition, as equipment consumed energy in terms of electricity or fuel to maintain operations, restricting equipment involvement could bring both economic and environmental benefits, both of which were essential to the success of a green project. This was the situation where the just-in-time (JIT) concept could be applied with minimum modifications. By applying the JIT concept, in the pre-design stage, all the deliveries of the materials, construction and installation activities were taken into consideration, so that no unnecessary movements would happen on site. The right parts needed were delivered to site at the time they were needed and only in the amount needed (Ohno 1988). While the implementation of the JIT concept in traditional projects was not mandatory at the moment, it was always good practice to apply this in green projects, based on the knowledge that most rating systems did take minimum energy usage into consideration.
3. **Material.** Generally speaking, a material plan could be divided into three phases: selection, conservation and usage. When selecting the materials for the project, other than fulfilling the project requirements, there were some other considerations that could not be neglected. Prefabricated materials were widely adopted right now because of their easy installation, fire protection and so on. In addition, prefabricated materials and products were manufactured in factories within a closed environment where quality management could be better applied to fulfill the requirement of being environment-friendly. Location of the supplier was another consideration especially when all rating systems allocated a certain amount of credits to regionally extracted, processed and manufactured materials. However, when using regionally manufactured materials was not viable, it was always a good practice to try to combine possible deliveries together to reduce delivery cost, which was often referred to as the Rim Delivery System. As stated previously, if the JIT concept was applied at the construction stage, materials would be immediately used when they reached the site. It was not necessary to provide storage area for materials. However, as the Green Globes proposed, building materials made of organic matter, as well as steel should be protected on site and in transit.

2.3.4.4 Post-occupancy

Green building was better regarded as a process rather than a product due to the long post-construction operational duration. Being a process, operations and maintenance needed to be taken into consideration. Green building might be different for different operations and maintenance, contractual issues, defects control, etc. There would be daily, weekly, seasonal and annual routines (Halliday 2008). It was especially important for the building to be commissioned if the client was seeking green certification. According to Glavinich (2008), the purpose of commissioning was to ensure that the building systems were operated and could be maintained in accordance with the owner's project requirements as expressed in the facility program that defined the owner's operational requirements. These included:

- Building systems have been installed properly and operate correctly.
- Each building system being commissioned interacts well with other building systems.
- Complete equipment and system documentation, including operations and maintenance, is provided to the client and final users.
- Client and final users have been properly trained in order to perform reliable, efficient and sustainable building operations.

In fact, operations and maintenance was the one area that green rating regulatory bodies have taken into consideration when revising the rating systems. BCA (2008) proposed the provision of a building users' guide that included details of the environmental friendly facilities and features within the building and their uses in achieving the intended environmental performance during building operations. GBI (2008) also mentioned the importance of operations and maintenance manuals to the success of the green building, which, like other products, had running problems.

As stated above, the demand for green buildings was growing. Yudelson (2008) forecasted green building growth rates in the double digits until 2007. However, this did not erase the fact that green building was in its infancy and was still a niche market. Continuous improvement based on previous experience was essential for the development of green buildings. The feedback review could be carried out by different parties in different stages. After the completion of the building, the project team could investigate the building performance to draw lessons. At the same time, feedback could be documented by final users in the following years after completion. Finally, the performance could be assessed by third parties to highlight both generic and technical lessons for further improvement. Although documentations for continuous improvement were not allocated with credits in green rating systems right now, there remained a good chance that this would happen in the near future because of the benefits this could bring to the construction industry.

2.4 Global Climate Change

Among all current environmental issues, climate change appeared to be the most significant one, which caused the most considerable threat to human development (Tang and Yeoh 2007). The worst-case predictions for rising sea levels in the Thames Estuary would see the level of the river rising by up to 4 m by 2100, which means that eventually, large parts of London – one of the world's business capitals – would be under water (Tang and Yeoh 2007).

Billions of people were exposed to natural disasters caused by global climate change, which took lives, damaged infrastructures and resources, disrupted economic activities and threatened social development (Pelling et al. 2004). The potential impact of climate change on the global economy could be enormous: re-insurance companies estimate that it could be of the order of hundreds of billions of dollars per year in the form of natural disasters and disruptions to agricultural cycles (Brown 2005). If actions were not taken to reduce greenhouse gas emissions, the overall costs and risks of climate change would be equivalent to losing at least 5 % of global GDP per year, now and forever (Stern 2007).

Climate change was mainly caused by an increase in greenhouse gases (GHGs) emissions from both natural and man-made sources. However, it was widely believed that man-made sources, such as human activities, were the factors which contributed the most GHGs emissions. The developed countries emitted half of the world's carbon dioxide and America was the largest emitter of GHGs. Without changing current consumption habits, it was estimated that the power generation and transport in developing economies could drive CO_2 emissions to 40 billion tons by 2030 (Tang and Yeoh 2007).

The construction industry played a significant role in economic growth everywhere around the world. In the UK economy, the construction industry accounted for 8 % of GDP and employed 1.5 million people (1 in 14 of the total working population) directly (Brown 2005). However, this industry was also the largest source of GHGs. According to the American Institute of Architects (AIA 2007), it was estimated that nearly 50 % of all the GHG emissions were generated by buildings and their construction in terms of energy used throughout the life cycle of construction projects. The main sources which caused carbon emissions in the construction life cycle could be identified as follows (Kruse 2004):

1. The cement sector alone accounted for 5 % of global man-made CO_2 emissions;
2. Mining/manufacturing of materials and chemicals had considerable impact on CO_2 emissions;
3. Transport of heavy materials such as cement was energy intensive, which consumed a large amount of energy and generate emissions;
4. On-site construction of building was currently not effective and was high energy consuming, which led to unnecessary emissions; and
5. The maintenance of buildings consumed significant energy, especially for heating, lighting and air conditioning.

The rise of the global focus on climate change led to the foundation of the United Nations Framework Convention on Climate Change (UNFCCC) in 1992 to deal with the environmental concerns that were brought about by global climate change. It aimed to introduce measures to control adverse climate changes caused by GHGs in both industrialized and developing countries (Yates 2007). Global environmental concerns about climate change also led to the establishment of the Kyoto Protocol, an international agreement that set binding targets for the reduction of GHG emissions by industrialized countries by the year 2012 (Yates 2007). The Kyoto Protocol was regarded as a small step towards stabilizing atmospheric concentrations of greenhouse gases at a level that would have acceptably manageable consequences on the global environment, human health, natural resources and physical infrastructure (Tang and Yeoh 2007). The objectives of the Kyoto Protocol included (Yates 2007):

1. Changing consumer patterns;
2. Protecting and promoting human health condition;
3. Promoting sustainable humane settlement development; and
4. Protecting the environment, air, water, and ecosystems and combating deforestation and managing wastes.

The Kyoto Protocol set specific target of GHGs reduction for each country, which has ratified the protocol, to meet by 2012 (Yates 2007). It was obvious that the Kyoto Protocol would affect the construction industry at several levels, including both the technical and managerial levels. It encouraged the construction industry to adopt more efficient and more environment-friendly materials and resources, as well as sustainable architecture design and urbanism, to cut energy consumption, thus reducing carbon emissions.

Generally speaking, the global recognition of climate change might affect the industry level at three levels. First, governments intended to impose limits on GHGs emissions. As the building industry generated almost 50 % of the total emissions, it was beneficial for the industry to act proactively. Secondly, global climate change would directly affect the business of the construction industry by rising sea levels, causing abnormal weather and geological conditions, etc. Thirdly, the construction industry was known as a 3D's industry, which was dirty, dangerous and demanding (Singapore Construction 21 Committee 1999). In addition, rising global recognition of climate change might have the potential to change the business behavior of the construction industry.

However, what has been done in the construction industry so far was mainly focused on the technical level. Implementing new carbon reducing technologies might not provide the best cost/performance solution, not to mention additional costs when applying these technologies to non-value adding activities, which should not have been allowed to take place in the first instance. This contradiction led to the research of applying lean production principles in the construction industry to reduce carbon emissions. It embraced a new way of thinking, which divided GHG emissions in the building and construction sector into two groups: embodied emissions and unnecessary emissions. Embodied emissions referred to the emissions that were caused by energy consumption during producing, delivering and installation, without which the production, delivery and installation processes could not be completed. Unnecessary emissions could be defined as emissions that were generated by non-value adding activities in building and construction processes, and could be significantly reduced if the processes are appropriately managed.

2.4.1 Global Climate Change and the Construction Companies

With the growing global awareness of carbon emissions and climate change, construction companies are now under increasing pressure to take environmental considerations in their daily decision making processes. In this context, construction companies primarily faced risks from the physical, regulatory and competitive levels (Tang and Yeoh 2007):

- **Physical risks.** As stated above, global climate change caused threat to construction activities. This required that the construction companies took the impacts of climate change into consideration when choosing the site, planning and scheduling. The physical impacts of climate change, including flooding, coastal erosion, etc., should be properly assessed. Otherwise, the project might not be profitable, or even viable in some cases.
- **Regulatory risks.** Regulatory bodies, both national and international, have kept pressure on construction companies to invest in carbon elimination techniques and environment-friendly techniques. Besides the UNFCCC and the Kyoto Protocol, countries around the world were developing various rating systems and regulations to achieve energy efficient and carbon neutral projects. For example, the LEED (Leadership in Energy and Environmental Design) Green Building Rating System and the U.S. Environmental Protection Agency's Energy Star program have been used to promote less energy consumption and less carbon emissions in various aspects during the project life cycle.
- **Competitive risks.** With the global concentration on climate change, construction companies needed to take carbon emissions into consideration for them to survive in the competitive construction market. The burden of corporate social responsibility kept falling onto construction companies. In some cases, the projects had to obtain a certain "carbon mark" before it could be approved, built and sold. Investing in carbon reduction would not only improve public perceptions of the company, but also bring about considerable benefits, as oil prices continued to rise recently. Over time, rising oil prices were likely to translate into higher energy costs, including higher electricity and gas prices. In addition, most cities were providing government incentives to projects with green impacts, especially when these were large developments with major infrastructure impacts (Yudelson 2008). By 2007, prolonged oil prices above

$50–$60 per barrel had changed the psychology of consumers and businesses for the first time in a generation (Yudelson 2008). During this time, both global awareness of carbon emissions and the rising oil prices would bring about a tremendous change to the management and operation of construction companies.

Besides the three categories of risks identified above, construction companies were facing other kinds of risks from the managerial level to the technical level. Whether or not the company was able to manage its carbon emissions level under the limit imposed by regulatory bodies might affect the survival of the company. A comprehensive strategic package to deal with global climate change was essential for the Construction companies to succeed, which might include (Tang and Yeoh 2007):

1. A systematic analysis of the company's situation, which included:
   (a) Financial status. The financial status was related to choices that the company might take to reduce carbon emissions. Technical improvements took less time to validate the effect, but would usually cost a lot more. On the other hand, managerial improvements, which could also contribute to reduce carbon emissions, might take a longer time to validate. Managerial improvements aimed to change people's behavior and it was not a job of a one-off nature.
   (b) The company's exposure to climate change. Manufacturers, suppliers and contractors with high energy consumption and significant carbon emissions needed to assess their exposure to respond to pressure from the regulatory bodies. A high exposure often required that strong proactive measurements be taken to ensure the stability and continuity of the business.
2. Companies must both mitigate the risks and seize the opportunities brought by climate change. By adopting new corporate policies to reduce carbon emissions, companies should be able to seize new market opportunities. Other actions included setting GHGs approachable reduction targets which could be achieved by purchasing carbon reducing technologies and improving current energy efficiency. Companies should also participate in seminars, dialogues and research relating to climate change to get themselves updated on current issues. These could both help to reduce the risk of being caught totally unprepared and to seize the emerging market opportunities.
3. Consulting, sharing and discussing climate change strategies with other stakeholders. First of all, this highlighted the importance of employee involvement to reduce carbon emissions. In the construction industry, front-line employees were directly responsible for carbon emissions by activities such as delivery and installation. The importance of employee involvement could be explained by the fact that organizational goals and personal goals could both be achieved if employees were treated with equity and respect. They should be involved with decision making, provided with meaningful jobs and given the opportunity to learn (Stendel and Desruelle 1992). Secondly, carbon reducing technology was a

complex concept with lots of subdivisions. It was impossible for a company to be familiar with all the carbon reducing technologies, not to mention the ones which would perfectly fit in well with the company's present situation. Consulting, sharing and discussing with other professionals from the industry would help to reduce the time and resources consumed to identify feasible strategies for the company.

4. It should be noted that adapting to climate change was not a job that can be done once and for all. It was a long term improvement rather than a short term endeavour. Although adopting low carbon emissions technologies might produce instantaneous results, this would still require the company to monitor the process and to make periodic records upon which evaluations and improvements were based on.

2.4.2 Measuring Carbon Emissions – The Building Research Establishment (BRE) Methodology

This section reviewed guidelines on how carbon emissions in the construction industry could be measured. However, it should be noted that the literature relating to the evaluation process was limited, especially with respect to a complete construction life cycle. The methodology presented in this section served as a framework rather than implementation strategies to carry out the research. Details relating to how the evaluation process was designed in this study would be explained in Chap. 6 – Research Methodology.

In line with the aim of this study, which was to reduce carbon emissions by applying lean construction principles in the construction industry, a detailed evaluation process relating to how carbon emissions was measured in the construction life cycle was necessary. The *BRE methodology for environmental profiles of construction materials, components and buildings*, according to Howard et al. (1999), was useful by providing reliable and independent environmental information about building materials and components. The BRE methodology was developed by the Building Research Establishment (BRE), which was a world leading research organization delivering sustainability and innovation across the built environment and beyond (BRE 2008). It standardized the way of identifying and assessing the environmental effects of building materials over their entire life cycle, through their extraction, processing, construction, use and maintenance and their eventual demolition and disposal (Howard et al. 1999). Although the methodology was mostly applied to evaluate the environmental impacts of construction materials in their life cycle, this did not restrict the methodology from being applied in the construction life cycle to measure the level of carbon emissions. According to the BRE methodology, the following steps were necessary to carry out the measurements (Howard et al. 1999):

1. **Boundaries.** The BRE methodology advocates a cradle to grave measurement of environmental impacts to include all of the processes from the procurement of raw materials through to production, use, reuse and recycling until eventual disposal. However, the cradle to grave measurement is not always quantifiable, because of the long timescale of a building. Therefore, a few assumptions would need to be made before the measurement process can be carried out. For example, the environmental impacts of insulation materials in their operational stage are far more important than the ones in their production stage, given the consideration that the operation stage usually occupies a longer timescale than the production stage. In addition, before carrying out the measurement from cradle to grave, assumptions should be made relating to the schedules of repair, maintenance and replacement, which are usually not predictable in real life cases.

   The boundaries of the assessment vary with the aim and objectives of the research. The BRE methodology aims to provide guidance when research relating to investigate environmental profiles of construction materials, components and building is carried out. Hence, this can include all major environmental impacts, such as climate change, acid deposition, ozone depletion, water extraction, waste disposal, etc. However, according to the aim and objectives of this study, only factors relate to climate change would be investigated in this study.

   In the production stage, the available per tonne data is adopted, tracing all raw materials back to their extraction, describing the mode of transport and distance traveled to the processing site and the processing activities carried out therein. However, the per tonne data is not suitable for transport and construction. The methodology designed for transport and construction will be explained later in this chapter.
2. **Inventory data collection.** In order to define which kind of data is required to fulfill the aim and objectives of this study, the process tree needs to be defined, including any major transportation stages with a clearly marked system boundary to indicate the included from the excluded processes (Howard et al. 1999). This will be discussed further in the section of "Decision Tree". The process tree is necessary to define the inputs and outputs of the process. Defining inputs and outputs of the process vary with the aim and objectives of the research. The BRE methodology is restricted to environmental impacts from: Energy, Minerals and Waster consumption, Waste, Air and Water emissions. Thus, the inputs are limited to materials associated with the manufacture of a product and also the consumption of fuel and water, which is classified into the following domains:

   (a) **Materials.** These include inputs associated with the production process of a product, including ingredients, packaging materials and consumable items. In addition, it is not always realistic to gather all the information, since a large number of materials are used in small quantities. Thus, the methodology takes two sources of data into consideration, which are materials and substances used in large quantities and those which are used in small quantities but have a significant contribution to the process.

(b) **Transport of materials to the plant.** Transportation information from each supplier must be recorded, including the type of travel, distance traveled, delivery weight and all other information which is required to calculate the fuel consumption.
(c) **Direct consumption of fuel.** In the BRE methodology, each type of fuel consumption must be recorded to ensure that all the fuel consumption is taken into consideration. This direct consumption of fuel is calculated only for the construction stage. Offsite transport activities will be calculated separately.
(d) **Water use and capital equipment.** If the aim of the research is to measure the environmental impacts relating to water pollution, the water brought into the plant should be recorded. In addition, the contribution of capital equipment to the environmental impacts is not considered. According to the BRE methodology, although the capital equipment is a form of indirect energy input in the production process, its contribution to the environmental profile is not normally considered. In addition, as the maintenance schedule is not predictable in the future, maintenance activities, such as using of lubricants, are also not included in the life cycle assessment.

3. **Inventory data analysis.** Raw data collected from the manufacturing yard or on site must be modified before identifying the environmental profiles. The processes relating to calculating emissions include:

   (a) Convert data to standard units, e.g. MJ for energy, Tonne for inputs.
   (b) Transport figures should be calculated to obtain data on product delivery to site.
   (c) The data in the production stage should be normalized to per tonne levels.
   (d) Apply conversion figures for fuel consumption to translation the consumption into emissions.

The BRE methodology provides a general guideline to assess the environmental impacts of construction materials, components and buildings. It should be noted that although the methodology is not designed to address the evaluation process for carbon emissions, it provides useful guidance, especially the data collection and data analysis processes. More importantly, it offers the information on how many potential considerations should be taken into account when assessing the carbon emissions levels.

2.4.3 Measuring Carbon Emissions – The IPCC Methodology

The BRE methodology offers a framework to develop environmental profiles for construction materials, components and buildings. However, it does not provide a general methodology to calculate carbon emissions in the construction process.

Carbon emissions, which are largely driven by the combustion of fuels, can be divided into two categories: combustion and fugitive emissions, or escape without combustion. However, fugitive emissions are closely related to the process of coal mining and handling, which is not the focus of this study. Therefore, the following sections aim to review the process of calculating the carbon emissions from combustion sources. The *2006 IPCC Guidelines for National Greenhouse Gas Inventories* provides the general methodology and world-wide average emissions factors to calculate carbon emissions.

2.4.3.1 Tiers

There are three tiers presented in the 2006 IPCC Guidelines for estimating emissions from fossil fuel combustion (IPCC 2008). The Tier 1 method is used to calculate the emissions based on the quantities of the fuel combusted, using average emission factors. However, other emissions that are related to climate change, including nitrous oxide and methane will largely depend on the combustion technology and prevailing operating conditions. Hence, for calculating these emissions, the Tier 3 method that takes combustion technology and prevailing operating conditions into consideration is preferred.

In the Tier 2 method, emissions from combustion are estimated from similar fuel statistics, as used in the Tier 1 method, but country-specific emission factors are used in place of the Tier 1 defaults (IPCC 2008). The Tier 2 method is applied when country-specific emission factors are available.

With the availability of detailed emission models, measurements and data, the Tier 3 method is applied to better estimate the GHG emissions levels, at the cost of more detailed information and effort (IPCC 2008). By applying the official standard measurement methods and field tests, accurate GHG emissions can be obtained. However, according to the methodology, calculating CO_2 emissions using the Tier 3 method is often unnecessary because CO_2 emissions do not depend on combustion technology. The only situation where technology specific CO_2 emissions data is required is the possibilities of emissions trading.

2.4.3.2 Decision Tree

In order to decide which method from Tier 1 to Tier 3 is to be applied in the calculation process, a decision tree would be helpful for evaluation. The selection of the method will largely depend on the aim of the study, availability of resources in terms of time, work force, sophisticated models and budget (IPCC 2008).

As illustrated in Fig. 2.1, the decision process is largely dependent on the availability of measurement models. For calculating CO_2 in general, a Tier 1 method will often suffice.

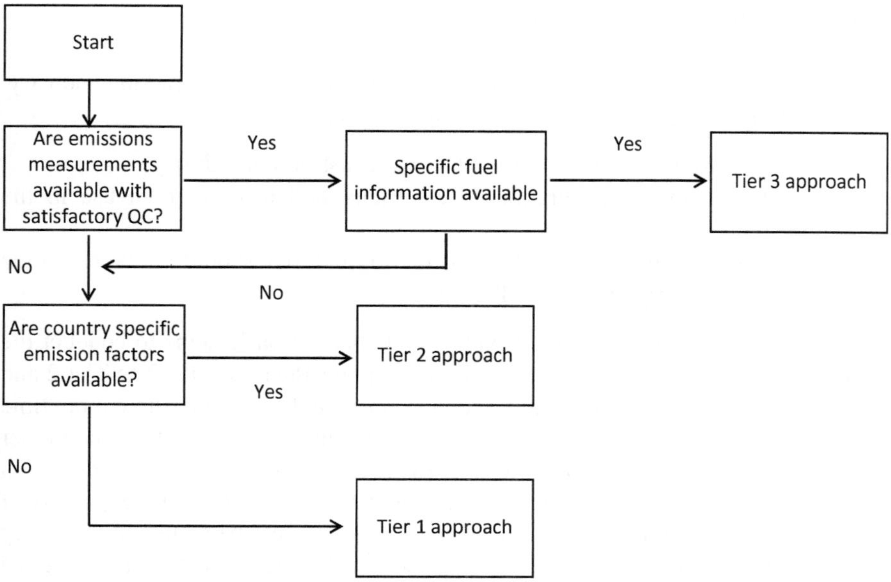

Fig. 2.1 Generalized decision tree for estimating emissions from fuel combustion (Adapted from: IPCC 2008)

However, where country-specific emission factors are available, a Tier 2 method is able to help regulatory bodies from that country to conduct a CO_2 inventory analysis that will benefit the country by controlling carbon emission levels.

According to IPCC (2008), emission models and technology-specific methods for road transport might be based on vehicular kilometres travelled rather than on the fuel used. However, this does not contradict the fact that the activity data applied in such models and higher tier methods should be consistent with the fuel sales data.

This means that the three methods do not contradict one another. It is however, at the abstraction level that they differ. The Tier 1 method can be applied to test the accuracy of the results generated from the Tier 3 method, and vice versa. However, it should be noted that where a discrepancy between fuel sales and vehicular kilometres travelled is detected, the activity data, used in the technology-specific method should be adjusted to match fuel sales statistics, unless it can be shown that the fuel sales statistics are inaccurate (IPCC 2008). This means that the Tier 1 method has a high priority over the Tier 3 method, as long as the fuel consumption data is accurate. This is based on the fact that the embodied carbon in a certain kind of fuel is determined by its quantity. No matter what kind of measurement method is used, the carbon emissions level cannot exceed the level which is determined by the embodied carbon. If the level do exceed, there must be a mistake in measuring the carbon emissions in the higher tier method.

2.4.3.3 Stationary Combustion

According to IPCC (2008), the emissions sources in the construction industry, which belong to stationary combustion, can be grouped as follows:

(a) Emissions from combustion of fuels in the construction industry;
(b) Combustion for the generation of electricity and heat for own use in the construction industry; and
(c) Energy used for transport in the construction industry should not be classified under this category, but under Transport.

As explained above, the Tier 1 method sets out the basic steps to conduct the simplest calculations. It is the method which requires the least data. The Tier 2 and Tier 3 methods require more detailed country-specific data and consume more time and resources to conduct the calculations. When appropriately applied, the higher tiers methods have more accurate results. However, when there is a huge deviation between the results obtained from the Tier 1 and Tier 3 method, and the quantity of the fuel consumption is accurate, the Tier 1 method should be adopted. According to IPCC (2008), the equations which are used to calculate emissions are explained as follows:

Tier 1 Approach

The following equation is used to calculate the estimated emissions:

GHG emissions from stationary combustion:

$$\text{Emissions}_{\text{GHG,fuel}} = \text{Fuel Consumption}_{\text{fuel}} * \text{Emission Factor}_{\text{GHG,fuel}} \tag{2.1}$$

Where:

$\text{Emissions}_{\text{GHG, fuel}}$ = emissions of a given GHG by type of fuel (kg GHG)
$\text{Fuel Consumption}_{\text{fuel}}$ = amount of the fuel combusted (TJ)
$\text{Emission Factor}_{\text{GHG, fuel}}$ = default emission factor of a given GHG by type of fuel (kg gas/TJ)

Tier 2 Approach

The equation used in the Tier 2 approach to calculate emissions is similar to the Tier 1 method. The only difference between these two is that the Tier 1 default emission factors are replaced by country-specific emission factors.

Tier 3 Approach

The Tier 1 and 2 methods use average emission factors as the basis of the calculations. However, in reality, emissions depend on several factors, including fuel type, combustion technology, operating conditions, control technology, quality of maintenance, and age of the equipment used to burn the fuel.

The equation used to calculate emissions by the Tier 3 method is:

GHG emissions by technology

$$\text{Emissions}_{\text{GHG, fuel, technology}} = \text{Fuel Consumption}_{\text{fuel, technology}} * \text{Emission Factor}_{\text{GHG, fuel, technology}} \quad (2.2)$$

Where:

$\text{Emissions}_{\text{GHG, fuel, technology}}$ = emissions of a given GHG by type of fuel and technology (kg GHG)

$\text{Fuel Consumption}_{\text{fuel, technology}}$ = amount of the fuel combusted per type of technology (TJ)

$\text{Emission Factor}_{\text{GHG, fuel, technology}}$ = emission factor of a given GHG by fuel and technology type (kg GHG/TJ)

To calculate the overall emissions from the Tier 3 category, it would only involve a simple process of summing up the results from Eq. 2.2.

2.4.3.4 Mobile Combustion

Mobile combustion is related to major transport activities, including road, off-road, air, railways, and water-borne navigation. In a typical off-shore construction project, road and off-road transport are normally considered as the mobile combustion sources for carbon emissions. However, for a country like Singapore, which is surrounded by sea from three sides and relies heavily on the imports of materials and human resources from overseas, air and water-borne navigation must not be neglected.

1. Road Transportation

   The mobile source category of road transportation includes all types of light-duty vehicles such as automobiles and light trucks, and heavy-duty vehicles such as tractor trailers and buses, and on-road motorcycles, including mopeds, scooters, and three-wheelers (IPCC 2008). Carbon emissions from road transport can be calculated by the fuel consumption or the traveling distance of the vehicles. However, according to IPCC (2008), fuel assumption is more appropriate for measuring carbon emissions. The equation used in this methodology is similar to Eq. 2.1. In addition, whether or not the Tier 2 methodology will be applied is largely dependent on the availability of the country-specific emission factors. It should be noted that if this category of emissions is a key category which occupies a significant proportion of the country's total emissions, the country-specific emissions factors should be collected by the researcher, even if they are not available from the regulatory bodies.

2. Off-road Transportation
   The off-road transportation includes vehicles and mobile machineries used within agriculture, forestry, industry (including construction and maintenance), residential and sectors such as airport ground support equipment, agricultural tractors, chain saws, forklifts, and snowmobiles (IPCC 2008). The preferred methodology to calculate the off-road transportation emissions is to multiply fuel consumption by the country-specific emission factors. However, difficulties relating to the diversity of equipment types, locations and usage patterns will largely affect the accuracy of the results. The equations used for the Tier 1 and Tier 2 method are quite similar to Eq. 2.1. The equation which is applied in the Tier 3 method is shown below:

   Tier 3 emissions estimate:

$$\text{Emission} = \text{N} * \text{H} * \text{P} * \text{LF} * \text{EF} \tag{2.3}$$

   Where:

   Emission = emission in kg
   N = source population
   H = hours of use of vehicle (h)
   P = average rated power of vehicle (kW)
   LF = typical load factor of vehicle (fraction between 0 and 1)
   EF = average emission factor for use of fuel in the vehicle (kg/kWh)

3. Railways/Water-borne navigation/Civil aviation
   According to IPCC (2008), CO_2 emissions from railway vehicles are estimated on the basis of the full carbon content of the fuel, which means that the equation used in this calculation is similar to Eq. 2.1. The reason for listing these three types of emissions together is based on their similarities in the calculation process. However, the carbon footprint of the workers is significantly different from the calculation process. One simple explanation is that it is not realistic for a flight to carry one single passenger. The carbon footprint for human resources can be measured by per return flight per person. The data relating to the carbon footprint per return flight is largely dependent on the distance between the destination country and the country where the flight begins. On the other hand, the carbon footprint for materials can be measured by per return flight per kg. The detailed assumptions and design of the measuring process of these three categories will be explained in the methodology chapter.

2.5 Summary

This chapter reviews the concept of sustainable development from three dimensions: economic, social and environmental sustainability. Although these three areas form the general sustainable concept, this study focuses on the connection between environmental sustainability and the construction industry. One important reason for this is the global recognition of climate change, which might bring considerable changes to the construction industry. Both the construction industry and construction companies should be fully prepared for the increasing pressures from the physical, regulatory and competitive aspects. Institutions have provided methodologies to calculate carbon emissions for both building materials and construction activities. These include the "BRE methodology for environmental profiles of construction materials, components and buildings" and the "2006 IPCC guidelines for national greenhouse gas inventories", which would serve as the methodology to compute carbon emissions in this study.

Sustainable construction incorporates both technical and managerial activities. These have been verified by green building rating systems that placed project management, including both practices and processes, in their revised versions. In accordance with this trend, this study aims to investigate management-related improvements in the construction industry that can help to reduce carbon emissions. These improvements should have less investment costs than the technical improvements, which are likely to be more expensive.

References

Adler A et al (2006) Green building: project planning and cost estimating, 2nd edn. R.S. Means, Kingston

American Institute of Architects (2007) Architects and climate change. [Online] Available at: http://www.aia.org. Accessed 20 Feb 2008

ASTM (2006) Standard terminology for sustainability relative to the performance of buildings. [Online] (ASTM Standard E 2114-06a) Available at: http://www.astm.org. Accessed 5 Aug 2008

Athena Sustainable Materials Institute (2002) LEED Canada adaptation and BREEAM/green leaf harmonization studies: part III BREEAM green LEaft harmonization study. ECD Energy and Environment Canada Ltd., Toronto

BCA (2008) BCA green mark scheme. [Online] Available at: http://www.bca.gov.sg/GreenMark/green_mark_buildings.html. Accessed 12 Aug 2008

Bohle HG, Downing TE, Watts MJ (1994) Climate change and social vulnerability: toward a sociology and geography of food insecurity. Glob Environ Chang 4(1):37–48

BRE (2008) BRE mainwebsite. [Online] Available at: http://www.bre.co.uk. Accessed 16 Mar 2009

Bright U (2005) Comparison of the LEED and Green Globes green building rating systems. Backgrounder report of the Natural Resources Defense Council. Unpublished review document

Brown CS (2005) The sustainable enterprise: profiting from best practice. Kogan Page, London

Business Impact (2000) Winning the integrity: a guide to social responsibility. Business in the Community, London

Carpenter TG (2001) Environment, construction and sustainable development: the environmental impact of construction, vol 1. Wiley, Chichester

Cassidy R (2003) White paper on sustainability: a report on the green building movement. A supplement to building design and construction. A Reed Business Information Publication, Oak Brook

DETR (1999) A better quality of life: a strategy for sustainable development. HMSO, London
EPA (2004) Exhaust and crankcase emission factors for nonroad engine modelling – compression-ignition. [Online] Available at: http://www.epa.gov/otaq. Accessed 16 Aug 2007
Forum for the Future (2008) Formal definitions of sustainable development. [Online] Available at: http://www.forumforthefuture.org/node/327. Accessed 15 Aug 2008
GBI (2008) The Green Globes systems. [Online] Available at: http://www.thegbi.org/commercial. Accessed 29 Aug 2008
Glavinich TE (2008) Contractor's guide to green building construction: management, project delivery, documentation and risk reduction. Wiley, Hoboken
Halliday S (2008) Sustainable construction. Butterworth-Heinemann, Oxford
Haselbach L (2008) The engineering guide to LEED-new construction: sustainable construction for engineers. McGraw-Hill, New York
Howard N, Edwards S, Anderson J (1999) BRE methodology for environmental profiles of construction materials, components and buildings. CRC, London
IPCC (2001) Climate change 2001: the scientific basis. Cambridge University Press, Cambridge
IPCC (2008) 2006 IPCC guidelines for national greenhouse gas inventories, volume 2: energy. [Online] Available at: http://www.ipcc-nggip.iges.or.jp/public/2006gl/vol2.html. Accessed 15 Jan 2009
Kibert C (2008) Sustainable construction: green building design and delivery, 2nd edn. Wiley, Hoboken
Kruse C (2004) IIGCC briefing note: climate change and the construction sector. [Online] Available at: http://www.iigcc.org/docs/PDF/Public/ConstructionSector_2004.pdf. Accessed 17 June 2008
Liddell H (2006) Eco-minimalism- less is more. Green Building Press. [Online] Available at: http://www.seda2.org/articles/Ecominimalism.html. Accessed 29 Aug 2008
Moir L (2001) What do we mean by corporate social responsibility. Corp Gov 1(2):16–22
Moody-Stuart M (1999) Forword (supplement) – responsible business. Financial Times, June, p 2
Munasinghe M, Swart R (2005) Primer on climate change and sustainable development facts, policy analysis and applications. Cambridge University Press, Cambridge
OECD (2001) Sustainable development: critical issues. Organisation for Economic Co-operation and Development, Paris
Ohno T (1988) Toyota production system: beyond large-scale production. Diamond, Inc., Tokyo
Pearce DW (1988) Optional prices for sustainable development. In: Collard D, Pearce DW, Ulph D (eds) Economics, growth and sustainable environment. MacMillan Press, London
Pearce DW, Anil M, Edward BB (1989) Blueprint for a green economy. Earthscan Publications, London
Pelling M, Maskrey A, Ruiz P, Hall L (2004) Reducing disaster risk: a challenge for development. United Nations Development Bank, Bureau for Crisis Prevention and Recovery, New York
Pezzey J (1992) Sustainable development concepts: an economic analysis, World Bank environment paper number 2. The World Bank, Washington, DC
Rees WE (1989) Planning for sustainable development: a resource book. UBC Centre for Human Settlements, Vancouver
Repetto RC (1986) World enough and time: successful strategies for resource management. Yale University Press, New Haven
Rogers PP, Jalal KF, Boyd JA (2008) An introduction to sustainable development. Glen Educational Foundation, Inc., London
Schubert U, Stormer E (2007) Sustainable development in Europe: concepts, evaluation, and applications. Edward Elgar, Cheltenham
Singapore Construction 21 Committee (1999) Construction 21: reinventing construction. Ministry of Manpower and Ministry of National Development, Singapore
Smith TM, Fischlein M, Suh S, Huelman P (2006) Green building rating systems: a comparison of the LEED and Green Globes systems in the US. [Online] Available at: http://www.usgbc.org/docs/LEEDdocs/LEED_RS_v2-1.pdf. Accessed 12 Aug 2008

St. John A (1992) The sourcebook for sustainable design: a guide to environmentally responsible building materials and processes. Architects for Social Responsibility, Boston

Stendel HJ, Desruelle P (1992) Manufacturing in the nineties: how to become a mean, lean, world-class competitor. Van Nostrand Reinhold, New York

Stern N (2007) The economics of climate change: the Stern review. Cambridge University Press, New York

Tang K, Yeoh R (2007) Cut carbon, grow profits: business strategies for managing climate change and sustainability. Middlesex University Press, London

UNEP (2003) Scientific assessment of ozone depletion: 2002: executive summary. Global ozone research and monitoring project – report no 47. UNEP, Geneva

UNEP (2007) Global Environment Outlook – GEO 4: environment for development. Progress Press Ltd., Valletta

USGBC (2008) LEED for new construction. [Online] Available at: http://www.usgbc.org/DisplayPage.aspx?CMSPageID=220. Accessed 18 Aug 2008

WBCSD (1999) CSR: meeting changing expectations. [Online] Available at: http://www.wbcsd.org/DocRoot/hbdf19Txhmk3kDxBQDWW/CSRmeeting.pdf. Accessed 3 Mar 2010

WCED (1987) Our common future. Brudtland report. Oxford University Press, Oxford

Wood DJ (1991) Corporate social performance revisited. Acad Manag Rev 16(4):691–718

WRI, IUCN and UNEP (1992) Global biodiversity strategy. WRI, IUCN and UNEP, Washington, DC/Gland

Wu P, Low SP (2010) Project management and green buildings: lessons from the rating systems. J Prof Issues Eng Educ Pract 136(2):64–70

Yates JK (2007) Global engineering and construction. Wiley, Hoboken

Yudelson J (2008) The green building revolution. Island Press, Washington, DC

Chapter 3
Modelling the Lean Production Philosophy

3.1 Introduction

As noted in Chap. 2, there is an increasing recognition on the importance of global climate change. The construction industry which consumes much energy and generates nearly 50 % of the greenhouse gases emissions is largely affected by this trend. Global efforts have been made to control carbon emissions, from design, materials to techniques. These efforts include increased energy efficiency, environment-friendly materials, passive design and other technical innovations.

Previous studies (Lee et al. 1999; Alwi et al. 2002) have shown that construction activities are not always efficient. There are still many non-value adding activities in the construction life cycle, from production, delivery to installation. The essence of the lean production philosophy is straightforward. It concentrates on simplifying production and organization system and eliminating waste by changing the attitude of the employees. It is only recently that the lean production principles are seen to be related to sustainable development. Following the aim of this study, which is to create a bridge between the lean principles and reduction in carbon emissions, an investigation into the history, principles and applicability of lean production principles is necessary.

This chapter investigates the origin, and principles of lean production, and its applicability in reducing carbon emissions. In addition, a detailed comparison between lean principles and just-in-time principles is presented in this chapter as justification because the lean and just-in-time principles have often been used interchangeably. The literature review presented in this chapter aims to answer the following questions:

1. What is lean production and lean construction?
2. What are the principles of lean production and lean construction?
3. What are the differences between the lean philosophy and just-in-time philosophy?
4. Why is lean production chosen as the research focus?

P. Wu and S.P. Low, *Lean and Cleaner Production: Applications in Prefabrication to Reduce Carbon Emissions*, DOI 10.1007/978-3-642-42062-7_3,

3.2 History – The Toyota Production System

It is believed that the lean principles have their origin in the Toyota Production System. According to Howell (1999), lean production was developed in Toyota by Engineer Ohno. The term "lean" was coined by the research team working on international automobile production to reflect both the waste reduction nature of the Toyota Production System and to contrast it with the craft and mass forms of production (Womack et al. 1990). One of the first studies, aimed at understanding and applying the lean principles in the construction industry, was carried out by Lauri Koskela in 1992 (Koskela 1992). By applying these principles to the construction industry, Koskela identified the basis of the "new production philosophy", which is also known in western countries as "lean production" (Isatto and Formoso 1998).

The Toyota Production System was developed by Toyota to deal with the problems arising from the oil crisis. The main purpose of the system was to achieve cost reduction. However, this cannot be achieved without fulfilling three sub-goals, which include (Monden 1993):

1. Quantity control, which enables the system to adapt to daily and monthly fluctuations in demand of quantity and variety.
2. Quality assurance, which assures that each process will supply only good units to subsequent processes.
3. Respect for humanity, which must be cultivated while the system utilizes human resources to attain its cost objectives.

According to Monden (1993), it is a special feature of the Toyota Production System that without fulfilling the three sub-goals, the primary goal, which is cost reduction, cannot be achieved and vice versa. By adopting the new production system, Toyota could still maintain a profit even in periods of slow growth, by cutting unnecessary costs generated from non-value adding activities, including maintaining inventory and work force.

The basis of the Toyota Production System is to achieve profit growth by reducing costs through completely eliminating waste such as excessive stocks or work force, which is supported by the following two pillars (Ohno 1988):

1. Just-in-time (JIT).
2. Autonomation ("Jidoka" in Japanese), or autonomous defects control in a loose interpretation.

In the automobile industry, JIT means that the rights parts needed in production reach the assembly line at the time they are needed and only in the amount needed in a flow process (Ohno 1988). However, this is not always the case. Products are made of thousands of parts, which means that any interruption in the flow process will affect the achievement of "just-in-time". Thus, quality control must coexist with the JIT operations (Monden 1993). Autonomation means a mechanism which can prevent defects, through autonomous checking of abnormalities in machines or in product lines (Ohno 1988).

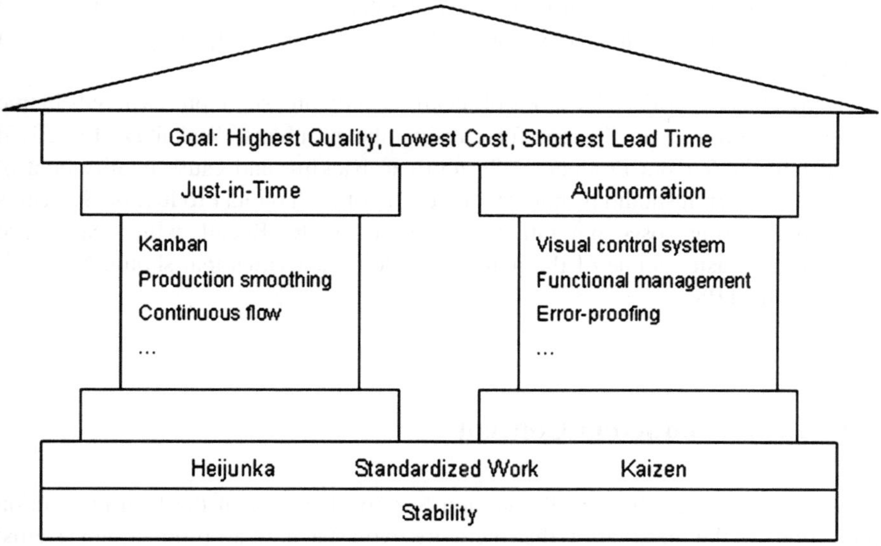

Fig. 3.1 House of the Toyota Production System (Adapted from: Ohno 1988)

Two other concepts are also key to the Toyota Production System. These include the flexible work force ("Shojinka" in Japanese), which means varying the number of workers to demand changes, and creative thinking or inventive ideas ("Soikufu" in Japanese), which means capitalizing on worker suggestions (Monden 1993). In order to achieve these four concepts mentioned above, namely the two pillars and two less important concepts, Toyota has established the following subsystems and methods for implementation (Monden 1993):

1. "Kanban system" to maintain JIT production.
2. "Production smoothing method" to adapt to demand changes.
3. "Shortening of the setup time" for reducing production lead time.
4. "Standardization of operations" to attain line balancing.
5. "Machine layout" and "multi-function workers" for the flexible work force concept.
6. "Improvement activities by small groups and suggestion system" to reduce the work force size and to increase worker morale.
7. "Visual control system" to achieve the Autonomation concept.
8. "Functional management system" to promote company-wide quality control.

It is acknowledged that all the subsystems may seem overwhelming and difficult to implement at first sight. The House of the Toyota Production System is a simple explanation of the Toyota Production System, including all the subsystems. This is illustrated in Fig. 3.1. The house is supported by two pillars: JIT and Autonomation, which are then supported by several fundamentals, including production smoothing ("Heijunka" in Japanese), standardized work, continuous improvement ("Kaizen" in Japanese) as well as stability. When working with little inventory, which can be achieved by applying the JIT concept, defects, interruptions or problems that stop production and cause instability to the process can be eliminated. This explains the importance of stability to achieve the ultimate goals: Highest Quality, Lowest Cost and Shortest Lead Time.

Toyota has experienced a series of recall recently and some professionals blame the Toyota Production System for causing the problems. Wakabayashi (2010) stated that the Toyota Production System can backfire and cause quality control issues. However, it should be noted that the core of the Toyota Production System is not about cutting costs, but rather eliminating waste. Recall, which can cause tremendous waste, is one of the non-value adding activities that should be eliminated by the TPS.

3.3 Lean Production Concept

Originating from the Toyota Production System, the core of the lean production philosophy is the observation that there are two aspects in all production systems: conversions and flows (Koskela 1992). Conversion activities refer to those which actually add value to the product or process. Flow activities refer to non-value adding activities, which consume time, costs and resources but do not add value to the product or process. Traditional management improvement, through technology innovation, efficiency improvement and so on, is not fully aware of the existence of these non-value adding activities, which lead to efficiency being lost during production processes. For example, when technology innovations are applied to the process, these are applied to both the value adding activities and the non-value adding activities.

However, only the value adding activities are essential to the production process. It is a waste of time and resource to apply improvements on flow activities, which do not add value. Thus, the lean production philosophy aims to create an environment where conversion and flow activities are separated and treated differently. Conversion activities are improved while flow activities are eliminated at the same time.

Lean production is not a totally new concept. It is a combination of existing management philosophies in a new way. Among the methodologies for attaining lean production, the most important ones are (Koskela 1992):

1. **Just-in-Time (JIT)**: The JIT concept was firstly initiated by Ohno and Shingo in Toyota to reduce and eliminate inventories. In order to achieve less inventories, other techniques were added to the concept, including continuous improvement, uninterrupted work flow, top management commitment and so on. Other than the inventories which are normally regarded as waste, the non-value adding activities in the production process include: overproduction, moving time, inspection time, waiting time, idling equipment, inexperienced employees, etc.
2. **Total quality control (TQC):** The term "total" refers to three extensions (Shingo 1988): expanding quality control from production to all departments;

expanding quality control from workers to management; and expanding the notion of quality to cover all operations in the company. By ensuring that every part of the production process is in good quality, poor quality materials will not occur, thus eliminating both time and money to fix any problem.
3. **Time based competition**. Time based competition refers to compressing time throughout the organization for competitive benefit (Koskela 1992). This concept is also mentioned in the JIT philosophy as simplifying the work process, reducing the process set-up time and decreasing inventory levels.
4. **Concurrent engineering**. Concurrent engineering refers to an improved design process characterized by rigorous upfront requirements analysis, incorporating the constraints of subsequent phases into the conceptual phase, and tightening of change control towards the end of the design process (Koskela 1992).
5. **Value based management**. Value based management refers to "conceptualized and clearly articulated value as the basis for competing" (Carothers and Adams 1991). The concept is also mentioned in JIT as the pull system, which focuses on customer demand.
6. **Visual management**. Visual management is concerned about visual control in production, quality and workplace organization (Greif 1991, p. 281). By setting up a standard, any deviation from it will be recognized immediately. Thus, appropriate actions will be taken before damage happens.
7. **Total productive maintenance (TPM)**. Total productive maintenance refers to the autonomous maintenance of production machinery by small groups of multi-skilled operators (Nakajima 1988, p. 166).
8. **Employee involvement**. The importance of employee involvement can be explained by the fact that organizational goals and personal goals can both be achieved if employees are treated with equity and respect in terms of being involved with decision making, being provided with meaningful jobs and being given the opportunity to learn (Stendel and Desruelle 1992).

Originating from the Toyota Production System, the lean production philosophy was developed as a new way of thinking which advocates reducing or eliminating non-value adding activities as well as improving the efficiency of value adding ones at the same time. The lean philosophy can be considered as a new way to design and make things that are differentiated from mass and craft forms of production through the objectives and techniques applied on the shop floor, in design and along supply chains (Howell 1999). Eleven important principles are essential to the lean philosophy, including (Koskela 1992):

1. Reduce the share of non-value adding activities (also called waste);
2. Increase output value through systematic consideration of customer requirements;
3. Reduce variability;
4. Reduce cycle time;
5. Simplify by minimizing the number of steps, parts and linkages;
6. Increase output flexibility;
7. Increase process transparency;
8. Focus control on the complete process;

9. Building continuous improvement into the process;
10. Balance flow improvement with conversion improvement; and
11. Benchmark.

Although the principles may seem overwhelming at first sight, these can be reorganized at two levels to achieve better understanding. First of all, lean production aims to use minimum amount of resources, including materials, labors and time, to achieve defect-free products with no inventory. Secondly, any activities or processes which do not add value to help achieve customer requirements are regarded as non-value adding activities and should be eliminated. Thus, there is another way to interpret lean production principles:

1. Reduce non-value adding activities, such as maintaining inventory, multi-handling and multi-delivery, reproducing processes due to product defects, etc. The interpretation of non-value adding activities is based on the processes that are being carried out and the stages where the construction projects are.
2. Produce according to customer requirements to prevent deviation costs.
3. Continually improve rather than regard the improvement processes as implementations that can be done once and for all.
4. Respect for humanity, which must be cultivated while the production system utilizes human resources to attain its cost objectives (Monden 1993).

According to Howell and Ballard (1998), in order to achieve the three objectives identified above, there are five techniques that can be applied to support the goal:

1. Stopping the line.
2. Pulling production forward.
3. One-piece flow.
4. Synchronize and align.
5. Transparency.

Traditional construction activities are controlled by planning that determines the sequence and time of the activities. Cost reduction can be achieved by accelerating activities. However, under lean thinking, waste also includes costs that could have been avoided, if appropriate planning is carried out at the beginning. Cost reduction is not only achieved by accelerating activities, but also meeting customer requirements in zero time with nothing in stores (Howell and Ballard 1998). The tenets of lean thinking and lean production are drawn by Womack and Jones (1996). While lean production includes five principles discussed above, lean thinking is supported by:

1. Specify value by product. According to Howell and Ballard (1998), specifying value by product shapes all actions around customer requirements. Although this principle is currently applied in the construction industry, the application seems to lack the "pursuit of perfection" goal. As will be discussed in the next section, one core area in the Toyota Production System is attacking fundamental problems, simply specifying value by product before design is not lean thinking.

2. Identify the value stream. This tenet focuses on the time and method that the values established in tenet one can be realized. In fact, a value stream map is usually adopted which help project managers to identify the hidden issues that will hinder the flow of the activities.
3. Make production flow. Making production flow means that the production process should not be interrupted. In fact, if the tenet is applied to the construction industry, delivery and construction processes should not be interrupted as well. The products that have been produced in factories should be in constant motion without stopping (Womack and Jones 1996). However, according to Ballard and Howell (1998), current construction planning systems are unable to produce a reliable flow of work, therefore leading to buffers or wasted capacity.
4. At the pull of the customer. This concept is closely related to the "pull" system, which is derived from the Toyota Production System. The term "customer" used in this tenet can be extended to a wider background. It is usually referred to the ultimate users of the product. In a wider background, each work station can be identified as a "customer" and its requirement can flow back to previous work stations to regulate the activities.
5. While pursuing perfection, custom product, zero time delivery and nothing in stores. This tenet has been discussed with 1–4 and will not be discusses separately.

It should be noted that although lean production is a new thinking to produce based on orders and eliminate non-value adding activities, partial implementation of lean thinking, particularly organizational or relational aspects, have been developed and used on complex, uncertain and quick projects (Howell and Ballard 1998). According to Howell and Ballard (1998), these projects are likely to fail when only traditional approaches are used while applying lean thinking. Although these tenets or principles are likely to be helpful, they do not fix the underlying problem alone. This study therefore focuses on implementing the overall lean thinking in the precast concrete industry to achieve low carbon emissions.

3.4 Linking Lean Production with the JIT Concept

Conventional management methodology focuses on continuous improvement by implementing new technology to all activities. However, the lean philosophy observes two different aspects in all production activities: value-adding and non-value adding activities, namely conversions and flows. While all activities consume time and resources, only conversions will add value to the products. Thus, the core idea of the lean philosophy is to eliminate flow activities and make conversion activities more efficient at the same time (Koskela 1992).

The JIT management philosophy focuses on creating the right environment for effective operations, from both the strategic and tactical level (Low and Chan 1997). In the strategic level, it aims to identify the major and fundamental issues

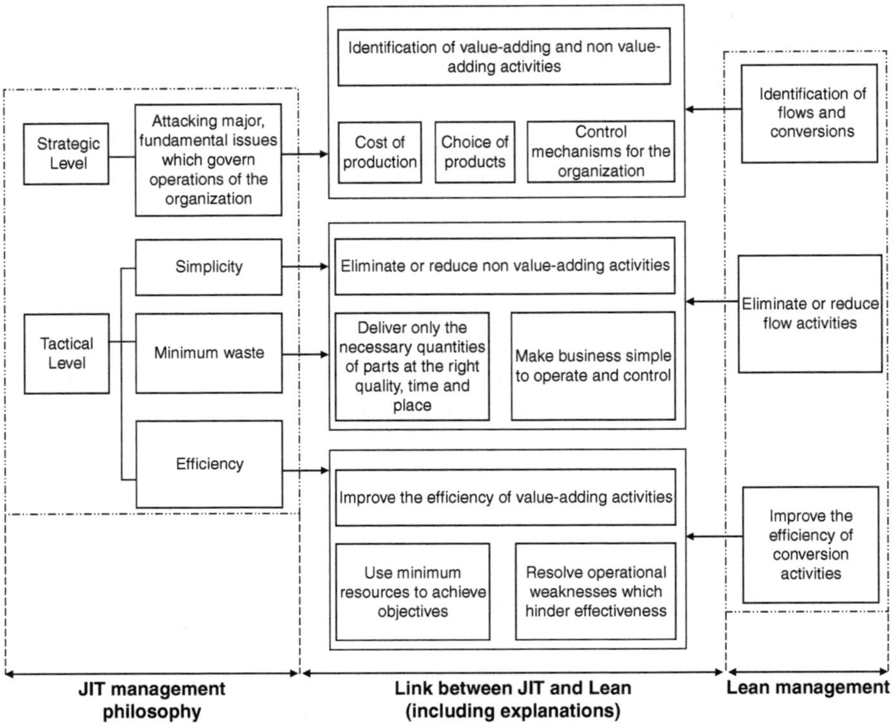

Fig. 3.2 Comparison of lean and JIT management philosophy

that will influence operations of the organization. At the tactical level, JIT is concerned with less important actions and decisions which can improve operations of the organization from three levels: simplicity, efficiency and minimum waste. Conventional management methodology focuses more on the tactical level. However, the JIT philosophy covers both the strategic and tactical areas to bring about more effective management.

According to Tommelein and Li (1999), the term "JIT" is used to describe such a delivery that the materials that are brought to their location are installed immediately upon arrival without any delay due to storage. JIT originated from the Toyota Production System and it was later translated into the west as lean production system. In the JIT system, overproduction and delayed delivery can be eliminated by "Kanban". Therefore, when there is a shortage from the workstation, information will flow back to previous workstation to order the materials or products in the right amount.

As general management philosophies, it seems that Lean and JIT have much in common. Identification of non-value adding and value adding activities is at a more strategic level of thinking. Non value-adding activities, which are represented by flow and waste in Lean and JIT correspondingly, are expected to be reduced or eliminated. At the same time, improving the efficiency of value-adding activities is essential to both Lean and JIT to improve overall productivity. The relationship between the Lean and JIT philosophies is summarized in Fig. 3.2.

The similarities between these two management methodologies come from their same origin – the Toyota Production System. Although the JIT philosophy is considered to have been derived directly from the system, the term "lean" was coined by the research team working on international automobile production to reflect both the waste reduction nature of the Toyota Production System, and to contrast it with the craft and mass forms of production (Womack et al. 1990; Howell 1999). However, it would seem inappropriate to say that the lean and JIT philosophies are the same based on the same origin and the concepts that have many similarities. To achieve the objectives identified in the concept, both philosophies have principles, tools or techniques for implementation. These principles are the core areas of both philosophies. Thus, it is necessary to compare them to evaluate if Lean and JIT are the same, as these principles are fundamentals for implementations.

3.4.1 Principles

As stated in Sect. 3.3, there are eleven important principles are essential to the lean philosophy. Although these principles originated from the same concept which aims at reducing non value-adding activities, they are not on the same abstraction level (Koskela 1992). By examining these eleven principles more closely, their relationships with JIT principles can be identified.

3.4.1.1 Reduce the Share of Non-value Adding Activities

Reducing the share of non-value adding activities is the fundamental principle in lean (Koskela 1992). However, in order to identify these non-value adding activities, an in-depth investigation of the organization and production processes is needed, which is represented by attacking fundamental problems in JIT principles. At the same time, waste is defined as anything other than the absolute minimum resources of materials, machines and personnel required to add value to the product (Hay 1988). In other words, the meaning of reducing the share of non-value adding activities is represented by elimination of waste in JIT principles.

3.4.1.2 Increase Output Value Through Systematic Consideration of Customer Requirements

Consideration of customer requirements can be regarded as an application of the "Kanban" or "Pull system", which originated from the Toyota Production System. Most manufacturing organizations operate in a push system, which means that they plan what is going to be produced which is then pushed through the factory (Low and Chan 1997). In a broad sense, each work station in a production process can be identified as a customer. In a system where customer requirements can be systematically considered, these requirements flow backwards through the process whilst products or materials flow in the opposite direction.

3.4.1.3 Reduce Variability

The reasons why variability should be reduced can be explained at two levels. First, any deviation from a target value in the production causes a loss, which is a quadratic function of the deviation, to the user and wider society (Bendell et al. 1989). In addition, variability, especially in the duration of activities, increases the volume of non-value adding activities (Koskela 1992). Thus, reducing variability is merely a conformance to the two principles mentioned above, which are referred to as the pull system and elimination of waste in JIT principles.

3.4.1.4 Reduce the Cycle Time

In a production flow, the cycle time can be represented by the following equation (Koskela 1992):

$$\text{Cycle time} = \text{Processing time} + \text{inspection time} + \text{wait time} + \text{move time}$$

However, except processing which does add value to the production flow, inspection, move and wait time are considered as waste, or non-value adding activities. At the same time, progressively reducing the cycle time by eliminating non value-adding activities indicates another important issue: continuous improvement. Reducing the cycle time cannot be done once and for all. It is the long term improvement rather than the short term benefit that continuous improvement aims at, which can be illustrated in Fig. 3.3.

3.4.1.5 Simplify by Minimizing the Number of Steps and Parts/Increase Output Flexibility/Increase Process Transparency

These three principles are in fact closely related to create the environment where just-in-time production can be achieved with minimum interruption to the work flows. The concept is referred to as uninterrupted work flow in JIT principles. These are tools or techniques to achieve uninterrupted work flow rather than principles.

The costs of production increase with complexity, with other things being equal. Complex systems are less reliable than simple ones due to more complicated working procedures, prolonged information flow, etc. Simplification is essential to ensure that the production process is not interrupted at all.

The essence of increasing output flexibility can be illustrated through the use of a U-shaped layout in a production process. The U-shaped layout, as illustrated in Fig. 3.4, which assists communications, since workers on a particular flow line are

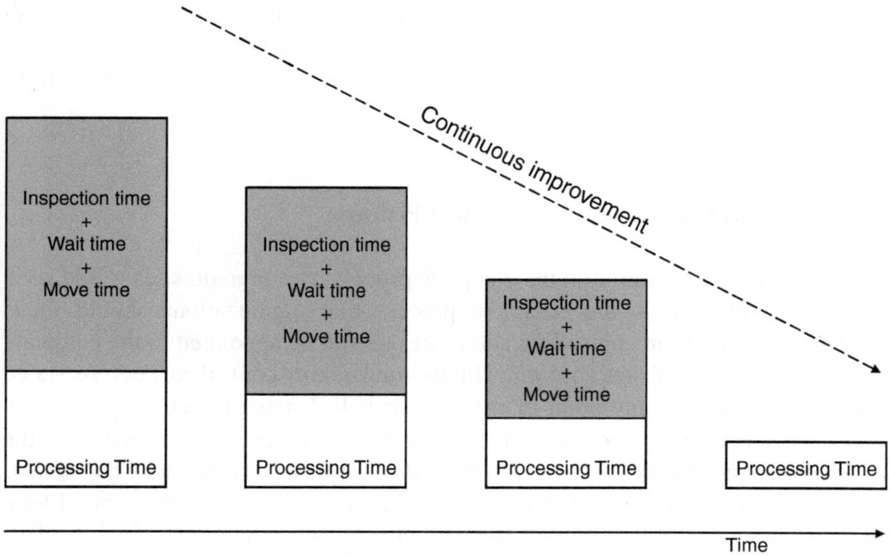

Fig. 3.3 Continuous improvement which progressively helps to eliminate non-value adding activities and improve the efficiency of value adding activities (Adapted from: Berliner and Brimson 1988, p. 253)

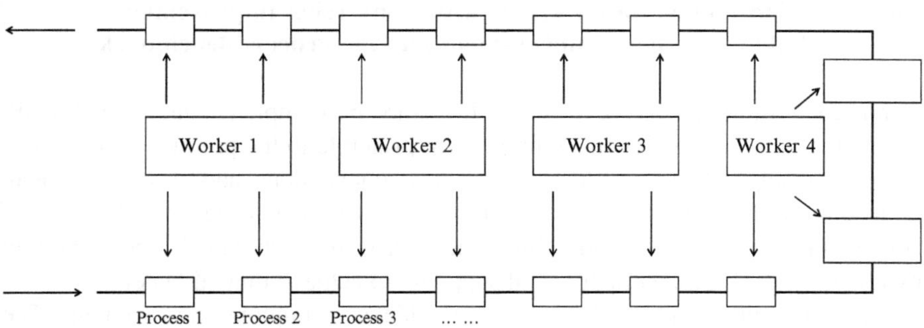

Fig. 3.4 A U-shaped line with multiple-function workers (Adapted from: Nakade and Ohno 1997)

physically closer to each other, enables workers to have access to a number of machines, thus increasing flexibility (Low and Chan 1997). This layout, of course, requires that the workers be trained as multi-skilled employees. The effectiveness of this layout is also related to the effort that workers put into the training, which leads to another important JIT aspect: employee involvement, meaning that organizational goals and personal goals can both be achieved if employees are treated with equity and respect in terms of being involved with decision making, being provided with meaningful jobs and being given the opportunity to learn (Stendel and Desruelle 1992).

Lack of process transparency can increase the propensity to err, reduce the visibility of errors and diminish motivation for improvement, thus exposing the workflow to interruptions (Koskela 1992). From the objective of process transparency, which is to make the main flow of operations from start to finish visible and comprehensible to all employees (Stalk and Hout 1990), it clearly addresses two embedded principles: uninterrupted work flow which lasts from start to end and employee involvement.

3.4.1.6 Focus Control on the Complete Process

To achieve focused control on the complete process, two prerequisites would need to be met. First, it is the complete process that organizations should focus on. Secondly, different controlling authorities should be appointed to the complete process. Both prerequisites conform with the total quality control concept. As stated above, disruptions in any point of the process will damage the entire production chain and it is therefore necessary for the workers, foremen, engineers and all the firm's employees to take primary responsibility for quality (Low and Chan 1997).

On the other hand, the complete process requires that organizations build long term collaborative relationships with suppliers and subcontractors, which is identified in JIT principles as supplier and client relations. The quality of the products can be guaranteed only with first-rate materials that are used to build the products.

3.4.1.7 Balance Flow Improvement with Conversion Improvement/ Build Continuous Improvement into the Process/Benchmark

Continuous improvement never ends. It cannot be completed once and for all. When applying the continuous improvement principle to the production system, it is actually applied at two different aspects: conversions and flows. The lean philosophy aims to reduce the flow activities while increasing the efficiency of conversion activities at the same time. The improvement should also be advocated by all employees, subcontractors and suppliers to achieve mutual benefit.

Benchmarking is a useful stimulus to achieve continuous improvement. The basic steps include investigating the process to identify the weaknesses and strengths, finding, understanding and comparing the best practices adopted by industry leaders and competitors and incorporating the best practices (Camp 1989).

It should be noted that most of the principles are intertwined with one another. Sometimes, it is difficult to analyze them separately. However, the comparison here is able to identify the primary and secondary JIT principles which are closely related to the Lean principles being analyzed. The links are shown in Table 3.1.

The elements of the JIT system may be slightly different in different interpretations. For example, in Voss and Robinson (1987), the elements are organized into 14 principles, which are:

1. Flow/layout. The factory layout should be arranged in such a way that the process is not interrupted. The value adding activities are therefore maximized and the non-value adding activities are eliminated.

Table 3.1 Major links between lean and JIT principles

	Lean principles	Primary link with JIT principles	Secondary link with JIT (if any)
1	Reduce the share of non value-adding activities	Elimination of waste	Attacking fundamental problems
2	Increase output value through systematic consideration of customer requirements	The "Kanban" or "Pull" system	
3	Reduce variability	Elimination of waste The "Kanban" or "Pull" system	
4	Reduce the cycle time	Elimination of waste	Continuous improvement
5	Simplify by minimizing the number of steps and parts	Uninterrupted work flow	Employee involvement
6	Increase output flexibility		
7	Increase process transparency		
8	Focus control on the complete process	Total quality control	Suppliers relation
9	Build continuous improvement into the process	Continuous improvement	Top management commitment and employee involvement
10	Balance flow improvement with conversion improvement		
11	Benchmark		

2. Smoothed line build rate. The JIT system advocates a smoothed line build rate, which is in consistent with principle 1.
3. Mixed modelling. Mixed modeling is related to the pull system, which is designed to match the production rate order demand as closely as possible. In other words, the production is driven by orders.
4. Set-up time reduction. Set-up time reduction is one technique that can be used to achieve uninterrupted workflow.
5. Work-in-progress reduction. In the JIT system, work-in-progress reduction can be achieved by minimizing inventory level.
6. Kanban. As has been discussed, Kanban is the "pull" system by organizing process in such a way that information flows backwards to regulate activities.
7. Quality. Materials with high standard are one prerequisite of the JIT system to avoid double delivery and double handling.
8. Product simplification. In the JIT concept, production simplification can help to achieve uninterrupted workflow by simplifying the work process.
9. Standardized containers. The JIT system advocates using small standardized containers to deliver products.
10. Preventive maintenance. Preventive maintenance helps to achieve uninterrupted workflow by taking different uncertainties into consideration.
11. Flexible workforce. Flexible workforce has been discussed in Sect. 3.4.1.5. For example, a U-shaped line is often adopted to achieve flexible workforce.

12. Organization in modules or cells. According to Voss and Robinson (1987), Many JIT factories are organized in small autonomous modules or cells. Each of the modules or sells is responsible for its own production.
13. Continuous improvement. As stated in Sect. 3.4.1.7, achieving a JIT environment is not a one-off effort, long-term commitment is necessary.
14. JIT purchasing. Quantity, quality and delivery schedule should be well defined before deliveries are actually made.

Although JIT principles can be interpreted by different ways, they all originated from the same origin – the Toyota Production System. It is the abstraction level that these interpretations differ. For example, flexible workforce, preventive maintenance and product simplification are the tools that can be taken to achieve uninterrupted workflow. These principles are all discussed under the heading of "uninterrupted workflow" rather than being discussed separately.

3.4.2 Similarities and Differences

The basic idea for applying the lean concept in the construction industry is very simple: to keep the production system and organization simple and avoid waste (Alarcón 1997). In fact, the most important instrument in lean construction is believed to be Kaizen (改善) or continuous improvement, which originated from the Toyota Production System. With Kaizen in mind, the construction industry can make the production process transparent with less waste and more efficient communication structure.

While the lean management methodology has developed into a philosophy with eleven principles mentioned above, the applications of the lean concept in the construction industry are mostly focused on implementations these principles separately. For example, when implementing multi-functional task groups on site, it was found that homes or houses should not be finished via a construction project based at a construction site and using specialized work crews (such as kitchen installers, electricians, plumbers) but rather this should be from a central yard using all round work crews (Melles and Wamelink 1993). This idea was also adopted by O'Grady (1988) to implement the JIT philosophy through the U-shaped flow line which can simplify control, allowing for the gradual reduction of inventory and work-in progress levels (Low and Chan 1997).

As can be seen from the House of the Toyota Production System, the applications of lean or JIT in the construction industry focus on Kaizen or continuous improvement of existing work processes to achieve the ultimate goal. It is the attitude rather than a series of techniques or tools that Lean or JIT brings about. If an organization would like to make the total production process transparent, to simplify the communication structure and to reduce the inventories in the production processes, the organization would need to think about the improvements in construction activities (Alarcón 1997).

While lean and JIT appear to overlap with one another, there are a few differences that should be noted, which include the following:

1. Fundamentally speaking, it seems that the lean principles do not really invoke new management techniques. The lean management philosophy seems to only combine existing principles in a new way (Alarcón 1997). Taking total quality management (TQM) for example, the concept of TQM combines quality assurance with quality improvement, while the latter is usually referred to as Kaizen in Japan.
2. Both methodologies embrace the ideas generated from the Toyota Production System, but they are not on the same abstraction level. Some JIT principles can be detailedly interpreted by a combination of several lean principles.
3. While most lean principles overlap with each other, they can be implemented as stand-alone applications. For example, it is possible to implement just-in-time deliveries in a construction site without using simultaneous engineering or multifunctional task groups (Alarcón 1997).

   However, JIT principles, on the other hand, seem to have tighter bonds with one another. For example, elimination of waste cannot be achieved without coordination with other principles. "The pull system" is necessary to reduce inventory, while "uninterrupted work flow" ensures that there are no waste of waiting time and waste of motion. In addition, "the total quality control concept" guarantees that there are no waste from product defects, while "employee involvement" makes sure that no waste due to human resources happen on site. Furthermore, "supplier relations" lead to less time spent on communications with suppliers and subcontractors as well as leading to higher quality standard materials. A JIT environment can therefore be created, as long as all the aspects above are considered on a "continuous improvement" basis. This difference between lean and JIT seems to show that lean is more application-oriented while JIT is more principle-based. By generalizing all the subsystems in the Toyota Production System, JIT has developed into a whole management methodology where these subsystems or principles cannot be separated. On the other hand, lean principles review and polish the subsystems into detailed applications.
4. The Toyota Production System was developed by Toyota within the context of the Japanese culture. However, western countries are culturally different with Japan from societal structure as well as economic structure, not to mention the different industrial background when applying the system. For example, in Japan, everybody is proud of and loyal to his company and his part of the production process, which forms the emotional links with the company, while in western countries, the attitude to one's own company is less emotional (Alarcón 1997). Hence, in a sense, the lean philosophy is a revision and reorganization of the JIT principles which originated from the Toyota Production System.

Liker (2004) stated that "lean production" is actually invented by Toyota. In fact, dozens of publications on lean are based on the Toyota Production System. JIT and lean production are all used as abbreviations for the Toyota Production System. It's the abstraction level that they differ.

3.5 Linking Lean Production with Green

The lean concept has proven to be effective in increasing environmental benefits by eliminating waste, preventing pollution and maximizing value to owners (Huovila and Koskela 1998; King and Lenox 2001; Ferng and Price 2005; Lapinski et al. 2006; Nahmens 2009; Miller et al. 2010). Huovila and Koskela (1998) identified the contribution of the lean concept to sustainable development, which included:

- Waste activities were identified. Elimination of the waste activities directly contributed to the sustainable development.
- The customer concept should be re-examined. Owners, end users, contractors and the environmental were all customers of the production and construction activities, as shown in Fig. 3.5.

EPA (2003) found that lean produced an operational and cultural environment that is highly conducive to waste minimization and pollution prevention, and that lean provides an excellent platform for environmental management tools such as life cycle assessment and design for environment. Luo et al. (2005) applied the lean concept to prefabrication and stated that lean could contribute to improve quality

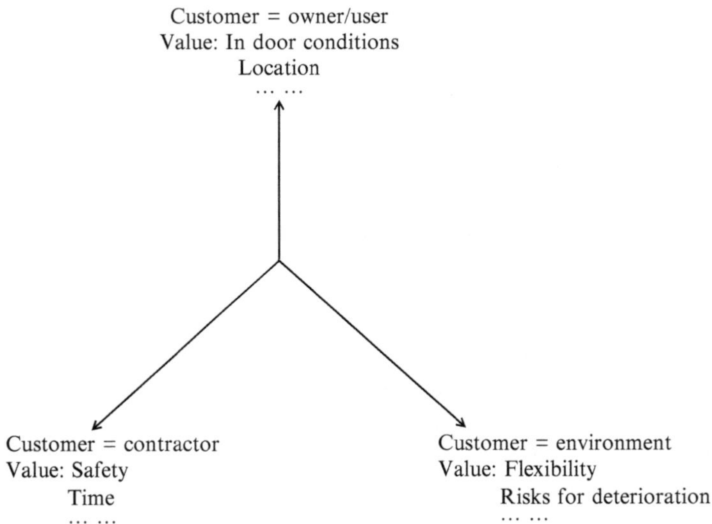

Fig. 3.5 Classified requirements for a facility and their possible priorities for different customers (Adapted from: Huovila and Koskela 1998)

and supply chain and reduce waste. Bae and Kim (2007) found that different lean applications might have different results on the three pillars of sustainable development (i.e. economic, social and environmental sustainability). For example, lean supply might have influence on economic and environmental impacts rather than social impacts. Nahmens (2009) stated that it is a natural extension to apply the lean concept to achieve green production and construction. By applying the lean concept to a production line, 9 to 6.5 people (labor waste), 12 % space (equipment waste) and 10 % wallboard (material waste) can be reduced (Nahmens 2009). Miller et al. (2010) applied the lean principles to a small furniture production company and found that the lean principles can help the company meet every increasing customer demands while preserving valuable resources. In these studies, wastes, environmental burdens, and environmental deterioration were commonly used as the contributions that can be achieved by applying the lean concept. In these studies, wastes, environmental burdens, and environmental deterioration were commonly used as the contributions that can be achieved by applying the lean concept.

However, there were a few studies which argued that lean may show a negative impact on environmental performance (Cusumano 1994; Rothenberg et al. 2001). Helper et al. (1997) also showed that there was no appreciable relationship between Lean and green in the manufacturing industry.

It should be noted that the evaluation of the lean concept in achieving environmental sustainability could only be examined when the environmental sustainability was set at the target at the very start. The contribution of the lean concept to green could not be fully assessed when reducing initial costs and eliminating waste were set at the targets.

Although the lean concept and principles were believed to have many benefits for environmental sustainability, the impacts should be quantified. According to Bae and Kim (2007), the difficulty in understanding and quantifying the impacts of the lean principles was perhaps on reason that stakeholders hesitated to use the lean principles to be green. The quantification of the lean principles in achieving environmental benefits, such as low carbon, low emissions, low waste, should be conducted.

3.6 Summary

Originated from the Toyota Production System, the lean production philosophy is a management philosophy which focuses on eliminating non-value adding activities and improving the efficiency of value adding activities at the same time, when delivering quality products and services. Lean construction is the application of the lean production philosophy in the construction industry. It seems that the JIT philosophy, which also originated from the Toyota Production System, has many similarities with the lean production system, but not at the same abstraction level. Over the years, JIT has developed into a management philosophy, rather than a sub-system that was adopted in the Toyota Production System. The JIT concept focuses on investigating more deeply into the system to attack fundamental problems. Some of the JIT principles may incorporate several applications. For example, uninterrupted work flow can be achieved by grouping related activities, simplifying the production process, reducing the set-up times of process and implementing total preventive maintenance. However, on the other hand, the lean principles are defined to address specific problems and are more application-oriented.

The decision to adopt the lean principles in this study is based on the research aim. As a management tool-kit will be provided at the end of the research, the lean principles fit in more closely with the research objectives. More importantly, the concept of lean has been widely adopted among professionals whose backgrounds are grounded in the application of the Toyota Production System. It would therefore ease the communication process when the findings of this current study are shared with these professionals.

References

Alarcón L (1997) Lean construction. Balkema, Rotterdam

Alwi S, Hampson K, Mohamed S (2002) Waste in the Indonesian construction project. In: Proceedings of the 1st international conference of CIB W107 – creating a sustainable construction industry in developing countries, Spier, South Africa

Bae JW, Kim YW (2007) Sustainable value on construction project and application of lean construction methods. In: Pasquire C, Tzortzopoulos P (eds) Proceedings of IGLC-15, East Lansing, MI

Ballard G, Howell G (1998) Shielding production: an essential step in production control. ASCE J Constr Eng Manag 124(1):18–24

Bendell A, Disney J, Pridmore WA (1989) Taguchi methods: applications in world industry. IFS Publications/Springer, Bedford

Berliner C, Brimson JA (1988) Cost management for today's advanced manufacturing. Harvard Business School Press, Boston

Camp RC (1989) Benchmarking: the search for industry best practices that lead to superior performance. ASQC Quality Press, Milwaukee

Carothers GHJ, Adams M (1991) Competitive advantage through customer value: the role of value-based strategies. In: Stahl MJ, Bounds GM (eds) Competing globally through customer value. Ouorum Books, New York, pp 32–66

Cusumano MA (1994) The limits of lean. Sloan Manag Rev 35(4):27–32

EPA (2003) Lean manufacturing and environment. [Online] Available at: http://www.epa.gov/lean/performance/index.htm. Accessed 14 Nov 2010

Ferng J, Price ADF (2005) An exploration of the synergies between Six Sigma, total quality management, lean construction and sustainable construction. Int J Six Sigma Compet Advant 1 (2):167–187

Greif M (1991) The visual factory. Productivity Press, Cambridge

Hay EJ (1988) The just-in-time breakthrough – implementing the new manufacturing basic. Wiley, New York

Helper S, Clifford PG, Rozwadowski H (1997) Can green be lean. A paper submitted to the Academy of Management annual meeting, Organizations and the Natural Environment

Howell G (1999) What is lean construction. In: Tommelein ID (ed) Proceedings of IGLC-7, U.C. Berkeley, CA
Howell G, Ballard G (1998) Implementing lean construction: understanding and action. In: Formoso CT (ed) Proceedings of IGLC-6, Guaruja, Brazil
Huovila P, Koskela L (1998) Contribution of the principles of lean construction to meet the challenges of sustainable development. In: Formoso CT (ed) Proceedings of IGLC-6, Guaruja, Brazil
Isatto EL, Formoso CT (1998) Design and production interface in lean production: a performance improvement criteria proposition. In: Formoso CT (ed) Proceedings of IGLC-6, Guaruja, Brazil
King AA, Lenox MJ (2001) Lean and green? An empirical examination of the relationship between lean production and environmental performance. Prod Oper Manag 10(3):244–257
Koskela L (1992) Application of the new production philosophy to construction. CIFE technical report#72, Stanford University, Stanford
Lapinski AR, Horman MJ, Riley DR (2006) Lean process for sustainable project delivery. J Constr Eng Manag 132(10):1083–1091
Lee SH, Diekmann JE, Songer AD, Brown H (1999) Identifying waste application of construction process analysis. In: Tommelein ID (ed) Proceedings of IGLC-7, U.C. Berkeley, CA
Liker JK (2004) The Toyota way: 14 management principles from the world's greatest manufacturer. McGraw-Hill, New York
Low SP, Chan YM (1997) Managing productivity in construction: JIT operations and measurements. Ashgate Publishing, Brookfield
Luo Y, Riley D, Horman MJ (2005). Lean principles for prefabrication in Green Design-Build (GDB) projects. In: Russell K (ed) Proceedings of IGLC-13, Sydney
Melles B, Wamelink JWF (1993) Production control in construction: different approaches to control, use of information and automated data processing. Delft University Press, Delft
Miller G, Pawloski J, Standridge C (2010) A case study of lean, sustainable manufacturing. J Ind Eng Manag 3(1):11–32
Monden Y (1993) Toyota production system: an integrated approach to just-in-time. Industrial Engineering and Management Press, Norcross
Nahmens I (2009) From lean to green construction: a natural extension. In: Ariaratnam ST, Rojas EM (eds) Proceedings of the 2009 construction research congress, Seattle, pp 1058–1067
Nakade K, Ohno K (1997) Stochastic analysis of a U-shaped production line with multiple workers. Comput Ind Eng 33(3–4):809–812
Nakajima S (1988) Introduction to TPM. Productivity Press, Cambridge, MA
O'Grady PJ (1988) Putting the just-in-time philosophy into practice: a strategy for production managers. Kogan Page Ltd., London
Ohno T (1988) Toyota production system: beyond large-scale production. Diamond, Inc., Tokyo
Rothenberg S, Pil FK, Maxwell J (2001) Lean, green and the quest for superior performance. J Prod Oper Manag 10(3):228–243
Shingo S (1988) Non-stock production. Productivity Press, Cambridge, MA
Stalk GJ, Hout TM (1990) Competing against time. Free Press, New York
Stendel HJ, Desruelle P (1992) Manufacturing in the nineties: how to become a mean, lean, world-class competitor. Van Nostrand Reinhold, New York
Tommelein ID, Li EY (1999) Just-in-time concrete delivery mapping alternatives for vertical supply chain integration. In: Tommelein ID (ed) Proceedings of IGLC-7, Berkeley, CA
Voss CA, Robinson SJ (1987) Application of just-in-time manufacturing techniques in the United Kingdom. Int J Oper Prod Manag 7(4):46–52
Wakabayashi D (2010) How lean manufacturing can backfire. Available from: http://online.wsj.com/news/articles/SB10001424052748704343104575032910217257240. Cited 29 Nov 2013
Womack JP, Jones DT (1996) Lean thinking. Simon & Schuster, New York
Womack JP, Jones DT, Roos D (1990) The machine that changed the world: the story of lean production, 1st edn. Harper Perennial, New York

Chapter 4
The Precast Concrete Industry

4.1 Introduction

Precast concrete is widely adopted in the construction industry. In the UK, precast concrete represents about 25 % of the market for cementitious products, including a wide range of products such as blocks, pavings, suspended floors, precast concretes, cast stone and architectural claddings (Glass 2000). The precast concrete industry is especially important for the construction industry in Singapore due to its large demand from public housing projects. The Housing and Development Board (HDB) was formed in 1960 to bring clean and decent homes in public housing estates to the population as quickly as possible, which constitute one of the main engines of Singapore's economic and social stability (Tan 1999).

One of the solutions to achieve speedy construction would be precast concrete components which are able to provide a cost-effective way of carrying out "system building" types of construction projects. According to BCA (2011), construction time for the HDB projects in the industrialization programme was reduced from an average of 18 months for conventional designs to 8–14 months for projects of similar size.

In addition, there are similarities between the precast concrete industry and the manufacturing sector, where the lean principles are firstly applied. It means that lean principles can be applied with minimum modifications to the former, and thereby to test the applicability in reducing carbon emissions.

This chapter aims to investigate the production process of precast concrete components. In addition, the applicability of lean principles to reduce carbon emissions in the precast concrete industry is examined from both the precasters' and contractors' point of view. The literature review in this chapter aims to answer the following questions:

1. What are the processes for producing typical precast concrete components?
2. What are the benefits to adopt precast concrete components in construction projects?
3. How will the lean principles help to reduce carbon emissions in the precast concrete industry?

P. Wu and S.P. Low, *Lean and Cleaner Production: Applications in Prefabrication to Reduce Carbon Emissions*, DOI 10.1007/978-3-642-42062-7_4,

4.2 Production Considerations

The precast concrete industry, associated with prefabrication, is a supporting industry, which usually involves the design and manufacture of units or modules, and their installation to form the permanent works at the work site (Gibb 1999). This section aims to investigate the production processes of precast concrete components. In addition, the handling equipment are highlighted with a view to reduce carbon emissions.

4.2.1 Production Processes

According to Richardson (1973), the production processes in the precast concrete industry can be divided into three categories. These are:

1. **Job or jobbing production.** The job or jobbing production refers to the production process that: (1) units are cast daily from specific moulds and alterations are carried out between casting operations; and (2) a small number of units are required. Job production methods are adopted in factories on a project basis. As the products required by projects are different, the production centre will re-tool itself accordingly. Hence, there is little likelihood of repetition work.
2. **Batch production.** When manufacturing large quantities of products, batch production methods are usually adopted. In batch production, groups of components are manufactured at a rate in relation to, but sometimes in excess of, site requirements (Richardson 1973). Precast concrete products including structural frames, bridge beams and staircases usually fall into the category of batch production.
3. **Flowline production.** Where the manufacture involves repetitive casting, for example, the casting of panels for large panel construction or the production of flooring components by proprietary manufacturers, the process lends itself to the adoption of flowline techniques.

The production process in a precast concrete factory will largely depend on the nature of the precast concrete components. If the production process would frequently vary with the contracts, there is a high possibility that jobbing production will be adopted. On the other hand, if the precaster is in a monopolistic position in the precast concrete market, the batch production will usually be applied.

4.2.2 Concrete

The ingredients of concrete include: cements, aggregates, water, admixtures, additives and pigments. The selection of the ingredients is largely dependent on the design criteria and economic considerations. While some precasters would like to produce concrete by themselves in the factories, there are a few who would like to purchase concrete due to reasons such as site restrictions, economic considerations, etc. Producing concrete in precast factories will help achieve just-in-time supplies. However, it will complicate the scheduling and planning process. Some other factors such as the supply of concrete materials should then be taken into consideration. Ordering concrete from outside suppliers due to site restrictions or economic considerations can help precasters to reduce their investments in capitals that are related to producing concrete, but additional factors resulting from transportation, including costs and traffic conditions should be taken into consideration.

The ingredients in the concrete would largely affect the energy consumption and carbon emissions during its life cycle. Cement is the main source of carbon emissions in the concrete production, although it only makes up 10–15 % in quantity. According to Vares and Häkkinen (1998), the carbon emissions of producing 1 t of a specific concrete product with a cement content of 280 kg/m^3 is 190 kg, while the figure would increase to 240 kg with the cement content rising up to 350 kg/m^3. In addition, due to the temperature needed during cement production, cement would account for over 60 % of the energy used in concrete production. This highlights the fact that energy consumption and carbon emissions arising from concrete production will largely dependent on component choice and concrete design (Vares and Häkkinen 1998).

4.2.3 Reinforcement

This section addresses the considerations when using steel reinforcement bars to product precast concrete products. According to Levitt (2008), reinforcing steel is used in precast concrete components for the following reasons:

1. To enable the product to take loads mainly in the tensile region of the product.
2. To enable the product to withstand handling where a region concerned would otherwise be in compression in the works.
3. To enable a region of the product normally in compression to withstand shrinkage.

When using reinforcement steel to produce precast concrete products, the amount, type and configuration are the three main points of interest (Levitt 2008). Similar to concrete, the configuration can be completed at the precast factories. It can also be completed by the suppliers before the reinforcements are delivered to the precast factories. The environmental concern for reinforcements is largely dependent on the type of steel and the amount of steel which is used in producing reinforcement. For example, utilizing 1.3 % pre-stressing steel in precast concrete slab production accounts for 9 % of the total energy used and 4 % of the total carbon emissions (Vares and Häkkinen 1998).

4.2.4 Moulds

The materials that can be used to produce moulds can vary from steel, timber, concrete to other composite materials. However, no matter what materials are chosen, all moulds in precast concrete production should fulfill the following requirements (Levitt 2008):

1. Maintaining a required geometry.
2. Maintaining the required tolerances.

The selection of the moulds is largely dependent on the tolerances required in specific geometry as well as surface appearance. Steel mould is commonly adopted in precast concrete factories due to its high strength, abrasion and high temperature resistance. More importantly, the lifetime for steel mould is usually very long, as long as appropriate maintenance is conducted regularly. Timber can also be used as mould materials. However, its application as mould materials is very limited due to the nature of timber, which can be easily affected by moisture content, releasing agents, etc. The number of uses of timber mould generally varies from 20 to 100, largely dependent on the paint used to protect the surface (Levitt 2008). Other materials such as concrete, plastics, aluminum and composite can also be used as mould materials. However, due to their limited applications in large scale production of precast concrete components, their applications would not be elaborated in this study.

4.2.5 Demoulding and Stacking

The finished precast concrete products can be easily damaged during the demoulding process, especially for those freshly demoulded products that require special attention. In the stacking process, the commonly adopted method to prevent damage in precast concrete factories is cast stone. The problems in demoulding hardened concrete generally relate to arris damage due to poor compaction or inadequate curing (Levitt 2008). Other than technical solutions such as using low polymer content mortar, the employee skills are also of vital importance to prevent such damage from happening.

Precasters may face the same problems during stacking to prevent damages when handling products. It is important for the operators involved in the stacking process to be aware of the safety requirements. Another important consideration for precasters is the stacking method. The commonly adopted stacking method on site is in the vertical form where products are stacked one on top of the other. However, this method can lead to problems caused by the singling out process, even though precasters may have marked their manufacturing dates. This is because the newly manufactured products will be placed on the top level while the old ones will remain at the bottom of the stack.

4.2.6 Equipment in Precast Concrete Production

The equipment used in precast concrete production can be divided into several groups, according to the nature of the production process. For example, delivering large quantities of freshly mixed concrete to the point of placing usually requires a conveyor belt. Cranage and hoists are commonly adopted for the lifting and singling out process of precast concrete units. Handling precast concrete components will not only require well-maintained equipment, but also an appropriate handling arrangement, including the size and mass of the units, availability of spaces, rate of movement, etc. Generally speaking, the equipment used in precast concrete production can be classified into the following:

1. Uncoupling activities. The equipment in this category are related to deliver freshly mixed concrete to the point of placing. Uncoupling activities can be achieved by using forklift, crane, truck and trolley; mechanical or manually operated barrow; and conveyor belt (Richardson 1973). The selection of the equipment for uncoupling activities is largely dependent on the production capacities of the precast concrete factories and also the nature of the precast concrete components. While the conveyor belt is suitable for large quantities of concrete production, its application requires careful selection due to the constant contact with grout and abrasive pastes (Richardson 1973). Skips and barrows are usually adopted for limited production of precast concrete components.
2. Cranage and hoists. Cranage and hoists are widely adopted in precast concrete factories for the lifting process. The driver is well placed above the factory floor to allow him a clear view of all parts of the factory. Hoists on gantries are adopted where operations on a shop floor are being carried out with equal intensities, or alternatively monorail hoists can be used (Richardson 1973).
3. Forklift trucks. When producing smaller precast concrete units or smaller quantities of precast concrete components, the forklift truck can be used to handle and transport these in the precast concrete factories. Using the forklift truck as handling and transporting equipment requires that the operator be adequately trained; otherwise, the precast concrete units can be damaged easily.
4. Mobile cranes. Mobile cranes can be used in the precast concrete factories where operations are spread over a large area and where there is a considerable distance between the casting yard and the storage yard (Richardson 1973). Mobile cranes are usually adopted in the absence of a gantry lifting system. Operating mobile cranes would require the drivers to be adequately trained due to the safety problems. Drivers who intend to hoist a load that is too heavy for the equipment can cause the crane to topple over, as well as those who attempt to reach out too far.
5. Derrick cranes. Derrick cranes are able to operate with large lifting capacities over a very wide radius. These are usually adopted to handle structural components.
6. Roller conveyors. Roller conveyors require higher investment costs than the other equipment, but these offer an opportunity to free up the units from the cranes in the casting yard early. The roller conveyor can also be used to transport goods and components (Richardson 1973). Roller conveyors require less maintenance than the other equipment mentioned above. Consequently, the roller conveyors are able to transport and operate with minimum effort.

4.3 Transportation and Erection Considerations

In order to ensure that the precast concrete components which are delivered to the construction site are in good condition, transportation from the factory to the site needs to be appropriately scheduled. The considerations during the transportation stage would include:

1. Vehicles. Appropriate frames and supports should be provided on the transportation vehicles to support the concrete units. Screw-jacks and fastenings with a positive locking and anti-rotational device would be useful to minimize movements of the concrete products during transportation.
2. Drivers. The drivers should be fully trained in loading and unloading techniques. This is especially important when placing the concrete units in a corner, where the units can be easily damaged during the unloading process.
3. Schedules. Many features on the transportation routes should be surveyed, including buildings, traffic condition and obstacles before arranging the delivery. The survey is necessary to ensure that the precast concrete products are delivered just-in-time, with no damage caused during transportation and with minimum delivery time.

The processes after the products are delivered from the precast concrete factories are intertwined with each other so much that it is difficult to clearly distinguish between them. In this study, the unloading process is investigated under the category of erection process. However, as the aim and objectives can vary with different research, the processes can be reorganized depending on the nature of the research.

The operations during the unloading processes are quite similar with the loading processes. A skilled banksman is responsible for directing the crane movement. In fact, damages during the unloading process, especially nib and corner damages, are very common if the crane hook is not correctly positioned.

The first consideration relating to the erection activities would be whether or not the details of the precast concrete components are provided to the contractors. Contractors should be provided with details by the designers or precasters concerning lifting, handling and erecting. The erection process can therefore be carried out as long as the deliveries reach the construction site. Other considerations during the erection process would include the following (Richardson 1973):

1. Accurate operation. The sockets and bracing bolts must be accurately positioned so that the site erector can fit in the components, thus reducing the time needed for erection. Accurate operations can be achieved by paying attention to a series of simple matters, such as the cleanliness of threads, clearance of holes and sockets, etc.
2. Site conditions. Site conditions, including access roads, place of cranages, office facilities and storage areas will inevitably influence the erection time of the precast concrete components. The site condition is also of vital importance for the components to be erected on a "just-in-time" basis.
3. Human resources. Like the other processes which need supervision, it is essential that all erection processes should be carried out under supervision from experienced and trained supervisors. Otherwise, performance and productivity would be affected.

4.4 Benefits of Precast Concrete Components

The three traditional construction objectives are time, cost and quality. Much has been discussed about the benefits from these three objectives through the use of precast concrete components in the construction industry. By combining off-site precast production with on-site construction activities, the overall duration of construction projects can be reduced. According to Gibb (1999), the overall construction time has fallen from several months to less than 2 weeks for McDonald's drive-through restaurants. Cost savings by using precast concrete components also lead to reduced construction time. On the other hand, according to Nutt-Powell (1985), the use of off-site fabrication and standardized components allows manufacturers to hire semi-skilled rather than skilled labour, thus reducing costs. In relation to the benefit of quality improvement, precast concrete components are produced in factories within a closed environment where quality control can be better applied. The quality of the construction project can therefore be improved. However, as transportation is required when using precast concrete components, costs introduced by transportation cannot be underestimated, as well as the damages which may happen during the transportation stage. According to Gibb (1999), other benefits by using precast concrete components in the construction industry include:

1. Predictability and reliability. While the precast concrete components have been tested in the factories, the uncertainties and unknowns in construction projects can be reduced, thus increasing the overall predictability. In addition, precast concrete units are produced in factories within a closed environment where quality control can be better applied.
2. Productivity. The manufacturing environment enables better improvement in productivity. As noted previously, semi-skilled labour rather than skilled labour are more likely to be adopted in precast concrete production. According to Gibb (1999), semi-skilled workers perform well-defined tasks at a given work-station, thus increasing productivity. In addition, the closed environment will enable techniques and management skills to be better applied to achieve productivity improvement.
3. Safety, health and the environment. On-site work at a height is always hazardous due to the potential risks. However, by adopting precast concrete construction, the on-site work at a height can be reduced to a minimum, thus reducing the exposure to hazards. Other benefits relating to safety, health and the environment include (Gibb 1999):

   (a) Reduction of on-site work which tends to be less environmentally sensitive.
   (b) Reduction of material wastage.
   (c) Less noise, dust, etc.
   (d) Better controls on atmospheric pollution.
   (e) Usually less energy use in transportation and on-site works.
   (f) Recycling of materials and supplies is easier in an off-site environment.

4. Interfaces and coordination. There are three types of interfaces during the construction life cycle, namely the physical, organizational and managerial (sometimes referred to as contractual) interfaces. The physical interface can be defined as the interfaces between different elements of the building or structure. Generally speaking, adopting precast concrete components in construction projects will reduce the amount of on-site work, which will simplify the process, thus reducing organizational and managerial interfaces by minimizing the number of parties who are involved in the process.

There are many benefits to adopt precast concrete components in construction projects. The benefits discussed above are only a few that are commonly acknowledged by the professionals. Other benefits include fire resistance, thermal performance, sound insulation, etc. For example, according to Glass (2000), concrete has an established advantage over metal and timer solutions in terms of fire performance. The specific advantages of adopting precast concrete products include speed and quality, which come along with the benefits of high productivity and predictability. The precast concrete industry plays an important role to connect the manufacturing industry with the construction industry. Innovations and management philosophies that originated from the production and manufacturing industry can be tested in the precast concrete industry before they are applied in the construction industry. This study aims to identify the application of lean production philosophy in the precast concrete industry to serve as a preliminary study on the general application of the philosophy in the construction industry. With this connection between manufacturing and the precast concrete industry, it would appear that the lean philosophy can be applied with minimum modifications.

Precast concrete may face several disadvantages due to its characteristics. For example, as the precast concrete products are produced on a standard basis, using precast concrete may affect the building design. In addition, changes in the specifications of the project may be hard to deal with as production in precast concrete factories is usually fixed. Any change in the order quantity or in type of the precast concrete products may actually affect the delivery of the project. In addition, adopting precast concrete products will usually come with investments in skilled workman and cranes to carry out the installation activities. Despite the disadvantages, precast concrete can offer high quality product, especially in "system" building and when the clients require a shorter delivery time of the construction projects.

4.5 Applicability of the Lean Principles to Reduce Carbon Emissions

From the lean perspective, carbon emissions in the construction industry can be divided into two groups: effective and ineffective emissions. The lean production principles focus on ineffective emissions which are generated by non-value adding activities and can be considerably reduced if appropriately planned. The following section aims to identify the applicability of the lean principles in three construction stages: production, delivery and installation. The analysis at the production stage is based on the precasters' point of view, while the investigation at the delivery and installation stages in construction sites are from the contractors' perspective.

4.5.1 *Precasters*

The objective of this section is to provide information on how carbon emissions can be reduced in the production processes of precast concrete components. In the precast concrete factory, energy consumption is of critical importance in the production process, including setting up moulds, installing reinforcement bars, spacers and lifting hooks, pouring concrete and demoulding. However, there are some activities which are considered unnecessary which do not add value to the production process. These include maintaining inventory, multi-handling and multi-delivery, inappropriate operations and producing replacements due to product defects. While reducing non value-adding activities has always been recognized as a way to improve productivity, it is also useful to reduce carbon emissions, especially in the construction industry where most carbon emissions are generated by energy consumption, or fuel consumption in other words.

4.5.1.1 Production

In a precast concrete factory, the bending and assembly yards are normally not located together, which leads to unnecessary movements. Cutting, bending and fixing operation sites should be located together to reduce the movements of forklift trucks, mobile cranes and lorries. At the same time, internal batching plants are isolated from production beds. Precasters have to invest in machines like mobile and tower cranes to transfer concrete from the batching plants to the production beds. These unnecessary movements can generate carbon emissions, not to mention the potential interruptions these may cause to other activities such as interruptions to general access and transportation of steel bars and so on. When the process of de-moulding is completed, the components have to be transferred to the finishing trade yard for the finishing trades, including tiling works, surface treatment works, rough concrete surfaces and other treatments. According to Low and Chan (1997), the finishing trade area and production beds are separated. Mobile cranes are often needed for transportation of the products. Needless to say, this will cause unnecessary carbon emissions.

4.5.1.2 Storage

According to Low and Choong (2001), precast concrete components can be divided into normal-weight and light-weight components. Light weight panels refer to the

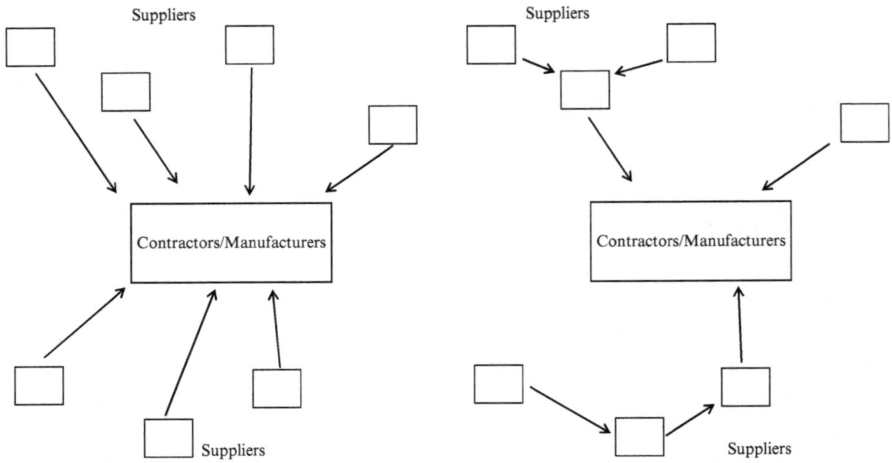

Fig. 4.1 The Spoke Delivery System and the Rim Delivery System (Adapted from: Low and Chan 1997)

standardized panels which are used for partitions and cement bricks. Other precast components which do not fit in the light-weight category are known as normal-weight components. Normal-weight components are stored in the open yard, while light-weight components are stored in the indoor yard, both by way of stacking (Low and Choong 2001). The specific concrete components ordered by contractors have to be singled out from the huge stockpiles before these are delivered to the construction site. This singling out process is defined as waste in the project life cycle, not to mention the potential damage to the components when using handling equipment, such as forklift trucks and dumpers, for transferring these components.

4.5.1.3 Relationships with the Suppliers

Long-term relationship based on trust is essential to reduce unnecessary energy consumption. It can help reduce carbon emissions at two levels. First, problems concerning the quality of raw materials can be avoided. Deliveries will be carried out at the right time with the right materials. By doing this, the chances of double delivery and handling due to undesirable quality would be very small. Secondly, production processes using precast concrete components involve the deliveries of different raw materials, including concrete, steel and painting materials. In the traditional delivery method, each material supplier is required to deliver directly to the plant, which is called the Spoke Delivery System, as shown in Fig. 4.1. In the Spoke Delivery System, each supplier operates independently with its own fleet delivery goods to customers (Liu et al. 2003). According to Liu et al. (2003), this method should be utilized when the lead-time requirement is tight, the goods need to be isolated, or the shipment is large. However, the new production philosophy proposes a more efficient delivery method: the Rim Delivery System, which organizes the deliveries in such a way that each supplier takes turns in delivering to the factory as well as calling at other materials suppliers along the way. The Rim Delivery System is sometimes referred to as direct shipment with milk runs. The transportation costs under the milk runs can be reduced by having each delivery vehicle visit several customer locations, provided that the total quantity of goods to be delivered does not exceed the vehicle capacity (Liu et al. 2003). The Rim Delivery System can significantly reduce carbon emissions caused by transportation by combining deliveries together, especially when a wide range of raw materials is required.

4.5.1.4 Other Applications

According to Low and Chan (1997), few precasters in the construction industry tried to reduce set-up times in their production lines. Constructability problems, repetitive mistakes and formwork de-moulding problems were the key problems which caused delays in manufacturing set-up times and create additional works that should not have had happened in the first place. To reduce energy consumption embedded in additional works due to different kinds of problems, precasters must ensure complete quality control over the production life cycle of the components. In addition, training is essential to raise awareness of sustainable construction and reduce carbon emissions. According to EPA (2004), if driven carelessly, even a perfectly maintained vehicle will pollute more than necessary. It is therefore necessary for the operator to understand and follow the standard operating procedures. For example, if the operator expects to idle for 30 s, it is better to turn off the engine and restart again. By doing so, gas consumption is reduced, thus eliminating carbon emissions.

In summary, the lean philosophy can be applied in the production stage to reduce carbon emissions. As can be seen in Fig. 4.2, most lean applications applied in the production stage are focused on reducing unnecessary energy consumption, in line with its concentration on eliminating non value-adding activities. By applying the new production philosophy, precasters can benefit from saving fuel consumption caused by unnecessary storage, carrying out additional works due to product defects and inappropriate operations, especially under such circumstances when oil prices have increased rapidly, as seen in the first half of 2008.

Take a typical prefabrication factory for precast concrete bridge beams for example. The reinforcement steel bar bending yards and assembly yards were generally not located adjacent to the production beds. Additional transportation needs to be arranged between the assembly yard and the production beds. Not only does the additional transportation consume energy and cause more carbon emissions, it also makes the factory production floor congested by introducing unnecessary mobile cranes, lorries, etc.

However, according to the lean philosophy, these unnecessary movements are considered as non-value adding activities that should not be allowed to take place in the first instance. By appropriate rearrangements, the production site layout can be

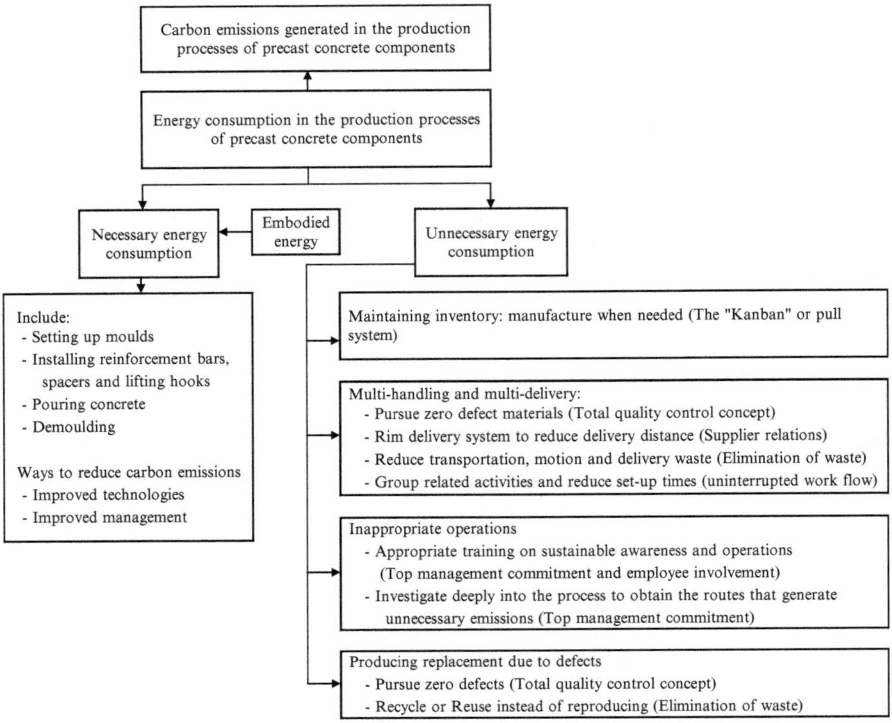

Fig. 4.2 Lean applications in the production stages of precast concrete components

improved so that unnecessary movements can be reduced to the minimum, leading to less fuel consumption and less carbon emissions. A concise and lean movement plan will lead to less obstruction in the production process. The possibility that precast concrete components are damaged during handling and transportation is therefore reduced to the minimum as well.

4.5.2 Contractors

The objective of this section is to investigate the applicability of the lean production philosophy from the contractors' point of view to reduce carbon emissions. Contractors are mainly involved in two kinds of tasks: dealing with the deliveries and scheduling the installation. Contractors must order enough quantities of the precast concrete products to be delivered at the right time to ensure smooth work processes. On the other hand, precasters try to deliver the right amount at the right time according to the contractors' needs. However, the balance may not always be achieved due to inappropriate scheduling or ineffective information flow, which requires the contractors to pay close attention to the following issues:

1. **Be accurate on the order quantities.** Although contractors prefer to order the exact amount of products which are required on site, this is easier said than done. CO_2 emissions will be caused by the additional deliveries, if the deliveries exceed the actual amount used on site. On the other hand, if there is a shortage on site and a short notice has been given before the on-site materials run out, additional deliveries need to be arranged. In order to achieve accuracy in the order quantities, prompt communications, as well as veracious predictions about the delivery conditions, including available vehicles and traffic conditions, need to be obtained. This is one of the prerequisites for reducing carbon emissions from the contractors' point of view. In addition, the Rim Delivery System is also feasible for the contractors when they are ordering a wide range of precast concrete products from different precasters. Of course, the Rim Delivery System can be better applied if long-term relationships between the contractors and the precasters have been established.
2. **Install immediately when deliveries arrive.** Contractors must be fully prepared that when the exact amount of precast concrete components is delivered to site at the right time, immediate installation is carried out to reduce the size of inventory. If successfully implemented, less site storage space is needed, thus reducing energy embedded in maintaining inventory. This is usually referred to as a JIT deliver system, according to Tommelein and Li (1999). It should be noted that contractors have a tendency to keep some level of inventory on site to deal with uncertainties that cannot be predicted during the installation stage. However, keeping inventories on site does lead to unnecessary CO_2 emissions generated by singling out processes, multi-handling and multi-delivery. How well the contractors are able to manage the uncertainties indirectly reflects their ability to control carbon emissions. The lean production philosophy enables the contractors to build up a strong capability to manage these uncertainties.
3. **Eliminate non value-adding activities.** Before examining how carbon emissions can be reduced by eliminating non value-adding activities, different types of non-value adding activities should be identified:

   (a) **Emissions during waiting.** Handling and installation activities have been given an exact time frame before the project begins, including weekly schedule, start and finish dates, exact timing (accurate to the hour), etc. The relevant persons in charge are responsible for scheduling the equipment to make sure that these activities are carried out at the right time in the right place with the right quantities. With these measures in place, installation equipment will not be left idling on site to wait for deliveries to arrive. At the same time, delivery vehicles will not be left idling to wait for the installation cranes to be made available. Should any changes be made to the pre-determined schedule, other procedures should be changed accordingly to avoid equipment being left idling.

   (b) **Emissions during transportation and motion.** The delivery of precast concrete materials for installation on site involves two procedures: delivering the components to the storage area and then transporting to the point of use (Low and Chan 1997). However, in line with the lean's core concept of eliminating non value-adding activities would advocate a single procedure: direct delivery for instant usage. If pre-installation treatments are needed on site, related activities should be grouped together to reduce multi-deliveries.

(c) **Emissions when re-producing.** Re-production due to product defects is another important factor which contributes to more carbon emissions. However, achieving a zero waste working environment may not always be possible due to management and technology restrictions. It is therefore better to apply the lean production philosophy with waste management plans, which would have a significant influence on controlling CO_2 emissions. According to (Gielen and Moriguchi 2002), the CO_2 emissions from steel production based on 100 % scrap is around 0.3 kg/t crude steel, compared to 2 t/t crude steel for blast furnace. More steel recycling will therefore reduce the CO_2 emissions.

Precast concrete components are widely used as construction materials because of their easy installation, fire protection properties and so on. Based on the benefit of easy installation, a smooth work flow with low equipment involvement would help the contractors to achieve low carbon emissions, compared with producing construction components on site such as cast in-situ concrete. In addition, the precast concrete components are produced in factories within a closed environment where quality management can be better applied. In order to smooth the work flow process and reduce carbon emissions that are caused by non-value adding activities, contractors need to:

1. Group related activities to achieve minimum movement.
2. Maintain installation equipment in good working conditions through regular checks to avoid breakdown.
3. Appropriately plan the site layout to make sure that the construction processes are not interrupted.

Although the lean principles can be applied individually, most of the principles are intertwined with one another. When evaluating the applications of lean principles from the contractors' point of view, it is clear that individual and grouped applications have a lot in common. No matter what principles the contractors apply on site to reduce carbon emissions, "always get the work done right the first time" and "cut down the non-value adding activities" are key to the application of the lean concept.

4.6 Pilot Studies

This study focuses on applying the lean principles in the construction industry to reduce carbon emissions. It is driven by the increasing pressure from the industry to address the problem of global climate change. This section aims to investigate current precast concrete production and construction practices to identify the non-value adding activities that consume energy and generate carbon emissions.

According to the aim and objectives of this study, two precasters were selected for the pilot studies. The two precasters occupied large market share of the precast concrete market and produced based on different production types. Therefore, their experiences in the precast concrete production would reflect the general production practices in the Singapore precast concrete industry.

This section aims to address the following questions:

1. Obtain information on production and construction processes to examine if some of the lean principles have already been applied on site. If not, what are the further improvements that can be recommended?
2. Communicate with practitioners in the industry to identify the non-value adding activities from the precasters' and the contractors' point of view.
3. Identify management improvements to achieve low carbon emissions. Traditional carbon reducing techniques focus on applying new environment-friendly materials and technology. These usually entail long research duration and possibly high implementation costs.

4.6.1 Background of Pilot Studies

The aim of this section is to investigate the production process of precast concrete components to identify the potential sources of waste, as well as to validate the applicability of lean production principles in the precast concrete industry. The questionnaire which was adopted for the pilot study is attached in Appendix 1.

Data collection was carried out by semi-structured interviews using a questionnaire with six sections: production type, quality control during production, delivery system, non-value adding activities during production, training system, recycle and reuse during production.

- **Production type.** This category is concerned with the production type of the precast concrete factory. The production type can be divided into the pull and push system. While the push system is usually referred to as mass production, the pull system adopts manufacturing only when needed. In addition, the handling of raw materials and finished products is addressed in this section.
- **Quality control.** This section aims to identify the quality control process during production if it is applied in the precast concrete factories. It should be noted that although the exact percentage of savings may not be available from the pilot studies, an estimate can still be obtained to measure the effect of quality control during the production process.
- **Delivery system.** The delivery system is concerned about two aspects of deliveries: raw materials and finished products.

- **Non-value adding activities.** As the site layout is provided by the precasters, non-value adding activities due to inappropriate site layout will be investigated, if there is any during the production process.
- **Training system.** Employee involvement is of vital importance for the construction industry to achieve the green objective, as most of the equipment and vehicles that consume energy and generate carbon emissions are handled by the operators. Employees who are directly involved in facilities management have the greatest potential and ability to save energy and reduce carbon emissions.
- **Recycle and reuse.** Statistics have shown that the CO_2 emissions in steel production can be significantly reduced by steel recycling (Gielen and Moriguchi 2002). This section aimed to obtain the current recycling and reusing practices in the precast concrete production process.

The two precasters are referred to as Precaster A and Precaster B to preserve their anonymity. In order to investigate the sources of waste in the precast concrete production, it should be noted that these two precasters have different production processes. Precaster A, which holds a dominant position in Singapore's precast concrete industry, produced based on mass production due to the large quantities of orders received. On the other hand, Precaster B, which only serves the public housing projects, produced based on order.

4.6.2 Results from the Pilot Study: Precaster A

4.6.2.1 Production Type

As stated above, Precaster A, who was a market leader in the Singapore's precast concrete industry, especially in precast concrete column production, has to mass produce due to the large quantities of order received, which led to a huge stock yard. The site layout of Precaster A is shown in Fig. 4.3. As can be seen from Fig. 4.3, the stockyard took up almost 30 % of the total land area, which leads to two problems in the factory:

- More time and resources would be consumed to select the products which were required by different projects, not to mention the potential damages during the singling out process. According to the project manager interviewed, although damages during the singling out process did not always happen, proactive measures should be taken, especially for the connection between the crane hook and the concrete components.
- As newly produced units would be stacked on top of old ones, it could take several months before the old ones which were placed at the bottom of the stack could be delivered. To make matters worse, according to the project manager, most of the clients preferred those which were freshly produced. Precast concrete units which have been produced for more than 40 days were usually not acceptable by the clients.

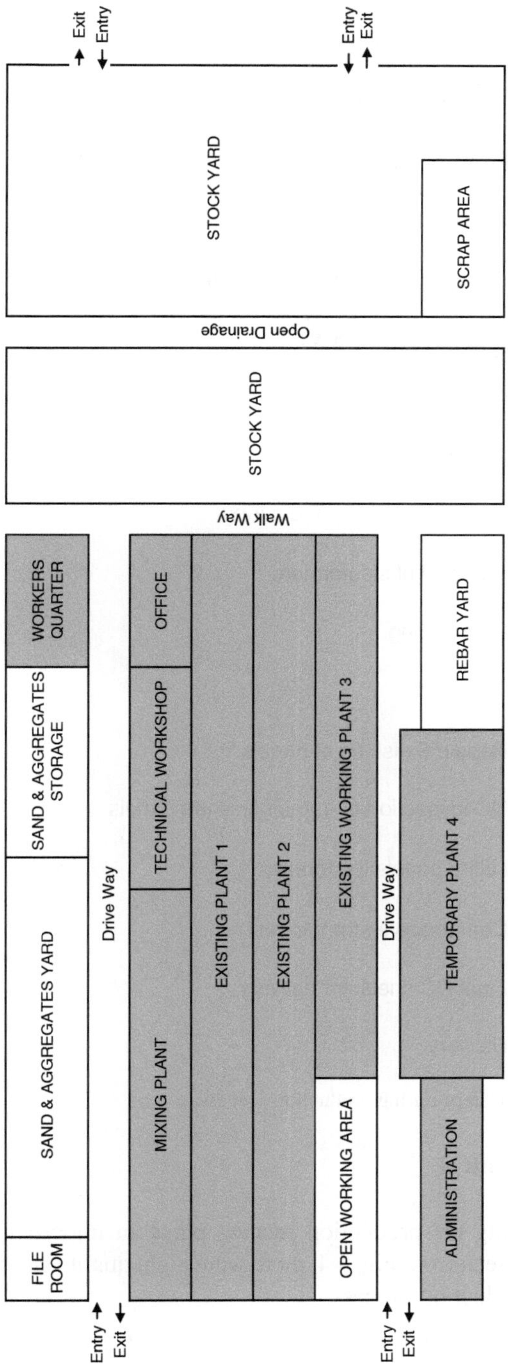

Fig. 4.3 Factory layout of Precaster A

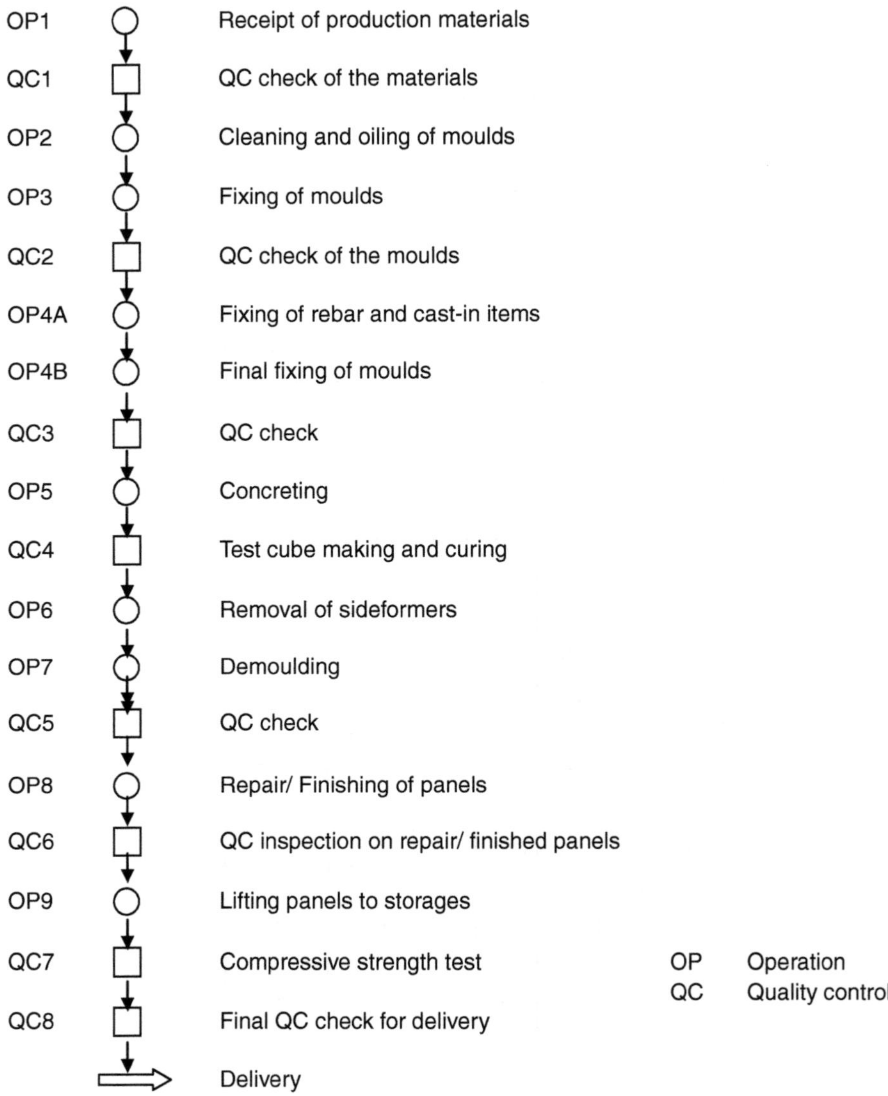

Fig. 4.4 Quality control steps during production processes

4.6.2.2 Quality Control

Quality control during the production process plays an important role to reduce damages. As can be seen from Fig. 4.4, there were eight quality control (QC) checks during the entire production process:

- **QC1.** QC1 referred to the quality control check that was carried out on the specifications of the raw materials. Raw materials which did not meet the specifications would be returned immediately.

- **QC2 and QC3.** QC2 and QC3 were carried out to check the cleanliness of the moulds. The cleanliness of the moulds was of vital importance to precast concrete production, as failure to maintain a clean mould face could cause visual defects on the surface of the concrete components. In addition, honeycombing on the face of the finished components due to badly cleaned joints was not uncommon during production and should be prevented from happening.
- **QC4, QC5, QC6 and QC7.** During the production process, there might be some components which were below the specified quality standard. Lack of compaction, misplacement of the reinforcement bars and damages during handling might lead to defective units. QC4 to QC7 were to ensure that the units below the specified quality standard were sent to other recycling or reusing process before these were totally rejected. As can be seen from Fig. 4.3, there was a scrap area in the factory to deal with units that were produced below the specified quality standard.
- **QC8.** This was the final quality control check before the precast concrete components were selected and delivered to the construction site. As explained earlier, damages during loading and unloading were very common. It was therefore necessary to make sure that the components delivered to the site were of good quality and would not be rejected by the contractors.

4.6.2.3 Delivery System

The materials which were used in the production process of Precaster A included: cements, aggregates, water, admixtures, additives, etc. According to the project manager interviewed, the concrete used in the production process was produced in-house. The decision to produce concrete in the factory was based on two levels of considerations. First of all, it was cost-effective for the business. Long term relationships between Precaster A and the various suppliers would be formed due to the large quantities of order every year. The raw materials could therefore be obtained at reasonable and acceptable prices. Secondly, as the delivery environment, including the availability of the ready-mixed suppliers, the traffic conditions and so on, was not predictable, relying solely on precast concrete suppliers was not entirely credible. In addition, producing concrete in the factory helped to achieve just-in-time production. This ensured that the production process would not be interrupted by the non-availability of concrete.

Although the Rim Delivery System was proposed to reduce the delivery cost and energy consumption during transportation, it was not adopted right now because raw materials like steel, cement, sand and aggregates required very different specialized delivery vehicles.

4.6.2.4 Non-value Adding Activities

Non-value adding activities in Precaster A's production process occurred largely because of the production type and its huge stockpile. Storage area was necessary because Precaster A adopted the mass production method. Without raw materials, the whole production process would be damaged. Although just-in-time supply could be applied to reduce the inventory of raw materials, Precaster A could not afford the loss if things went wrong with the delivery schedule, which involved a large number of uncertainties that could not be predicted. When raw materials were delivered to the factory, some amount of them went into the storage area. This practice led to further energy consumption when they were transferred from the storage area to the production place.

As noted earlier, due to the huge stockpile, the singling out process to select the units required by the project consumed energy and caused potential damage to the components. When damages actually happened in the precast concrete factory, the units would either be totally rejected, or sent for repair, depending on the level of the damage. Damages to the components would significantly contribute to two sources of energy consumption. Replacements have to be produced if the components were totally rejected, and the energy consumption when producing replacements would contribute towards embodied energy. However, when the damaged components were sent for repair, the energy consumed during repair fell into the category of multi-handling.

4.6.2.5 Training

The importance of employee involvement in precast concrete production to achieve energy efficiency and low carbon emissions was at three levels: handling techniques, plant maintenance and the awareness of sustainability. Although the appropriate handling techniques, plant maintenance and training programmes were provided by Precaster A through internal and external seminars, brainstorming and other educational and learning activities as well as programmes that aim at raising the employee's awareness of sustainability, these programmes were limited. According to the project manager interviewed, most of the programmes that were related to sustainability focused only on top-level management staff. On the other hand, the operators who were closely related to facilities operations and had the greatest potential to reduce energy consumption and cut down carbon emissions were often overlooked.

4.6.3 Results from the Pilot Study: Precaster B

One significant difference between Precaster B and Precaster A was the production type. Precaster B supplied only to public housing projects in Singapore, which means that its production was based on order. In addition, the precast factory was a 4-storey building, which was much larger than the layout of Precaster A. Because Precaster B only produced for public housing projects in Singapore, most of the components were standardized, such as staircases, partitions and so on. Nevertheless, there were some

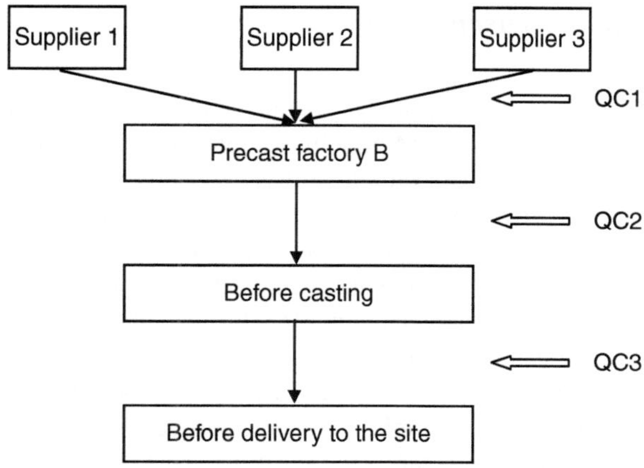

Fig. 4.5 Quality control processes of Precaster B

requirements which varied with the nature of the housing projects, including design, architecture and structure.

In accordance with the production type, the inventory level in the factory was much lower. First of all, the factory did not produce at its full capacity. The production process would not start until they have received the orders. In addition, precast units that were freshly produced would normally be scheduled for delivery in 3 days (5 days maximum).

The quality control process for precast production was almost the same as the procedure in Precaster A. As can be seen in Fig. 4.5, three main steps were carried out as quality control processes. It should be noted that although Precaster B was responsible for producing precast concrete components for public housing projects, the actual production work was subcontracted out. The properties, including the land, equipment and facilities were owned by Precaster B, but the subcontractor who was awarded the contract, was allowed to use the facilities for production. The decision to subcontract the production was based on economic considerations. In the quality control processes, all the QC checks would be carried out by both Precaster B and the selected subcontractor. Precaster B was involved in the quality control process as a third party who provided third-party evaluation that could significantly reduce the potential risks brought about by problems relating to quality, either from raw materials or finished products.

According to Precaster B, most of the training programmes were intended to reduce water consumption and electricity usage. Few programmes have been provided for other sustainability-related issues, such as eco-driving, eco-operation, etc. However, as noted above, staffs who were related to facilities management should have the greatest potential to reduce energy consumption in precast concrete factories. Hence, they should be the ones that the training programmes focus on, especially the programmes related to reducing energy consumption.

Precaster B also carried out internal briefing as an effort towards continuous improvement. However, as only management-level staffs were involved in the internal briefing, lower level employees might not have the proper channels to express their ideas.

4.6.4 Discussions

Based on the pilot studies conducted in the precast factories, it seems that there were some lean production principles applied to achieve better production at the time of the pilot studies. For example, it was very common that the total quality control concept and quality management were applied in precast factories. In addition, although both precasters were aware of the lean production concept, they were not familiar with the principles and how the lean production principles could be applied. The result is in line with previous findings that the lean production concept is now totally new. It is a combination of existing management philosophies in a new way (Howell 1999).

There were several types of non-value adding activities in the precast concrete factories. For example, the "push" production could lead to huge stockpile. While the stockpile was large, the selecting process might consume a considerable amount of energy and generate unnecessary carbon emissions, not to mention the potential damages during the selecting process. To address this problem, a better coordination between contractors and precasters should be formed. The delivery system might also cause ineffective operation. It seems that there were certain kinds of materials that could be grouped together to deliver, as long as the materials were not contaminated by each other during the delivery.

The coordination on site seems to be another problem that led to unnecessary carbon emissions. During the visit in the factories, when the finished products were placed on the lorry for delivery, the lorry was held idling for over 30 min. All the paper works, QC checks and fixing activities were conducted with the equipment being held idling.

There seems to be a gap between the management team and the employees who were actually conducting the production process. Few programs have been provided for the objective of reducing carbon emissions or other related issues. However, as the facilities operators had the greatest potential to reduce carbon emissions, they should be the ones that training programs should focus on.

4.7 Summary

Previous studies have shown that the lean production philosophy could be applied in the precast concrete industry to improve productivity and reduce waste. In this chapter, it was shown that the lean production philosophy had the potential to reduce carbon emissions from three aspects: eliminating non-value adding activities, improving the efficiency of value adding activities and raising the awareness of sustainability among the employees.

Compared with other carbon reduction technologies, such as research into environment-friendly materials, passive design, etc., lean production should provide a viable, affordable and beneficial solution to both the precasters and the contractors to achieve lower carbon emissions.

In addition, the lean production principles could be used as stand-alone applications, which means that the entry barrier was very low. Companies and organizations were able to choose one or several principles which they were good at to implement to achieve less energy consumption and lower carbon emissions. However, the literature review suggested that some of the improvement processes were based on a "continuous improvement" basis and might not produce instantaneous results. Without instantaneous results, companies and organizations might be reluctant to apply the lean production principles in their daily production activities. In addition, the organizational structure and decision-making process might have to change due to additional considerations which could affect short-term performance. Hence, before applying the lean production philosophy in their daily activities to achieve energy efficiency and cut carbon emissions, companies and organizations should be fully prepared for the forthcoming changes and challenges and keep in mind that on a "continuous improvement" basis, a lean environment could ultimately be created.

Reducing carbon emissions in the future could be achieved through technical improvements and innovations which professions and researchers were paying much attention to at the time of this study. However, it should be noted that management improvements should not be overlooked because human activities which consumed energy and generate carbon emissions were not always efficient. There remained many non-value adding activities which needed to be eliminated.

Based on two pilot case studies carried out in the precast concrete factories, it was observed production improvements related to production type, quality control, delivery systems, non-value adding activities, training, recycling and reusing could help precasters to cut carbon emissions. While both precasters have implemented several single applications, such as quality control and training, they were not aware of the whole package of applying the lean philosophy to reduce carbon emissions. This result, on the other hand, suggested that applying a single lean production principle was not totally new in the precast concrete industry; it was real challenge to implement the lean package.

References

BCA (2011) Annex B: Development of the precast concrete component industry in Singapore. [online] Available at: http://www.bca.gov.sg/publications/BuildabilitySeries/others/rscp_anxb.PDF. Accessed 20 July 2011

EPA (2004) Exhaust and crankcase emission factors for nonroad engine modelling – compression-ignition. [Online] Available at: http://www.epa.gov/otaq. Accessed 16 Aug 2007

Gibb AG (1999) Off-site fabrication: prefabrication, pre-assembly and modularisation. Whittles Publishing, Caithness

Gielen D, Moriguchi Y (2002) CO_2 in the iron and steel industry: an analysis of Japanese emission reduction potentials. Energy Policy 30(10):849–863

Glass J (2000) The future for precast concrete in low-rise housing. British Precast Concrete Federation, Leicester

Howell G (1999) What is lean construction. In: Tommelein ID (ed) Proceedings of IGLC-7. U.C. Berkeley, CA

Levitt M (2008) Precast concrete: materials, manufacture, properties and usage. Taylor & Francis, London

Liu JY, Li CL, Chan CY (2003) Mixed truck delivery systems with both hub-and-spoke and direct shipment. Logist Transp Rev 39(4):325–339

Low SP, Chan YM (1997) Managing productivity in construction: JIT operations and measurements. Ashgate Publishing, Brookfield

Low SP, Choong JC (2001) Just-in-time management of precast concrete components. ASCE J Constr Eng Manag 127(6):494–501

Nutt-Powell TE (1985) The house that machines built. Technol Rev 88(8):31–37

Richardson JG (1973) Precast concrete production. Cement and Concrete Association, London

Tan GC (1999) Major challenges for housing development in the next millennium. [Online] Available at: http://www.housingauthority.gov.hk/eng/events/conf/conferen/pdf/etan.pdf. Accessed 18 Dec 2008

Tommelein ID, Li EY (1999) Just-in-time concrete delivery mapping alternatives for vertical supply chain integration. In: Tommelein ID (ed) Proceedings of IGLC-7. Berkeley, CA

Vares S, Häkkinen T (1998) Environmental burdens of concrete and concrete products. [Online] (Nordic Concrete Research publication 21 1/98). Available at: http://www.itn.is/ncr/publications/doc-21-10.pdf. Accessed 15 Dec 2008

Chapter 5
Modelling for the Lean Production Philosophy

5.1 Introduction

This chapter aims to explain the theoretical support and form the conceptual framework of this study. The review of the theoretical support was carried out at four levels:

1. The theory of production. The theory of production was investigated as the basis to identify wastes in production and construction processes. There were mainly three concepts of the theory of production, namely transformation, flow and value. Each concept was investigated from the rational, conceptualization, principles and development domains.
2. Economic explanation of production. Microeconomic theories that were related to the production process were reviewed in this chapter, including demand theory, the theory of the firm and the cost of production theory of value.
3. Economic explanation of the environment. Economic theories that were adopted to explain and solve environmental problems were reviewed in this chapter, including the theory of public goods, the theory of externality, and the command and control theory.
4. Management theories adopted to solve environmental problems, including management science theory and organizational environment theory were reviewed in this chapter.

The theoretical support was then integrated into the conceptual framework of this study. This chapter aims to answer the following questions:

1. What are the mainstream theories in this study?
2. How would the mainstream theories be used in this study?
3. What are the contributions of this study to the mainstream theories?

P. Wu and S.P. Low, *Lean and Cleaner Production: Applications in Prefabrication to Reduce Carbon Emissions*, DOI 10.1007/978-3-642-42062-7_5,

5.2 Sustainability Science

Sustainability science was usually defined as a discipline that points out the way toward a sustainable society (Komiyama and Takeuchi 2006). This new science originated from the concept of sustainable development, which was usually defined as "development that meets the needs of the present without compromising the ability of future generations to meet their own needs" (WCED 1987, p. 43). However, although the concept of sustainable development was broadly accepted over the world, the application of sustainable development was far more complicated than what the concept indicated. Scientific and technological underpinnings of the concept remained unclear (Cohen et al. 1998). Research to analyze these scientific and technological underpinnings was initiated by the International Council for Science (ICSU) during the 1990s. There were three levels of systems in sustainability science, which were the global, social and human systems. All the three levels of system were of crucial for the survival of human beings and the environment. Therefore, by breaking sustainability science into these three systems, current sustainability issues could be analyzed.

1. Global system. The global system provided base for human survival, including the geosphere, atmosphere, hydrosphere and biosphere (Komiyama and Takeuchi 2006). However, the human activities have influenced and caused fluctuations to the global system. For example, global climate change and ozone deprecation were considered as man-made interruptions to the global system.
2. Social system. The social system provided the social base for human survival, including political, economic and other structures. Although the social system could be improved by economic growth and technology innovation and improvement, social problems which usually came with the innovation and improvement could not be overlooked, especially environmental problems. Environmental problems affected both the social system, as well as the global system mentioned above.
3. Human system. The human system referred to an aggregation of factors which were necessary for the survival of human beings. For example, physical health, mental health and security were all necessities for human survival. However, this human system was currently challenged by environmental problems, including global climate change, solid waste, ozone depletion, etc.

Sustainability science dealt with three basic systems, from global, social to human system, as well as the interaction among the three systems. For example, a typical interaction between the global and social system was global climate change, which required low carbon emissions from the society through technical and non-technical improvements. One problem generated from the interaction between the social and human system was waste. The construction industry therefore should be capable of sustainable production and consumption to deal with such problem (Sotherton et al. 2004).

According to Mihelcic et al. (2003), the boundaries of sustainability science was very wide that it must embrace the social and natural science. It was therefore difficult for an individual researcher to investigate all the three systems in sustainability science. Komiyama and Takeuchi (2006) proposed the following steps in sustainability science to address the problem:

1. A general framework, in which the three systems could be expressed by quantifiable criteria and indicators, should be provided. It was the criteria and indicators that the research should be based on. The criteria and indicators could be used to solve a single problem. However, it was extremely important to notice that these criteria and indicators were not supposed to offer a general solution to any particular type of problem because of the diversity of the earth. The solutions might offer useful applications in one area or district, but the applications might harm the sustainable development process in other areas.
2. One unique feature of sustainability science was the process from problem identification to problem solving. In conventional research, solutions were based on a full analysis of the potential problems. However, in sustainability science, solutions may have to be provided before those problems have been fully analyzed. Global climate change was one typical example of this contradiction. It was difficult to analyze various solutions to the problem of global climate change, but the research relating to the global climate change could not wait.
3. As indicated in step two, a precautionary approach was adopted in sustainability science research. However, the development process of the approach leading to the acceptance of the society was essential, as Komiyama and Takeuchi (2006, p. 5) stated "what is demanded of sustainability science is not only the development of scientifically sound models for predicting future scenarios and evaluating the effects of different countermeasures and solutions but also effective management of the process by which these forecasts and evaluations are accepted by society, to generate the social reforms necessary to ensure global sustainability".
4. Cooperation among different researchers in sustainability science was essential due to the complexity in sustainability science. This was especially important in addressing the problem of global climate change because solutions were often provided without being fully analyzed and testified.

Sustainability science provided new routes to analyze global climate change. According to Hay and Mimura (2006), the links between climate change and sustainability were highly complex, including both direct and indirect factors. Sustainability science was able to provide a systematic way to considerable the overall interaction between the global climate change and sustainability. All options, except the "do-nothing" response, had certain contribution to achieve sustainable development. More importantly, a transition process from reactive measure to proactive measures was important to achieve the balance among the global, social and human systems. The model of manufacturing – complex systems also embraced such complexity concept that the systems should be analyzed by a holistic approach, as will be discussed in Sect. 5.3.

5.3 Model of Manufacturing – Complex Systems

The dominant model of manufacturing was based on the automobile industry – culminating in lean production (Winch 2003). However, according to Winch (2003), this model of manufacturing was unlikely to be of broad relevance to the construction industry beyond housing. Large technical systems (Coutard 1999), or complex systems (Miller et al. 1995), which focused on the creation of infrastructure and related systems, were usually adopted as an alternative model of manufacturing. Winch (2003) argued that while the conventional model of manufacturing, such as mass production and lean production, could be applied to mass market commodity products, it was highly unlikely to apply to another important group of products and industries. These groups of products and industries were usually classified as large technical systems, or complex systems, which were large item, customized, engineering intensive goods that are seldom, if ever, mass produced, such as flight simulators, telecommunications, intelligent buildings and so on (Miller et al. 1995).

Complexity and complex systems were emerging new science, but there was not yet a general accepted and comprehensive definition (Horgan 1995). Williams (1999, p. 269) stated that "while many project managers use the term complex project, there is no clear definition about what is meant – beyond the general acceptance that it is something more than simply a 'big' project". According to Bertelsen (2003), complex systems were not a special class of systems, but a way of examining the system as opposite to the ordered viewpoint. In complex systems, the systems should be examined as a whole unit without simplification. By examining the systems with a holistic approach, new feature and behaviours could be found (Bertelsen 2003).

Lucas (2007) pointed out 18 characteristics of complex systems, 14 of which were chosen to analyze the interaction between complex systems and the construction industry. The 14 characteristics were divided into three groups, which were:

1. Autonomous agents. According to Bertelsen (2003), complex systems were composed of autonomous agents, each of which was non-uniform. It means that each agent could perform under different standards. However, they must co-evolve to respond to the changes in either internal or external environment. The systems were therefore self-modification systems by redesigning themselves over time with changes. In contrast with the upward causation in mass production system, a downward causation was adopted in complex systems. In addition, complex systems were able to self-reproduce by correcting errors in the systems to form a new system.

2. Undefined values. Complex systems had undefined values. The interaction between complex systems and the environment was not pre-determined but rather evolving from time to time. In addition, the self-organizing characteristic would enable the complex systems to evolve towards fitness, a situation that the errors in complex systems would be easily spotted. In addition, there was no single standard that was suitable for complex systems. Even two initially identical systems could be evolved into two totally different complex systems by dynamically self-organizing.
3. Non-linear. Complex systems are non-linear, which means that the output of the systems could not be interpreted by the equation $F(X + Y) = F(X) + F(Y)$. This was based on the mutual interference between different agents in complex systems. The overall complex systems were far more complicated than a simple aggregation of subsystems and should therefore be examined as a whole.

Miller et al. (1995) pointed out three basic characteristics of large technical systems, or complex systems, which are:

1. A high degree of customization. Due to the high degree of customization, complex systems usually came with high cost. Unlike the mass production system, much of the skill and knowledge needed to produce complex systems was embedded in people. There were many connections between the subsystems of complex systems, each of which was extremely complicated, customized and expensive.
2. Emerging properties and rising complexity with time. The emerging properties and rising complexity were usually driven by customer requirements on performance, capacity and reliability. For example, Arthur (1993) pointed out that in the flight industry, the jet designers added more and more subsystems in order to overcome extreme stress.
3. A high degree of user involvement in innovation process. In complex systems, users were heavily involved for their survival. Consequently, the users were involved in the life cycle of complex systems, including research and development, design and production methods, etc. One important difference between the mass production system and complex systems was that users were much more educated of the systems technologies in order to achieve profitability in their own business.

Large technical systems or complex systems were often treated as self-organizing systems. The characteristic of self-organizing could be expressed as responses to the needs of the external environment and large users in particular (Miller et al. 1995), or closed systems which were driven by the inner dynamic to achieve development (Salsbury 1991). This self-organizing feature explained the importance of end users in complex systems, who were directly engaged in the innovation process. For example, in the flight simulation industry, which was viewed as complex systems by Miller et al. (1995), the end users were involved in systems integrators, where their requirements could be integrated into the design of flight simulators. This is quite different from mass production where the customers usually took or bought what had been manufactured. Similar to the feature of "pull", the importance of customers validated the fact that some general "lean thinking" did apply to complex systems industry (Womack and Jones 1996).

The production industry and the construction industry had complexity characteristics from autonomous agents, undefined value to non-linearity that both industries could be viewed as complex systems. According to Bertelsen (2003), the similarities between the production system and complex systems were:

1. Autonomous agents. Parties participating in the production system acted as independent agents in the production process. It was the responsibility of project managers to bring these autonomous agents together (sometimes, control structure and leadership could emerge even if it was not established initially), according to Dam and Elsborg (2003). Typically, a temporary production process was designed at the beginning, including the autonomous agents. The development of the temporary production system could be viewed as a co-evolution of the subsystems and a self-modification of the whole system. Co-evolution could be brought about by customer requirements, technology innovation and changes in internal or external environment. The self-modification process of the production system was in fact an aggregation of the co-evolution processes of the autonomous agents. In addition, emergence and disappearance of production companies could be viewed as self-reproduction processes. This indicated that the production process did obtain some characteristics as an aggregation of autonomous agents.
2. Undefined value. Although initial values were pre-determined within the temporary production system, these values might change with time and environment. The changing environment made the production a self-organized structure in time and space.
3. Non linearity. The production system was not a linear system. According to Shingo (1988), the overall productivity of the production process could not be a simple add up of the productivity from each subsystem. In fact, the production system was very complex and dynamic that when investigating the production system, project management was often applied without knowledge of the project integration (Bertelsen 2003). It was the complexity and unpredictability of the subsystems that caused unpredictability of the whole system. When analyzing the whole production system, the complexity of the system should not be overlooked.

According to Bertelsen (2003), construction process could be viewed as complex systems as well from the three basic characteristics:

1. Autonomous agents. The subsystems and modules of buildings, including the interaction between them, could be viewed as autonomous agents. Although these agents were pre-determined within the design, they usually evolved in the operation stage of construction projects. The co-evolving process of these agents could be brought about by technology innovation, introduction of new building materials, new building methods, maintenance requirements, customer requirements, etc.

The aggregation of the co-evolving process of the agents was the self-modification of the whole construction process.

2. Undefined values. The agents discussed above also represented the values of the project in the initial design stage. Construction projects were often non-standard. It was possible that two construction projects obtained two identical subsystems, but no two construction projects would be identical in all manners.
3. Non linearity. The output of the construction process was related to the inputs of sub-processes. However, this relation is not linear. According to Bertelsen (2003), the relation must be expressed in a sort of fuzzy arithmetic. In fact, the non linearity of the construction process was in accordance with the lean production philosophy that increased efficiency in subsystems might not lead to increased efficiency in the overall system, as observed by Shingo (1988).

As can be seen from above, both the production process and the construction process had most of the complex systems' characteristics, from autonomous agents, undefined values and non linearity. Complex systems presented a challenge to the construction industry that the complexity of the construction process should never be ignored and the project management paradigms should therefore be re-examined (Bertelsen 2003). However, the rising recognition of complex systems did not restrain the adoption of mass production and lean production in the construction industry. The mass production had the possibility to be applied and has been applied in the construction industry to achieve better construction, especially in the housing projects (Winch 2003). The application of mass production model in the housing projects was based on highly mechanized and efficient uses of conventional building methods, together with extensive application of factory-made components (Herbert 1984, p. 216). However, the overall construction industry was more complicated than the housing industry that the mass production model might not apply. The construction industry was seen as flexible, responsive, and capable of working in networks (Winch 2003). On the other hand, the mass production model was seen as bureaucratic and inflexible (Miles and Snow 1986).

According to Winch (2003), there were three basic groups in the construction industry based on the connection of the industry with the manufacturing industry:

1. Private housing production. Private housing production was the industry which obtained the closest bond with the lean production. The potential volumes were high enough for almost similar customer requirements. According to Ball (1993), there were two major concentrations on housing production, which was production efficiency and development gain. When the development gain was the main focus, the production fell more to the make-to-forecast (MtF) basis, where the product was produced for stock and sold after it was manufactured or, sometimes, during manufacturing (Winch 2003). When production efficiency were separated from development gain and was the concentration of the production, as has been achieved in Japan, it would be more likely that the lean production would be applied.

2. Building. Building sector referred to the construction sector where the concentration was to meet the needs of clients for a wide variety of facilities with undemanding technical specifications.
3. Major projects. This sector included infrastructure development as well as highly engineered buildings, including hospitals and high-rise offices. According to Winch (2003), in the buildings and major project sectors, it was the complex systems which seem more appropriate. In building and major projects sectors, the volume was usually not high enough to allow mass production and lean production to be economically viable, not to mention the increasing diverse customer requirements.

Therefore, in order to match the appropriate model of manufacturing to the construction industry, the sector in the construction industry should be identified at the start. According to Winch (2003), although housing sector seemed to be the only one where lean production could be applied, it did not restrain the construction industry from learning from the manufacturing industry, especially in the area of project management. Winch (2003) stated that it was only project management that could have an entire overview about the value system of the life cycle of the project from conception to completion and this was the key step that lean production can be applied.

5.4 Model of Manufacturing – Lean Production System

According to Womack and Jones (1996), it was often thought that production consisted of three core phenomena: production development, order-delivery and production proper. A general theory of production should address all the three levels of thinking. Three generic actions from the management perspective could help to realize the production goals, which were (Koskela 2000):

1. Design of the production system.
2. Control of the production system in order to get the intended production realized.
3. Improvement of the production system.

Buffa (1961) mentioned cost, quality and delivery commitments as the production goals. Wild (1984) stated that two objectives, which were service and utilization, were the production goals. The primary goal was service to provide customers with goods of a given kind. The secondary goal was utilization to ensure that materials, machines and labors were utilized at agreed levels. Slack et al. (1995) listed five objectives as production goals, which were: quality, speed, dependability, flexibility and cost.

Koskela (2000) organized the production goals listed above into three kinds:

1. The goal of providing the intended products in general.
2. Goals related to the characteristics of the production itself, such as cost minimization and level of utilization.

3. Goals related to the needs of the customer, like functional performance, quality, dependability and flexibility.

The theory of production should contribute to the achievement of the three goals of production. It should be noted that the general concepts and principles of production would be focused in the literature review rather than particular types of production.

5.4.1 The Transformation Concept of Production

The transformation concept of production was the dominant model of production during the twentieth century. The transformation concept was based on the most prominent single element in modern scientific management: the task idea (Taylor 1913). According to Taylor (1913, p. 144), the task idea in scientific management was described as follows:

> Perhaps the most prominent single element in modern scientific management is the task idea. The work of every workman is fully planned out by the management at least one day in advance, and each man receives in most cases complete written instructions, describing in detail the task which he is to accomplish, as well the means to be used in doing the work. . . Scientific management consists very largely in preparing for and carrying out these tasks.

Kendall (1912) also argued that the task was necessary in production in order to make all the prerequisites of a task ready and to ensure that the best method was followed. The task idea in scientific management led to two production types, which were systematic and unsystematic production. In systematic production, the producer knew exactly what had to be done, decomposed the objective into different tasks and analyzed the best way to achieve the objective. On the other hand, in unsystematic production, the objective was described to the superintendent, who further explained it to the foreman. The achievement of the production objective in unsystematic production was largely dependent on the knowledge of the superintendent and the foreman. Unlike systematic production, unsystematic production lacked planning at the start, completed instructions over the production life cycle and might slow down the production process.

5.4.1.1 Principles

1. *Decomposition*
   According to Koskela (2000), the transformation process could be decomposed into subprocesses. The decomposition principle was quite similar to analytical reductionism, a well-known notion in the history of philosophy and had its origin in the second rule of Descartes, which was stated as (Checkland 1981, p. 330):

> The second to divide each of the difficulties that I was examining into as many parts as might be possible and necessary in order best to solve it.

The decomposition rule was then interpreted by Slack et al. (1995, p. 914) as:

> Look inside most operations and they will be made up of several units or departments, which themselves act as smaller versions of the whole operation of which they form a part.

The decomposition principle was highlighted in the "Walrasian production model" (Wortmann 1992), where production was regarded as transformation processes to transform production factors into complete products. In fact, the decomposition principle has been adopted in many practical applications in production planning and control, such as in inventory control (Browne et al. 1988) and computerized production control, etc. In these applications, the production objective was decomposed into specific tasks. By completing these specific tasks, the production objective could be achieved.

2. *Cost minimization*

Cost minimization was the core principle of the transformation concept of production, which was embodied in and operationalized by conventional accounting theory. The cost minimization principle was based on the following assumptions (Umble and Srikanth 1990):

(a) Total cost of the production process equaled the sum of the costs of each operation.
(b) Total cost of each operation was proportional to the cost of direct labor for that operation.

Based on these assumptions, it was stated that the cost of the total process could be minimized by minimizing the cost of each subprocess. Hence, cost management could be applied in each subprocess to reduce the overall cost. The key assumption in the cost minimization principle was that the production process could be divided into subprocesses. In addition, these subprocesses must be independent of each other so that their costs could be analyzed separately.

Preceding the cost minimization principle, specialization and scale were the main methods for productivity improvement espoused by the early eighteenth and nineteenth century economists (Istvan 1992). Smith (1986) argued that work was typically specialized by functions such as sales, marketing, engineering, manufacturing, purchasing, controlling, human resources and finance, which were further specialized within each function. This specialization paradigm deeply affected both economic and managerial thinking and led to the scale effect, by which fixed capital costs were substituted for variable labor costs and amortized over sufficient volume to achieve a net economic gain (Istvan 1992).

However, the cost minimization principle was based on the assumption that the subprocesses were independent of each other. When the subprocesses were inter-related, the calculation process needed to be revised.

3. *Buffering*
   According to Slack et al. (1995), it was advantageous to insulate the production process from the external environment through physical or organizational buffering.
   The buffering principle was derived from the economic order quantity model (Harris 1990), which was designed to calculate the level of inventory that could minimize the total inventory holding and ordering costs. The buffering principle was consistent with the cost minimization principle. The cost minimization principle was based on the assumption that all the subprocesses were independent of each other. When the subprocesses were not independent of each other, they needed to be made independent by buffering.
4. *Value of outputs*
   The value of output principle was similar to the cost minimization principle. According to the principle, the value of the outputs was related to the value of the inputs to the production process. The principle was supported by the idea of value added in economics. Christopher (1913) defined value added as the difference between the revenues from the sale of goods or services and the costs of purchase. The value of the outputs of the production process therefore equaled to the sum of the costs of purchase and the value added of the transformation activities.

5.4.1.2 Development of the Transformation Concept of Production

The transformation concept of production and its principles have seldom been consolidated and organized into a well-grounded theory, according to Koskela (2000). Two most fascinating outcomes from the transformation concept of production are the generalized as Walrasian production model (Wortmann 1992) and mass production. In the Walrasian production model, the production process was regarded as a transformation process which could be divided into subprocesses. The model was made up of technical coefficients that equaled the ratio of transformation between the amount of a certain production factor and the amount produced of a given product (Koskela 2000). Henry Ford was the pioneer of modern mass production, which originated from the transformation concept of production. He was also regarded as the first to establish this as a whole principle to reduce the price and increase the profit. The development of the transformation concept of production can be illustrated by Fig. 5.1.

5.4.2 The Flow Concept of Production

The flow concept of production emerged during the 1980s when the transformation concept of production was challenged by industrial practices. The transformation concept was challenged mostly by the representatives from the just-in-time (JIT)

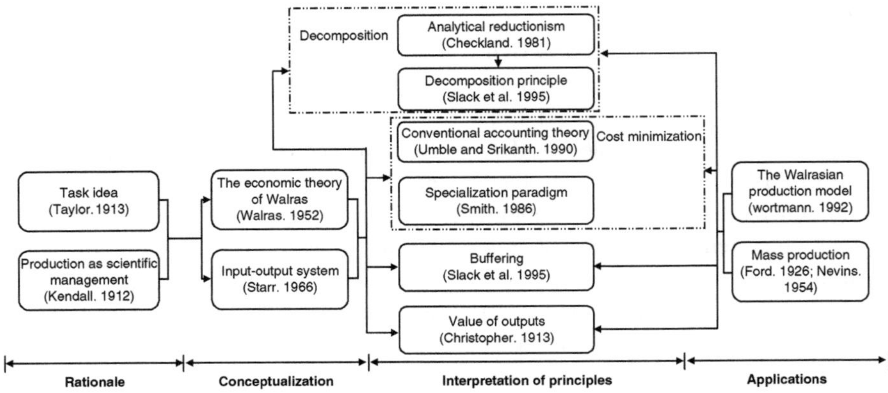

Fig. 5.1 The development of the transformation concept of production

camp (Koskela 2000). According to Shingo (1988, p. 454), the new insight in the production process was explained as follows:

> Process refers to an analysis of production in large units, and operation refers to an analysis of production in small units. Here apparently, processes and operations are considered only categories differing in size of units of analysis. Since processes and operations are perceived as phenomena that can be expressed on the same axis, there may be an unconscious assumption at work that improvements made in small-unit operations necessarily lead to improvements in collective processes.

In addition, according to Shingo (1988), production referred to the flow of products from one worker to another, while operation referred to the discrete stage at which a worker might work on different products. These two phenomena were different in nature. The improvements that were made to the different phenomena should therefore be different. Improvements made to operations should correspond to the nature of operations, while improvements made to processes should correspond to the nature of processes. However, according to Shingo (1988), the West misused these two terms and took for granted that improvements made to operations, the smallest units can lead to improvements to production as a whole.

The contribution of Shingo (1988) to the theory of production is at two levels. First of all, time was introduced to production which was perceived as a physical process rather than an economic abstraction in cost terms or in productivity terms (Koskela 2000). This level of contribution means that the production could be explained by appropriate models. For example, queueing theory could provide the basic physical model of production (Hopp and Spearman 1996). The theory of constraints could also be applied in factories to address a specific constraint of the factory (Goldratt and Cox 1984).

The second contribution of Shingo (1988) was the point that there were two activities in the production process, which were transformation activities and non-transformation activities. Improvements made to these two kinds of activities were totally different. While the transformation activities could be improved, non-transformation activities (also referred to as waste, non-value adding activities) should be eliminated.

5.4.2.1 Principles

1. *Waste*
   Waste was identified as activities that did not add value to the production process. Ohno (1988) identified seven categories of wastes: waste of production, waste of correction, waste of material movement, waste of processing, waste of inventory, waste of waiting and waste of motion. Waste existed in the production process because of three root causes: the structure of production system, the way that production was controlled and the nature of production (Koskela 2000). The products might be designed by one group, executed by another group and examined by another. Waste existed in terms of inspection, moving and waiting. In addition, waste seemed to be inevitable in production due to its nature. Defective materials, accidents, broken equipment and man-made mistakes seemed to be an inherent feature of all production activities (Koskela 2000). Eliminating these wastes was therefore of critical importance to improve the production process.
2. *Other principles*
   The other principles besides waste, which were to reduce the lead time, reduce variability, simplify, increase flexibility and increase transparency, have already been discussed in Sect. 3.4.1 and would therefore not be examined in details here.

5.4.2.2 Development of the Flow Concept of Production

The first application of the flow concept of production was the mass production system. According to Ford (1926), three principles underlay mass production:

1. The planned orderly progression of the commodity through the shop;
2. The delivery of the work instead of leaving it to the workman's initiative to find it; and
3. An analysis of operations into their constituent parts.

In fact, principle one was the feature of the flow concept of production. However, although the mass production system did have the "flow" feature, it was the transformation concept that the mass production system belongs to (Koskela 2000).

The breakthrough in the development of the flow concept of production was the Toyota Production System (also known as just-in-time) in the early 1950s. The idea of just-in-time started to be acknowledged and examined in Europe and America in the 1970s, mainly in the automobile industry (Koskela 2000). The flow concept in western countries was commonly known as lean production. The development of the flow concept of production can be seen in Fig. 5.2.

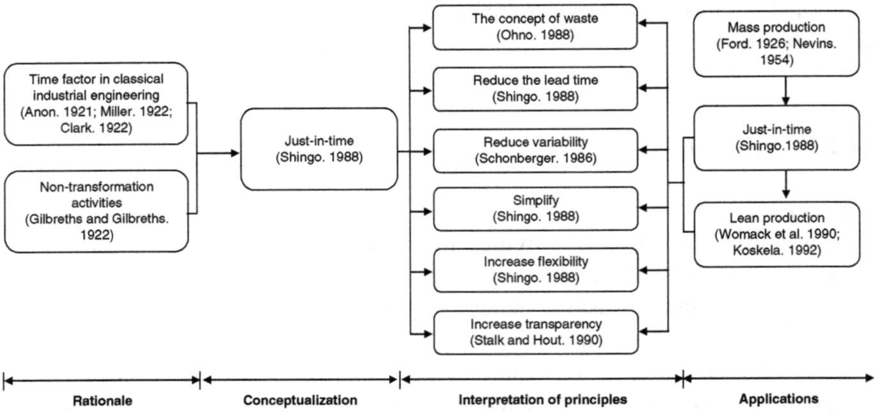

Fig. 5.2 The development of the flow concept of production

5.4.3 The Value Concept of Production

The value concept of production could be traced back to Levitt (1960, pp. 45–56), who stated that:

> Mass production industries are impelled by a great drive to produce all they can. The prospect of steeply declining unit costs as output rises is more than most companies can usually resist. The profit possibilities look spectacular. All effort focuses on production. The result is that marketing get neglected. . . .a truly marketing-minded firm tires to create value-satisfying goods and services that consumers want to buy.

Drucker (1989, p. 276) also stated the same point as follows:

> Finally the most important single thing to remember about any enterprise is that results exist only on the outside. The result of a business is a satisfied customer. . . .Inside an enterprise there are only costs.

Both Levitt (1960) and Drucker (1989) argued that the aim of the production process was to satisfy customer needs rather than to focus on the internal production matters. The problems associated with the value concept of production was that even though the customers provided requirements and the suppliers produced based on these requirements and generated value to the customers, how should these requirements and value be identified? The requirements of the customers could affect the production processes in two ways:

1. Karlsson et al. (1998) stated that the needs and wishes of a customer were condensed into a specification of the production and the specification controls the design function of the production.
2. Chase and Garvin (1989) argued that the customer could communicate directly with the production parties.

In addition, there were several ways to identify and measure the value that the production process generated:

1. Cook (1997) stated that the value could be viewed as prices attached to the product that the customer was willing to pay.
2. Barnard (1995) defined the value as the ratio of benefits/price.
3. Cook (1997) pointed out that the relative value of a production in comparison to its competitor could be viewed as another way to represent value.
4. Womack and Jones (1996) stated that the wrong production or service could be regarded as value loss. This provided one way of measuring value in a relative way.

5.4.3.1 Principles

The principles of the value concept of production included (Koskela 2000):

1. Requirements capture. Requirements capture was requisite to generate value. This principle was elaborated by Bergman and Klefsjo (1994) in the model of customer satisfaction.
2. Requirement flow-down. The requirements of the customer must be available in the production life cycle. The development of the principle was supported by Quality Function Deployment (Akao 1990), organizational means (Griffin et al. 1995) and interaction of employees (Ostroff and Smith 1992).
3. Comprehensive requirements. The comprehensive requirements principle meant that all deliverables to the customer should be taken into consideration in the requirement flow-down. The deliverables should include three levels of consideration: product, service and delivery (Kim and Mauborgne 1997).
4. Ensure the capability of the production system. This principle meant that even though the requirements of customer could be transferred in the production life cycle without value loss, the production system must obtain the capability to design, produce and deliver products as required by the customer. The capability of the production system included statistical control (Shewhart 1931), speed and dependability (Slack et al. 1995).
5. Measurement of value. Measuring value generated by the production process to the customer was of critical importance for further improvement (Whiteley 1991; Kordupleski et al. 1993).

5.4.3.2 Development of the Value Concept of Production

The applications of the value concept of production were at three levels, namely design, control and improvement. In the design stage, Schonberger (1996) proposed that the requirements of the customer could be integrated into the production process as a structuring principle. Quality Function Deployment (Akao 1990) as well as other methods to manage the value generation cycle was also generated from the value concept of production. In addition, the value concept of production has been applied to measure the value that the products could bring to the customer (Whiteley 1991). The development of the value concept of production was illustrated in Fig. 5.3.

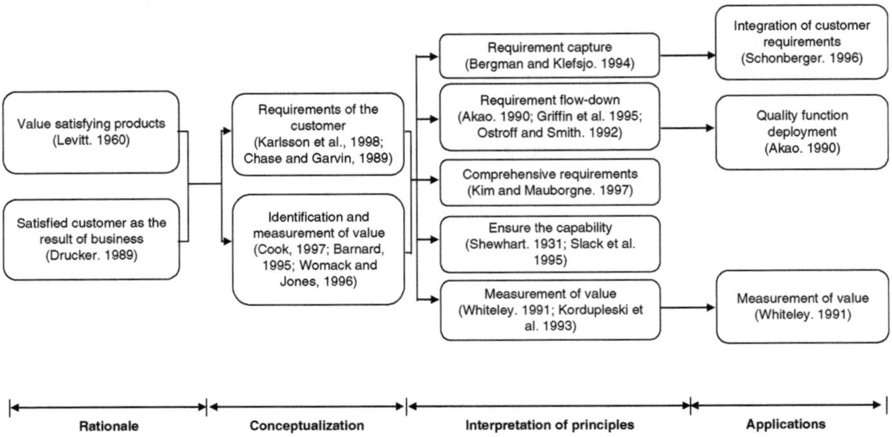

Fig. 5.3 The development of the value concept of production

5.4.4 *TFV (Transformation-Flow-Value) Framework of Production*

The TFV framework combined and balanced the three aspects, which were mentioned above as transformation, flow and value. The contribution of the TFV framework to the concept of production could be described as follows (Koskela 2000):

1. The goals of production could be achieved by "getting production realized efficiency", which was generated from the transformation view. These could also be achieved through the elimination of waste which originated from the flow view. In addition, elimination of value loss in the flow of value could help to achieve the goals of production. However, the three aspects of production should be balanced because their contributions to the production process were different.
2. Managerial-level actions could be taken from transformation, flow and value aspects, and these actions should have better performance than the actions that were taken based on a single perspective. From historical justification, each of the three principles has been proven to lead to major improvements, in terms of performance gains. Managerial-level actions from all three aspects were able to help achieve considerable improvement in production processes.
3. Previous studies relating to lean production failed to provide a theoretical support for lean production. Womack and Jones (1996) identified the philosophy of lean production into five principles. However, these principles were not able to cover the three areas fully. The TFV framework appeared to provide theoretical thinking rather than practices to support the philosophy of lean production.

According to Koskela (2000), the TFV framework can be summarized by Table 5.1. The TFV framework had three parts: conversion, flow and value. Seymour (1996)

Table 5.1 Transformation, flow and value generation concepts of design

	Transformation	Flow	Value
Principles	Decomposition	Elimination of waste	Requirement capture
	Cost minimization	Time reduction	Requirement flow-down
	Buffering	Simplification	Comprehensive requirements
	Value of outputs	Increase transparency	Ensure capability
Methods	Work breakdown structure (WBS); Critical path	Design structure matrix	Quality Function Deployment; Value engineering

Adapted from: Koskela (2000, p. 120)

argued that there was no single theory that could explain the lean production. Womack et al. (1990) argued that the dominant logic in the conversion concept of production was the Cartesian framework to state that the production process was conceived as a machine which converted raw materials into products. Miller and Rice's (1967) theoretical framework made distinctions between operating activities and maintenance activities. These aspects were in fact all included in the concept of the TFV framework. However, the boundaries of the TFV framework seemed to be limited. The actual value of the product was limited to the value that can be brought to the customer. In fact, there were certain values, for example, environmental values to the environment, which might seem unnecessary for the customer, but were truly important for the production process. This study will re-identify the value concept in the TFV framework.

5.5 Economic Explanation of Production

In microeconomics theories, production was regarded as activities that turned inputs into outputs. According to Samuelson and Nordhaus (1985, p. 579), production was defined as:

> The theory of production begins with specific engineering or technological information. If you have a certain amount of labor, a certain amount of land, and a certain prescribed amounts of other inputs such as machines or raw materials, how much output of a particular good can you get? The answer depends upon the state of technology: if someone makes a new invention or discovers a new industrial process, the obtainable output from a given factor inputs will go up. But, at any given time, there will be a maximum amount of output that can be produced with a given amount of factor inputs.

In order to examine the choices that the organization might take to accomplish the transformation goal, abstract models of production have been adopted by economists to avoid discussing many of the engineering intricacies (Nicholson 1998). By regarding production as transformation activities, there were several elements which needed to be addressed, which were the customers, the firm that carried out the transformation activities, and the costs associated with the transformation activities. Microeconomists have developed corresponding theories to address these elements.

5.5.1 Demand Theory (Consumer Choice Theory)

Consumer choice theory, or the theory of demand, was to explain consumer market behavior (Katzner 1970). The theory began with the assumption that demand curves relating to the amount of a particular commodity that a consumer was willing and able to purchase at various prices existed as phenomenal regularities in market economies (Mirowski and Hands 2006). According to Mirowski and Hands (2006), these regularities could be empirically observed by asking the individual how much he (or she) would be willing and able to purchase at various prices and/or by simply observing his (or her) actual consumption behaviour. The development of demand theory has passed through three stages (Hicks 1986):

1. The first stage was in Marshall's *Principles* (Marshall 1920). It was believed that although the statement that Marshall had made was a little different from subsequent theories, the core concept of demand theory was there.
2. The second stage was from the contribution of Pareto (1927). Pareto (1927) solved the set of quantities of the various commodities demanded depending on the set of prices.
3. The third stage of development of demand theory was from the contribution of Johnson (1913) and Slutsky (1915) to apply and improve Pareto's work.

The elementary form of demand theory was that the demand curve sloped downwards so that a fall in the price of a commodity tended to increase the quantity demanded, and that an increase in supply tended to lower price (Hicks 1986). According to Mirowski and Hands (2006), demand theory could offer a solution for the customer to choose the most preferred commodities within the constraint of limited money income. Changes in the price of the commodities would affect the affordability of the customers. The optimal consumption would therefore be affected. Demand theory could be applied to solve environmental problems that were brought by the consumption of products. By imposing a per unit tax on the products that had negative effects on the environment, the demand for the products could be reduced. The consumption would therefore be reduced. Detailed discussion of the market-based approach to solve environmental problems will be presented in Sect. 5.5.3.

5.5.2 The Theory of the Firm

According to Schmidt (2000), there is no single definition of the theory of the firm. There are three strands of literature which employ the same idea but differ on the underlying concepts and issues they addressed. These three strands of literature include (Schmidt 2000):

1. The traditional theory of the firm. The traditional theory of the firm was a part of the economic theory of markets which dealt with problems with identifying the

market prices. The fundamental assumption was that the firm or the organization could be viewed as a production function. Inputs from the markets were transformed into outputs through transformation activities by the production function. The firm was regarded as a profit-maximizing entity whose aims were to react to the changes of inputs and outputs of the market and achieved optimal production. Based on the traditional theory of the firm, the nature of the enterprise was not discussed at all. It was merely viewed as a production function which carried out transformation activities. In addition, the traditional theory of the firm did not provide a conceptual framework to analyze why all enterprises shared certain institutional features and why the precise nature of these features varied according to factors such as industry, firm size and legal form (Schmidt 2000).

2. The theory of the firm could also be viewed as an element in business administration. According to (Schmidt 2000), although economists in this strand did address the question of what constituted the nature of the firm, they failed to address:

   (a) The totality of those aspects and features which were important and necessary to the nature of the firm;
   (b) A clear understanding of the aspects and features which distinguished firms, or enterprises, from those which were not firms or enterprises; and
   (c) The aspects and features which all firms and enterprises had in common.

3. Economists in the traditional theory of the firm believed that the existence of the firms was a fact. It was therefore not important to ask why they existed. In addition, the firm must be profit-oriented and the internal organization of the firm was not relevant to a theory of the firm. These traditional notations were challenged by the modern theory of the firm. Coase (1937) raised the question of why firms existed as functions that were more complex than the firm of conventional pricing theory as indicated in the traditional theory of the firm. Coase's (1937) contribution to the modern theory of the firm was that it inspired researchers to investigate the existence of the firms by putting organizations into specific market problems. Williamson (1975, 1985) expanded Coase's contribution and developed the transaction cost theory, where the existence of firms was analyzed in economic terms – the transaction costs. To put it succinctly, both Coase (1937) and Williamson (1975, 1985) believed that a firm was an instrument for reducing transaction costs (Schmidt 2000). The second element in the modern theory of the firm was the theory of agency, which was clearly drawn by Jensen and Meckling (1976). The agency theory implied that in firms and enterprises, the ownership did not always equal to the management control. The firm was a legal entity which was at the center of a nexus of contracts (Jensen and Meckling 1976). These contracts were the agency between the ownership and the management control. The third building block in the modern theory of the firm was the theory of incomplete contracts. Based on the theory, Grossman and Hart (1986) analyzed the economic factors that had an influence on the decision making process but did not belong to the category of contracts.

5.5.3 The Cost of Production Theory of Value

The cost of production theory of value was one subdivision of Adam Smith's theory of value. According to Smith (1776), the money cost of production equaled to price under competition. Hence, the prices of the products were the sum of the prices of all the inputs, including wages, profits and rent. Dobb (1975, pp. 327–328) has stated:

> For a developed industrial society... Adam Smith advanced an expense-of-production, or 'components', theory of value... As for these components revenues themselves, these were determined by conditions of supply and demand prevailing respectively in the market for labour, for capital, and for land.

The cost of production theory of value could be applied with the demand theory and the theory of the firm. The money cost of production was the sum of the prices of all the inputs. Changes in the prices of the inputs would alter the price of the final product. The customer demand would therefore be affected. Based on the new customer demand, the firm could then analyze and find out the point where efficient allocation of resources could be achieved.

5.6 Economic Explanation of the Environment

Economics theories could be applied to analyze environmental problems – why they occurred and what could be done about them (Callan and Thomas 2000). Application of economic theories to solve environmental problems was not based on scientific knowledge, but on a production basis, the theory of which has been explained in previous sections. In fact, the applications had formed a field of study, which was called environmental economics. Environmental economics was a field of study concerned with solving environmental problems with economics theories. This section aims to explain why environmental problems cause a market failure and how companies should respond to these environmental problems.

5.6.1 The Theory of Public Goods

Paul A. Samuelson had traditionally been credited as the first economist to development the theory of public goods (Samuelson 1954). A public good had the following characteristics (Callan and Thomas 2000):

1. Nonrivalness. Nonrivalness meant that when the good was consumed by one individual, another person was not preempted from consuming it at the same time.
2. Nonexcludability. Nonexcludability meant that preventing others from sharing in the benefits of a public good's consumption was not possible.

According to Callan and Thomas (2000), public goods could lead to a market failure because the nonrivalness and nonexcludability characteristics prevented natural market incentives from yielding an allocatively efficient outcome. In a normal market, the demand was the aggregate value of the expected benefits associated with consumption. When the product was private (e.g. foods, clothes, etc.), the benefits to the customer could only be obtained through purchasing. However, when the product had the characteristics of nonrivalness and nonexcludability and became a public good (e.g. environmental quality), its benefit could be obtained even though it was consumed by other people. When the public good was environmental quality, the consequence could be a damaged ecology.

As the market was not able to allocate effectively when the product was a public good, the most commonly adopted method to solve the problems associated with public goods was intervention by the third party. The third party was usually the government to directly provide public goods. When the public goods fell within the environment-related issues, government agencies that would be activated in Singapore would include the National Environment Agency (NEA), Energy Market Authority (EMA), National Climate Change Committee, etc.

5.6.2 The Theory of Externality

The basic assumption in the theory of externality was that in the market where the production and consumption of goods generated environmental impacts, the effect of the environmental impacts could be identified as an externality. In microeconomics theories, the equilibrium price was the price where the marginal cost of the products equaled the marginal benefits that the products could bring to the customers. However, according to the theory of externality, this was not always the case in the real market. Sometimes, it was difficult to capture all the benefits and costs. If products were consumed by a third party and the benefits and costs associated with the consumption were difficult to define, the third-part effect was called an externality (Callan and Thomas 2000).

The effect of the third party could be either positive or negative. For example, according to Sankar (2000), a technological spillover was a positive externality, because the firm's invention not only benefited the firm but also the society through contributing to the pool of technological knowledge. On the other hand, pollution was regarded as a negative externality. Noise pollution and poor air quality associated with air transportation caused negative effect to the nearby properties. However, these negative effects to the nearby properties were not always captured in the price of the air ticket.

According to Sankar (2000), the property rights associated with environmental resources such as clean air, water in rivers and springs, oceans and atmosphere were not always well-defined. In fact, in most countries, these properties were identified as public domains that were free or unpaid. The prices associated with the use or consumption of these goods were therefore zero in most cases.

The theory of externality in microeconomics theories was the foundation to explain the factor of the environment in economic activities. Three contributions to the theory, from Pigou (1920), Coase (1960) and Baumol and Oates (1988), are briefly discussed as follows:

1. Pigou (1920) dealt with the smoke emissions given out by a factory. The smoke emissions generated negative effects to the neighbourhood. According to Pigou (1920), by imposing a per unit tax on the products of the factory, the demand of the products of the factory could be reduced. Hence, the environmental costs of the products could be realized in the price of the products. Pigou (1920, p. 194) also stated that "sometimes, when the interrelations of the various private persons affected are highly complex, the government may find it necessary to exercise some means of authoritative control".
2. Coase's contribution to the theory of externality was known as the Coase theorem (Coase 1960). Coase (1960) argued that when dealing with externalities with negative effects, the problem was not only about restraining those who were responsible for them, but also identifying the loss that would happen somewhere else as a result of the restraining.
3. Baumol and Oates (1988) worked on the per unit tax developed by Pigou (1920). The standards based on the availability of information and knowledge were identified first. The government could either force the standards by charging, or introducing tradable permits and quotas (Baumol and Oates 1988).

5.6.3 Economic Solution to Environmental Problems

Having discussed the reasons that could cause an environmental market failure, appropriate solutions could be provided. There were mainly two ways to solve environmental problems. One was the command and control approach. By adopting this approach, the government advocated pollution limits or technology-based restrictions to directly regulate polluting sources (Callan and Thomas 2000). However, the government and policy makers were usually not able to set the standards at an efficient level. There were extensive data which were necessary to measure the marginal cost and benefit in the demand and supply model, which made it difficult to set the standard at the point where the marginal cost of the standard equaled the marginal benefit. A market-based approach to solve the environmental problems was therefore necessary.

According to Callan and Thomas (2000), unlike the command and control approach, the market-based approach to solve environmental problems adopted economic incentives to achieve efficient allocation of resources. The major categories of the market-based approach included:

1. Pollution charge. The pollution charge was the fee imposed on the pollutants.
2. Abatement equipment subsidies. Government provided abatement equipment subsidies to reduce the cost of solving the environmental problems.

3. A deposit/refund system. In the system, the company was imposed an advance payment for the potential environmental problems that the production process might cause to the environment. The company was then refunded later by solving the environmental problems appropriately through proper disposal, recycling, etc.
4. The pollution permit trading system. This system involved the issuance of tradeable "rights to pollute" based on a given environmental objective. In fact, the Kyoto Protocol had already introduced this trading system for carbon trading in 1992.

The market-based approach forced the company to take the price of environmental problems into consideration and would therefore lead to a more efficient allocation of resources than the command and control approach would do. However, not all the market-based approaches were suitable to a specific situation. The selection of the appropriate market-based approach should be based on both the internal and external environment of the company.

5.7 Environmental Management

Concerns about the environment emerged during the 1960s (Garlauskas 1975). According to Beale (1975), the environment was perceived almost exclusively as a concern of the industrialized countries, rooted in the growing evidences that the processes of urbanization and industrialization which had produced benefits for these societies were also producing unexpected costs in terms of air and water pollution, destruction of natural resources and deterioration in the quality of urban life. According to Garlauskas (1975), the responses to the environment could be divided into four phases. The first phase evolved from the concerns about the environment, which were raised by scientific communities, supported by direct evidences and accepted by citizens, industries and the government during the 1960s. The second phase was a reaction to the concerns about the environment in the 1970s. The most important event in the second phase was the United Nations Conference on the Human Environment in Stockholm, Sweden, from June 5–16, 1972. The conference moved the environment issue into the centre of the world and established it as an important item on the agenda of governments throughout the world (Beale 1975). The second phase was followed by the organization phase when existing legislations, addition of action-forcing provisions and development of comprehensive compliance plans emerged. The fourth stage was characterized by implementing the reversal of environmental degradation at both national and global levels.

Environmental management was viewed as systematic approaches designed to deal with global, national and other scales of environmental problems through the control of disruptive human activities (Garlauskas 1975). The elements of environmental management will be discussed in the following sections.

5.7.1 Environmental Management: The Science of Ecology

According to Garlauskas (1975), environmental management could be defined as a systematic and scientific approach to assure a congruous interface between natural and man-made systems. In environmental management, the basic premise was the science of ecology, which was part of the biological sciences that dealt with the relationship between organisms and their surroundings (Odum 1971). The main objective of environmental management is to minimize the impact of man's activities on the physical and biological environment through an ecosystem perspective (Edmunds and Letey 1973).

According to Odum (1971), man was able to alter and disturb any ecosystem components, and thereby profoundly alter the environmental quality. Environmental management aimed to control man's activities to achieve the following objectives (Garlauskas 1975):

1. to prevent irreversible changes in ecology;
2. to assure man's survival and to progressively improve his quality of life; and
3. to direct man's processes as parts of evolving nature.

The actions that could be adopted to achieve the objectives listed above included (Garlauskas 1975):

1. visualize all processes in total perspective;
2. recognize and understand any processes or problems in the structure and its component interrelationships;
3. be able to manipulate or otherwise deal with the interdependencies characterizing the process and operation of the whole; and
4. be able to design, build and operate the management system which would serve as a means to manage any whole.

However, simply regarding environmental management as the science of ecology was not able to address the environmental problems that were caused by human beings. The activities which had negative effect on the environment were caused by man, who therefore was the subject of environmental management. This ecology view did not reveal the fact that man was the most important factor in environmental management. Environmental management was not "management of the environment"; it was about management of people and their activities within the constraints of the environment (Beale 1975).

5.7.2 Environmental Management: A Dynamic Equilibrium

Human beings should be the centre of environmental management. According to Garlauskas (1975), the essence of environmental management was that it allowed man to continue to evolve the technology without profoundly altering natural ecosystems through a systematic analysis, understanding and control. It could be

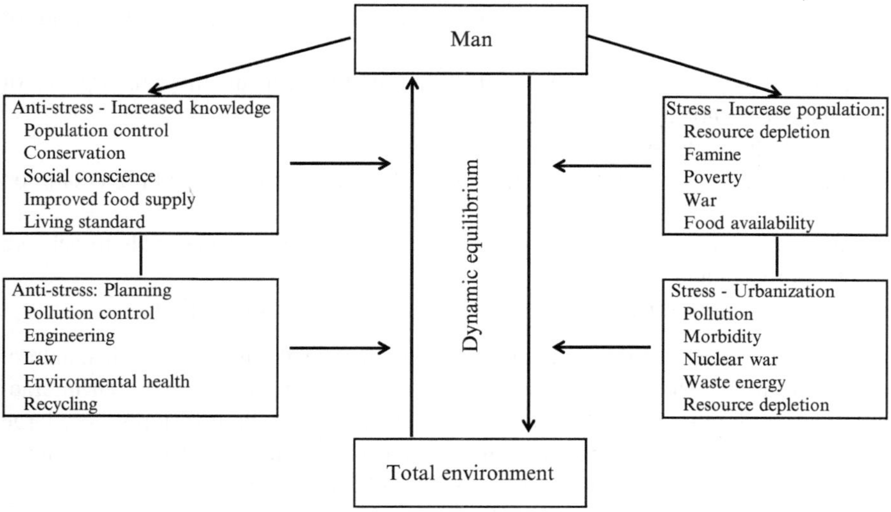

Fig. 5.4 The dynamic equilibrium between man and the environment (Adapted from: Garlauskas 1975)

explained by the dynamic equilibrium between man and the environment, which is described in Fig. 5.4.

As can be seen from Fig. 5.4, environmental management was about analyzing the current stress and anti-stress situation to achieve the dynamic equilibrium. The patterns that man could affect the dynamic equilibrium were divided into two groups, which were unbalancing and compensating activities. The unbalancing activities referred to the activities that imposed stresses on the dynamic equilibrium, such as increased population and the resulting demands (Ehrlich and Ehrlich 1972; Meadows et al. 1972). The compensating activities referred to the anti-stress activities which were brought about by improved knowledge.

As the anti-stress activities were developed according to the stress activities, there was a time lag in counteracting the equilibrium shift (Garlauskas 1975). A long time lag might result in the collapse of the dynamic equilibrium. In addition, a sudden crisis could also cause a partial or complete breakdown. The dynamic equilibrium was also affected by external stresses, such as earthquakes, radiation, floods and so on, which could periodically destroy the equilibrium between man and the environment (Tank 1973). Environmental management was a systematic tool which could be applied to manage the equilibrium between man and the environment. As can be seen from Fig. 5.4, environmental management was applied through anti-stress activities with the development of knowledge and technology.

5.7.3 Environmental Management: A Systems Concept

According to Miles Jr. (1973), the term "system" could be used to explain:

1. a complex unit formed of many often diverse parts subject to a common plan or serving a common purpose;

2. an aggregation or assemblage of objects joined in regular interaction or interdependence;
3. a set of units combined by nature in order to form an integral, organic or organized whole;
4. an orderly working totality;
5. a group of devices or artificial objects that form a network or used for a common purpose;
6. an organized and established procedure, method or set of materials to achieve the purpose; and
7. an organization for the collection and distribution of information.

Generally speaking, environmental management dealt with two systems: natural and man-made systems. These two systems were so complicated that their interaction with each other was difficult to define. In addition, as both systems were large in terms of input factors, functions and so on, each causal relationship between the natural and man-made environment was difficult to identify as well.

Brown (1970) therefore proposed a systems concept in environmental management, which includes:

1. Systems approach. The environmental management should be planned through a systems approach, by which the relationships between the natural and man-made environment as well as management techniques were fully taken into consideration.
2. Systems analysis. In systems analysis, it was the problems rather than the solutions that should be focused upon. Problem formulation, issue definition and past insight were the three main parts in systems analysis.
3. Systems management. Solutions to solve the problems identified in systems analysis were provided in systems management. The systems that should be considered in this stage included:

   (a) Environment. The science of ecology, which is explained in Sect. 5.7.1, might be applied to take the environment into consideration because management decisions should be made on the knowledge of the natural system. For example, when making management decisions related to reducing carbon emissions, the causes and impacts should be systematically analyzed. In addition, assessments should be conducted at both qualitative and quantitative levels through accepted and scientific procedures.
   (b) Disruptions. The stress activities in the dynamic equilibrium should be investigated as disruptions to the natural environment. Similar with the procedures in the environment section, both qualitative and quantitative assessments should be conducted to identify the scale, mode and effects of the disruptions.
   (c) Effects. The effects of the disruptions were fundamental to establish the solutions.
   (d) Human ecosystem. Public awareness and socioeconomic conditions might offer new insight to provided solutions to environmental problems other than technological improvements.

(e) Technological and engineering controls. This was the area where applied sciences were applied through artificial control to preserve and conserve the natural environment.
(f) Legal controls. According to Reitze Jr. (1972), many legal aspects were formulated without consideration or in-depth knowledge of the environment and technological capabilities to achieve the standards. Integration of scientific principles into legal concepts required a fundamental understanding of the related law and regulations (Barros and Johnston 1974).

5.7.4 Environmental Management and Management Theories in Production

Environmental management could be applied to investigate the relationship between companies and the environment. Project managers should then make appropriate management decisions based on the information from the analysis. In fact, the analysis was part of the systems theory. In an explanation of systems theory, Shafritz and Ott (1992, pp. 263–264) stated that:

> A system is any organized collection of parts united by prescribed interactions and designed for the accomplishment of specific goals or general purposes... Systems theory views an organization as a complex set of dynamically intertwined and interconnected elements, including its inputs, process, outputs, feedback loops, and the environment in which it operates and continually interacts... Organizations are not static, but... are adaptive systems that are an integral part of their environment.

How organizations were affected by the external environment originated from the work carried out by Daniel Katz, Robert Kahn and James Thompson, by viewing organizations as an open system (Katz and Kahn 1966; Thompson 1967). According to Katz and Kahn (1966), the organization was viewed as an open system with three states, namely the input, the conversion and the output stage. In the input stage, the organization acquired resources, such as materials, labor and equipment to produce goods or provide services. In the conversion stage, the inputs were transformed into outputs in the form of finished goods or services. In the output stage, the goods and services were released to the external environment.

Another important feature in environmental management was the adoption of management science theories. In management science theories, quantitative approaches were used to help managers to make decisions. In fact, management science theory was derived from the scientific management idea (Taylor 1913). If the negative environmental effects could be represented in the production process by imposing additional economic factors, the effects of these additional economic factors should be explained by management science theories, such as the theory of constraint, queueing theory, etc. These two management science theories will be discussed shortly.

The Goal (Goldratt and Cox 1984) firstly introduced the constraint concepts as "constraint management" and the "Theory of Constraints (TOC)". The TOC was an overall management philosophy which could help to achieve better management on a continuous basis. It had two broad viewpoints: that of the business system and of an ongoing improvement process itself (Gupta et al. 2002). When applying the TOC to the business system, it emphasized that organizations should take the following steps in order to "make more money now as well as in the future" without violating certain necessary conditions (Gupta 2003, p. 649):

1. Identify the system's constraint(s), whether physical or policy constraint.
2. Decide how to exploit the system's constraint(s), i.e. get the most possible from the limit of the current constraint(s); reduce the effects of the current constraint (s); and make everyone aware of the constraint(s) and its effects on the performance of the system.
3. Subordinate everything else to the above decision, i.e. avoid keeping non-constraint resources busy doing unneeded work.
4. Elevate the system's constraints(s), i.e. off-load some demand or expand capability.
5. If in the previous steps a constraint has been broken, go back to Step 1, but do not allow Inertia to cause a system constraint.

From the perspective of an ongoing improvement process, the TOC suggested that companies ask themselves three questions concerning change to accelerate their improvement process. These included (Gupta et al. 2002):

1. What to change, i.e. how do organizations identify the weakest link, i.e. the constraint(s)?
2. To what to change, i.e. once the weakest link is identified, how organizations should strengthen it by developing practical and good solutions?
3. How to cause the change, i.e. how organizations should implement the solutions?

Practical applications of the TOC in both production and manufacturing included production control, variation reduction and productivity improvement. The process to achieve these goals was similar, which is explained above at the organizational level and from the perspective of an ongoing improvement process.

The study of queues was a component of many courses in Operational Research (Bunday 1986). According to Bunday (1986), one main purpose of analyzing queues was to enable management to predict the outcomes of changes, including changes in the pattern of arrivals, reducing the service time, increasing the numbers of servers, etc. The basic characteristics of queueing processes were explored as follows (Gross and Harris 1974):

1. **Arrival pattern of customers.** The arrival pattern of customers (also referred to as inputs, if the queueing theory is applied to the production industry) was measured by the average number of arrivals per unit time or by the average time between successive arrivals. In addition, factors relating to simultaneous

arrivals and whether or not the arrivals would change with time should be taken into consideration.

2. **Service patterns of the servers.** The service patterns of the servers could be measured by the number of customers served per some unit of time (or the number of inputs which are processed in some unit of time). In the manufacturing and production industry, this was largely related to the system capacity, which was often seen due to the finite limit of the equipment.
3. **Queue discipline.** Queue discipline referred to the manner by which customers were selected for service. It can be first in, first out or last in, first out. One example in the manufacturing and production industry would be storage. Newly produced units were put on top of the old ones, which followed the last in, first out principle.
4. **Other considerations.** Other considerations such as the number of service channels and stages of services would have an impact on the result. It should be noted that these considerations were usually not sufficient to describe the processes that were being investigated. The correct model or at least the mode that best described the situations should be used for more accurate results.

Application of the queueing theory in the manufacturing industry was concerned about production flow and inventory control. Although mathematical models could be adopted in queueing theory, general recommendations using queueing theory should be provided, as not all problems in real life corresponded to a specific mathematical model. Applications of the queueing theory should go beyond the mathematical level and provide recommendations in making management decisions.

5.8 Conceptual Framework

The conceptual framework for this study was closely related to the aim and objectives of this study, which has been explained in Sect. 1.3. The conceptual framework is useful to outline the theoretical supports of this study and present an approach to identify new ideas and thoughts. In the concept of sustainable development and the construction industry, there were subsystems. For sustainable development, there were three domains, which were economic, social and environmental sustainability. For the construction industry, factors from the stakeholder level, business level, project level and so on should all be considered to deal with the potential changes brought about by sustainable development. In this study, the relationship between environmental sustainability and the project level in the construction industry was investigated.

In a similar way, the concept of environmental sustainability and the project level in the construction industry adopt a systems concept as well (Brown 1970). The relationship between global climate change and three main stages in the construction industry was investigated. According to the dynamic equilibrium, there should be a dynamic equilibrium between global climate change and the precast concrete industry as well. The dynamic equilibrium dealt with two concerns. Firstly, what were the potential changes that global climate change would bring to the precast concrete industry? For example, construction companies might face regulatory controls to reduce carbon emissions and may have to take the price of carbon emissions into consideration. Secondly, how should the precast concrete industry react to global climate change? Technical innovations to reduce carbon emissions were much focused upon to reduce carbon emissions in the construction industry at the time of this study. However, management improvements should not be overlooked because this could well be more cost effective.

As can be seen from Fig. 5.5, three different stages in the precast concrete life cycle were investigated in this study, which are the production, delivery and construction stages. Each stage had its own inputs and outputs. For example, in the production stage, the inputs were the various raw materials, human resources, equipment, etc. The outputs were the precast concrete components that served as inputs in the delivery and construction stages.

The three stages were organized into two open systems. In fact, these two open systems were related to the precasters and contractors in the precast concrete industry. Through the economic explanation of the environment, the impact of carbon emissions on the two open systems should be addressed. In traditional production theories, the production goals were about providing the intended products based on the needs of the customer. Environmental considerations were usually not included in the production goals. This study was concerned with taking the carbon emissions into consideration as one production goal in the precast concrete industry.

In addition, the dynamic equilibrium between global climate change and the precast concrete industry was currently managed under the command and control approach. However, the market-based approach should also be advocated because it could ensure an efficient allocation of resources. Government-mandated environmental standards were not likely to be set at an efficient level (Callan and Thomas 2000). In the near future, both the precasters and the contractors might have to adopt a cost-effective carbon reducing method. By transforming the negative effect of global climate change into economic consideration as one input into the production process, the traditional open-system view of the production process and construction companies should be re-investigated.

With the support of the transformation, flow and value concept of production, the traditional open-system view of the production process and construction companies were investigated by taking the economic factor of carbon emissions into consideration. A systems concept, which includes the environment, disruptions, effects, legal controls, technological and engineering controls and human ecosystems, was adopted in this study, as has been explained in Sect. 5.7.3.

Investigation of the environment and identification of the effects and disruptions of carbon emissions would be carried out based on the science of ecology. The two open systems would be re-investigated through the three production views. The transformation, flow and value concept of production were the fundamental of the

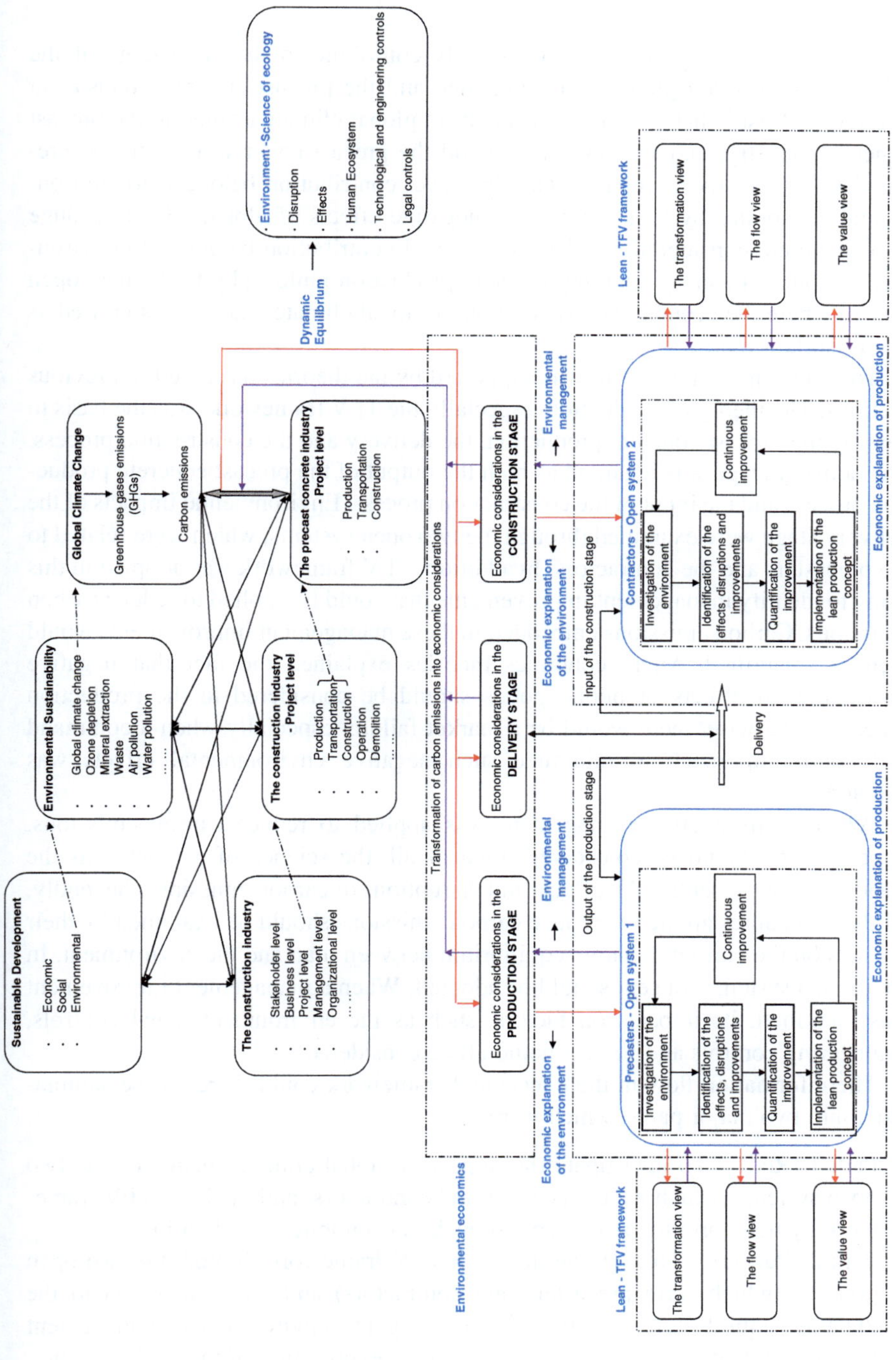

Fig. 5.5 Conceptual framework

lean production concept. The principles in each production view were the principles adopted to analyze the new open system where the costs of carbon emissions were considered.

The results of this study would directly contribute to the knowledge of the relationship between global climate change and the precast concrete industry at two levels. First, what were the impacts from global climate change in the precast concrete industry; and secondly how could the precast concrete industry be prepared to address these impacts. The first level contribution belonged to environmental economics by transforming the negative effects of global climate change into economic consideration. The second level contribution belonged to environmental management by applying the lean production philosophy to the new open system where the economic consideration of global climate change was viewed as one important input.

The conceptual framework was supported by the theories discussed in previous sections. The theory of production, especially the TFV framework, was the basis to analyze the precast concrete production, the delivery and the construction process. The delivery stage was regarded as both the output of the precast concrete production process and the input of the construction process. Environmental impacts of the delivery stage were examined through the two open systems, which were related to the precasters and the contractors. In addition, TFV framework was adopted in this study to identify management improvements that could be applied to reduce carbon emissions. Carbon emissions embedded in these management improvements would then be quantified. Microeconomics theories explained the fact that negative environmental effects of public goods should be considered in the production process. Otherwise, there would be a market failure, especially when a command and control approach to minimize such negative environmental effects was adopted.

When environmental management was applied to reduce carbon emissions, three aspects should be considered. First of all, the science of ecology was the basic premise to identify the effects and disruptions of carbon emissions. Secondly, all the procedures applied to reduce carbon emissions should be examined by their impacts on the current dynamic equilibrium between man and the environment. In addition, a systems concept should be adopted. When a management improvement was identified, its impacts on factors, such as the environment, legal controls, engineering controls and so on, should all be considered.

The information flow in the conceptual framework could therefore be summarized into two major paths, which were:

- Path 1 (Red) – To investigate the impact of global climate change on the two open systems (which were precasters and contractors) and the Lean TFV framework by transforming carbon emissions into economic considerations.
- Path 2 (Purple) – To examine the Lean TFV framework through the two open systems (which were precasters and contractors) and its contribution to the dynamic equilibrium in terms of improving the environmental management theory and practices, as well as the economic explanation of the environment.

5.9 Summary

This chapter investigated the theoretical framework of this study. General speaking, the TFV framework, the underlying theory of the lean production, was applied in this study. However, the boundaries of the TFV framework should be expanded in order to take the environmental impacts of the production process into consideration. In the transformation concept of the production process, when considering the raw materials, the economic interpretation of its environmental issues should also be taken into consideration. Their values should be re-examined by introducing the environment into the decision making process (research objective 3).

The theoretical framework was supported by microeconomic theories, especially the economic explanation of the production. In addition, the effects of global climate change due to carbon emissions were analyzed based on the science of ecology. Other aspects, including the legal issues, technological and engineering controls and so on, were all analyzed based on a systems approach.

This study would be carried out on two open systems, which were precasters and contractors. In each of the open system, its inputs and outputs were analyzed separately based on a systems concept. The environmental management process and the lean production concept were applied on both open systems and their effects on the open systems were measured by environmental factors. These two open systems are precasters (Chaps. 7 and 8) and contractors (Chaps. 9 and 10). The results would contribute to the dynamic equilibrium between the precast concrete industry and the global climate change by answering how the precast concrete industry should adapt to maintain the dynamic equilibrium (research objectives 1, 2 and 4). In addition, the impact of the lean philosophy on environmental evaluations (path 2) will be useful for the evaluation method to evolve (research objective 5).

References

Akao Y (1990) History of quality function deployment in Japan. In: Sinha MN (ed) The best on quality: targets, improvements, systems. ASQ Quality Press, Milwaukee, pp 183–196

Anon (1921) Report on elimination of waste in industry. Mechanical Engineering, September, pp 579–583

Arthur WB (1993) Pandora's marketplace. New Scientist Supplement. pp 6–8

Ball M (1993) Housing policy and economic power: the political economy of owner occupation. Methuen, London

Banard B (1995) Voice of the customer's impact on production inventory and process, resources management, and planning. In: Proceedings of the 38th APICS international conference and exhibition, Orlando

Barros J, Johnston DM (1974) The international law of pollution. Free Press, New York

Baumol WJ, Oates WE (1988) The theory of environmental policy. Cambridge University Press, New York

Beale JG (1975) Draft guide to environment protection and management: a basis for national environment initiatives. A report prepared for discussions with the UNEP, Nairobi, Kenya

Bergman B, Klefsjo B (1994) Quality from customer needs to customer satisfaction. McGraw-Hill, London
Bertelsen S (2003) Construction as a complex system. In: Formoso CT (ed) Proceedings of IGLC-11, Blacksburg
Brown MD (1970) Use of systems techniques in environmental quality management, Monograph no 7. National Sanitation Foundation, Ann Arbor
Browne J, Harhen J, Shivnan J (1988) Production management systems: a CIM perspective. Addison-Wesley, Wokingham
Buffa E (1961) Modern production management. Wiley, New York
Bunday BD (1986) Basic queueing theory. Edward Arnold Ltd., London
Callan SJ, Thomas JM (2000) Environmental economics and management: theory, policy and applications. Dryden Press, Fort Worth
Chase RB, Garvin DA (1989) The service factory. Harv Bus Rev 67(4):1
Checkland P (1981) Systems thinking, systems practice. Wiley, Chichester
Christopher W (1913) Economic productivity: a useful value added measure of business performance. In: Christopher W, Thor C (eds) Handbook for productivity measurement and improvement. Productivity Press, Portland
Clark W (1922) Relieving industry of burden. Mechanical Engineering, December, pp 819–820
Coase RH (1937) The nature of the firm. Economica 4(16):386–405
Coase R (1960) The problem of social cost. J Law Econ 3(1):1–44
Cohen S, Demeritt D, Robinson J, Rothman D (1998) Climate change and sustainable development: towards dialogue. Glob Environ Chang 8(4):341–371
Cook HE (1997) Production management – value, quality, cost, price, profit and organization. Chapman & Hall, London
Coutard O (ed) (1999) The governance of large technical systems. Routledge, London
Dam A, Elsborg S (2003) Entrepreneurship in the construction industry: a development and research project. In Proceedings of IGLC-12, A Danish Experiment on Cooperation in Construction (in Danish). Elsinore, Denmark
Dobb M (1975) Ricardo and Adam Smith. In: Skinner AS, Wilson T (eds) Essays on Adam Smith. Clarendon, Oxford, pp 324–335
Drucker PF (1989) The new realities: in government and politics, in economics and business, in society and world view, 1st edn. Harper & Row, New York
Edmunds S, Letey J (1973) Environmental administration. McGraw-Hill, New York
Ehrlich PE, Ehrlich A (1972) Populations, resources, environment. W.H. Freeman, San Francisco
Ford H (1926) Mass production. In: Encyclopedia Britannica, 13th edn. Encyclopedia Britannica, Chicago, pp 821–823
Garlauskas AB (1975) Conceptual framework of environmental management. J Environ Manag 3(3):185–203
Gilbreth FB, Gilbreth LM (1922) Process charts and their place in management. Mechanical Engineering, January, pp 38–41
Goldratt EM, Cox J (1984) The goal. North River, Croton-on-Hudson
Griffin A, Gleason G, Preiss R, Sevenaugh D (1995) Best practice for customer satisfaction in manufacturing firms. Sloan Manag Rev 36(2):87–98
Gross D, Harris CM (1974) Fundamentals of queueing theory. Wiley, New York
Grossman S, Hart O (1986) The costs and benefits of ownership: a theory of vertical and lateral integration. J Polit Econ 94(4):690–719
Gupta M (2003) Constraints management – recent advances and practices. Int J Prod Res 41 (4):647–659
Gupta M, Ko HJ, Min H (2002) TOC-based performance measures and five focusing steps in a job-shop manufacturing environment. Int J Prod Res 40(4):907–930
Harris FW (1990) How many parts to make at once. Oper Res 38(6):947–950
Hay J, Mimura N (2006) Supporting climate change vulnerability and adaptation assessments in the Asia-Pacific Region: an example of sustainability science. Sustain Sci 1(1):23–35

Herbert G (1984) The dream of the factory-made house. MIT Press, Cambridge, MA
Hicks J (1986) A revision of demand theory. Oxford University Press, New York
Hopp W, Spearman M (1996) Factory physics: foundations of manufacturing management. Irwin/McGraw-Hill, Boston
Horgan J (1995) From complexity to perplexity. Scientific American Magazine
Istvan RL (1992) A new productivity paradigm for competitive advantage. Strateg Manag J 13 (7):525–537
Jensen M, Meckling W (1976) Theory of the firm: managerial behavior, agency costs and ownership structure. J Financ Econ 3(4):305–360
Johnson WE (1913) The pure theory of utility curves. Econ J 23:483–513
Karlsson C, Rajesh N, Klas S (1998) Black box engineering: redefining the role of product specifications. J Prod Innov Manag 15(6):534–549
Katz D, Kahn R (1966) Organizations and the systems concept. In: Shafritz J, Ott J (eds) Classics of organization theory. Brooks/Cole, Pacific Grove, reprinted in 1992
Katzner D (1970) Static demand theory. Macmillan, London
Kendall H (1912) Systematized, unsystematized and scientifically managed plants. In: Gilbreth F (ed) Scientific management course: a landmark series of lectures given at the YMCA. Hive Publishing, Worcester
Kim W, Mauborgne R (1997) Value innovation: the strategic logic of high growth. Harv Bus Rev 75(1):103–112
Komiyama H, Takeuchi K (2006) Sustainability science: building a new discipline. Sustain Sci 1 (1):1–6
Kordupleski R, Rust R, Zanhorik A (1993) Why improving quality doesn't improve quality (or whatever happened to marketing?). Calif Manag Rev 35(3):82–95
Koskela L (1992) Application of the new production philosophy to construction. CIFE technical report#72, Stanford University, Stanford
Koskela L (2000) An exploration towards a production theory and its application to construction. Technical Research Centre of Finland, Espoo
Levitt T (1960) Marketing myopia. Harv Bus Rev 38:45–56
Lucas C (2007) The philosophy of complexity. Online available at: http://www.calresco.org/themes.htm. Accessed 4 Mar 2011
Marshall A (1920) Principles of economics: an introductory volume. Macmillan, London
Meadows DH, Meadows DL, Randers R, Behren WW III (1972) The limits of growth. Universe Books, New York
Mihelcic JR, Crittenden JC, Small MJ (2003) Sustainability science and engineering: the emergence of a new metadiscipline. Environ Sci Technol 37(23):5314–5324
Miles RF Jr (ed) (1973) Systems concepts. Wiley, New York
Miles RE, Snow CC (1986) Organizations: new concepts for new forms. Calif Manag Rev 28 (2):62–73
Miller FJ (1922) Prevention of wastes in industry. Mechanical Engineering, January, pp 9–10
Miller EJ, Rice AK (1967) Systems of organization. Tavistock Publications, London
Miller R, Hobday M, Leroux-Demers T, Olleros X (1995) Innovation in complex systems industries: the case of flight simulators. Ind Corp Chang 4(2):363–400
Mirowski P, Hands DW (2006) Introduction to agreement on demand: consumer theory in the twentieth century, History of political economy 38 (annual supplement). Duke University Press, Durham
Nevins A (1954) Ford: the times, the man, the company. Scribner, New York
Nicholson W (1998) Microeconomic theory: basic principles and extensions. Dryden Press, Fort Worth
Odum EP (1971) Fundamentals of ecology. W.B. Saunders, Philadelphia
Ohno T (1988) Toyota production system: beyond large-scale production. Diamond, Inc., Tokyo
Ostroff F, Smith D (1992) The horizontal organization. McKinsey Q 1:148–168
Pareto V (1927) Manual of political economy. Kelley, New York, Translated and printed in 1971

Pigou AC (1920) The economics of welfare. Macmillan, London

Reitze AW Jr (1972) Environmental law. North American International, Washington, DC

Salsbury S (1991) Facing the collapse of the Washington power supply system. In: La Porte TR (ed) Proceedings of the NATO advanced research workshop on social responses to large technical systems: regulation, management, or anticipation, Berkeley

Samuelson PA (1954) The pure theory of public expenditure. Rev Econ Stat 36(4):387–389

Samuelson PA, Nordhaus WD (1985) Economics, 12th edn. McGraw-Hill, New York

Sankar U (ed) (2000) Environmental economics. Oxford University Press, New Delhi

Schmidt RH (2000) Erich Gutenberg and the theory of the firm. In: Albach H (ed) Theory of the firm. Springer, Berlin, pp 3–39

Schonberger RJ (1986) World class manufacturing. The Free Press, New York

Schonberger RJ (1996) World class manufacturing: the next decade. The Free Press, New York

Seymour D (1996) Developing theory in lean construction. In: Proceedings of IGLC-4, Birmingham, England

Shafritz J, Ott J (1992) Classics of organization theory. Brooks/Cole, Pacific Grove

Shewhart WA (1931) Economic control of quality of manufactured product. Van Norstrand, New York

Shingo S (1988) Non-stock production. Productivity Press, Cambridge, MA

Slack N et al (1995) Operations management. Pitman, London

Slutsky EE (1915) Sulla teoria del bilancio del consonatore. In: Stigler GJ, Boulding KE (eds) Readings in prices theory. Homewood, Irwin, reprinted in 1952

Smith A (1776) The wealth of nations. Cosimo, Inc., New York

Smith A (1986) The wealth of nations. Viking Penguin, New York

Sotherton D, Chappells H, Van Vliet B (eds) (2004) Sustainable consumption: the implications of changing infrastructures of provision. Edward Elgar Publishing, Cheltenham

Stalk GJ, Hout TM (1990) Competing against time. Free Press, New York

Starr M (1966) Evolving concepts in production management. In: Buffa ES (ed) Readings in production and operations management. Wiley, New York

Tank RW (1973) Focus on environmental geology. Oxford University Press, New York

Taylor FW (1913) The principles of scientific management. Harpers & Brothers, New York

Thompson J (1967) Organizations in action. In: Shafritz J, Ott J (eds) Classics of organization theory. Brooks/Cole, Pacific Grove, reprinted in 1992

Umble MM, Srikanth ML (1990) Synchronous manufacturing. South-Western Publishing, Cincinnati

Walras L (1952) Elements of pure economics. The Econometric Society, Paris

WCED (1987) Our common future. Brudtland report. Oxford University Press, Oxford

Whiteley RC (1991) The customer-driven company. Addison-Wesley, Reading

Wild R (1984) Production and operations management, 3rd edn. Rinehart and Winston, London

Williams T (1999) The need for new paradigms for complex projects. Int J Proj Manag 17 (5):269–273

Williamson O (1975) Markets and hierarchies: analysis and antitrust implications. Free Press, New York

Williamson O (1985) The economic institutions of capitalism. Free Press, New York

Winch G (2003) Models of manufacturing and the construction process: the genesis of re-engineering construction. Build Res Inf 31(2):107–118

Womack JP, Jones DT (1996) Lean thinking. Simon & Schuster, New York

Womack JP, Jones DT, Roos D (1990) The machine that changed the world: the story of lean production, 1st edn. Harper Perennial, New York

Wortmann JC (1992) Factory of the future: towards an integrated theory for one-of-a-kind production. In: Hirsch BE, Thoben KD (eds) One-of-a-kind production: new approaches. Elsevier, Amsterdam, pp 37–74

Chapter 6
Research Design and Methodology

6.1 Introduction

This chapter aims to address the research methodology and operationalized measureables. Survey and case study research method are adopted in this study. Overall, the measureables were divided into two categories: one that was related to identify the non-value adding activities and to rank the relative importance of these activities; and the other category that was related to calculate the carbon emissions embedded in these activities. It should be noted that some of the measureables in the latter category should be carefully selected due to the calculation principle that was explained in Sect. 2.4.3. If the data required for a Tier 2 approach calculation (using country-specific emission factors) could be obtained, it was unnecessary to collect the data required for a Tier 1 approach calculation (using world-wide average emission factors).

6.2 Research Methodology

Some philosophical assumptions of this study, which are closely related to the research methodology, should be firstly examined. According to Burrell and Morgan (1979, p. 1), all social scientists approach their subject via explicit or implicit assumptions about the nature of the social world and the way in which it may be investigated.

According to Baptiste (2001), research is always an attempt to investigate something (i.e. reality) and ontology deals with the question of what is real. Ontology refers to normative commitments about the nature of reality, human nature, and the nature of human experience (Patterson and Williams 1998). The ontology of this study is to find whether there is such a link between lean philosophy and low carbon emissions.

P. Wu and S.P. Low, *Lean and Cleaner Production: Applications in Prefabrication to Reduce Carbon Emissions*, DOI 10.1007/978-3-642-42062-7_6,

Epistemology deals with the nature, sources and processes of knowledge and knowing (Baptiste 2001). Some general questions in epistemology that should be raised include (Baptiste 2001, p. 8):

1. In this project, what terms could be used to best describe what the research aims to find out or produce? According to the research objectives of this study, there are two terms that should be investigated, which are identification and quantification. While the identification can be obtained by literature review and survey, the quantification must be obtained by case study. Although the precast concrete production processes of different precasters share many similarities, there are differences regarding the location of suppliers, the site layout, etc.
2. In this project, what are the appropriate and inappropriate sources of knowledge? Previous research (e.g. Alwi et al. 2002) has discussed the wastes in the construction industry. However, there are many other wastes when examined by the lean philosophy, e.g. waiting time. A combination of literature review and observations are necessary to truly identify the non-value adding activities.

The epistemological part of this study defines the process of knowing, data collection and analysis in order to reach the findings, validate the results and ensure that the research can be duplicated by someone else using a similar research design and methodology. Two epistemological commitments of this study should be focused on as these are closely related to the research methods:

1. Dualism. The researcher is separated from the phenomenon observed, that is to say, the scientific observation of this study is an act of description. The non-value adding activities exist no matter who the observers are.
2. Hermeneutic circle. The observations and explanations may express the understanding at the time of this study. However, it should be noted that such observations and explanations are subjective to revision. Lean philosophy is continually evolving and the application of lean is therefore time bound.

Axiology is the third paradigmatic component of Laudan's model (Laudan 1984). Axiology, according to Patterson and Williams (1998), refers to goals underlying a particular approach to science. There are two types of goals, which are terminal and instrumental goals. While terminal goals refer to the ultimate aims of a specific paradigm (e.g., universal laws of human functioning, predictive explanation and understanding), instrumental goals refer to the criteria by which specific research efforts will be evaluated as good or bad science (Patterson and Williams 1998). The instrumental goals may include generalizability, reliability, predictive validity, persuasiveness and insightfulness (Patterson and Williams 1998). The axiological commitment of this study is related to how the result can be applied by precasters and contractors. It is also related to the continuing effort to reduce the effects of carbon emissions on global climate change through the more readily accessible management approach (and not the more expensive technology-driven approach). This is part of corporate social responsibility for companies that add value to industry.

This study was concerned with identifying the non-value adding activities in the precast concrete industry and assessing the carbon emissions in these activities.

Fig. 6.1 The selection of research methods in accordance with the research objectives

Type of research design		Choice
Experimental research		×
Quasi-experimental research		×
Non-experimental research	Case studies	✓
	Surveys	✓
	Comparisons	×
	Historical designs	×
	Correlation or regression	×

In accordance with the research objectives of this study, surveys and case studies are used as the research methodology of this study, which is shown in Fig. 6.1.

The selection of research is closely related to the research aim of this study. This study aims to apply the lean philosophy to the production cycle and installation cycle of precast concrete products. The application process involves a direct observation of current implementations to identify the improvement. It also involves innovative applications of the lean philosophy, e.g. JIT sourcing, grouping technology to both production and installation cycles. Action research, which involves practitioners in the research process from the initial design of the research project through data gathering and analysis to final conclusions and actions arising out of the research, can therefore be adopted in this study (Whyte 1991).

The research methodology and research design was divided into two sections, as shown in Fig. 6.2. Four stages were included in the research design, which were:

- Stage one: Literature review
- Stage two: Survey
- Stage three: Case studies
- Stage four: Findings, conclusions and results

6.2.1 Identifying Non-value Adding Activities

6.2.1.1 Data Collection Method

The research methodology adopted in this stage included:

Desk Research

Previous studies (Lee et al. 1999; Alwi et al. 2002) have provided information of the non-value adding activities in the construction industry, which could be adopted in

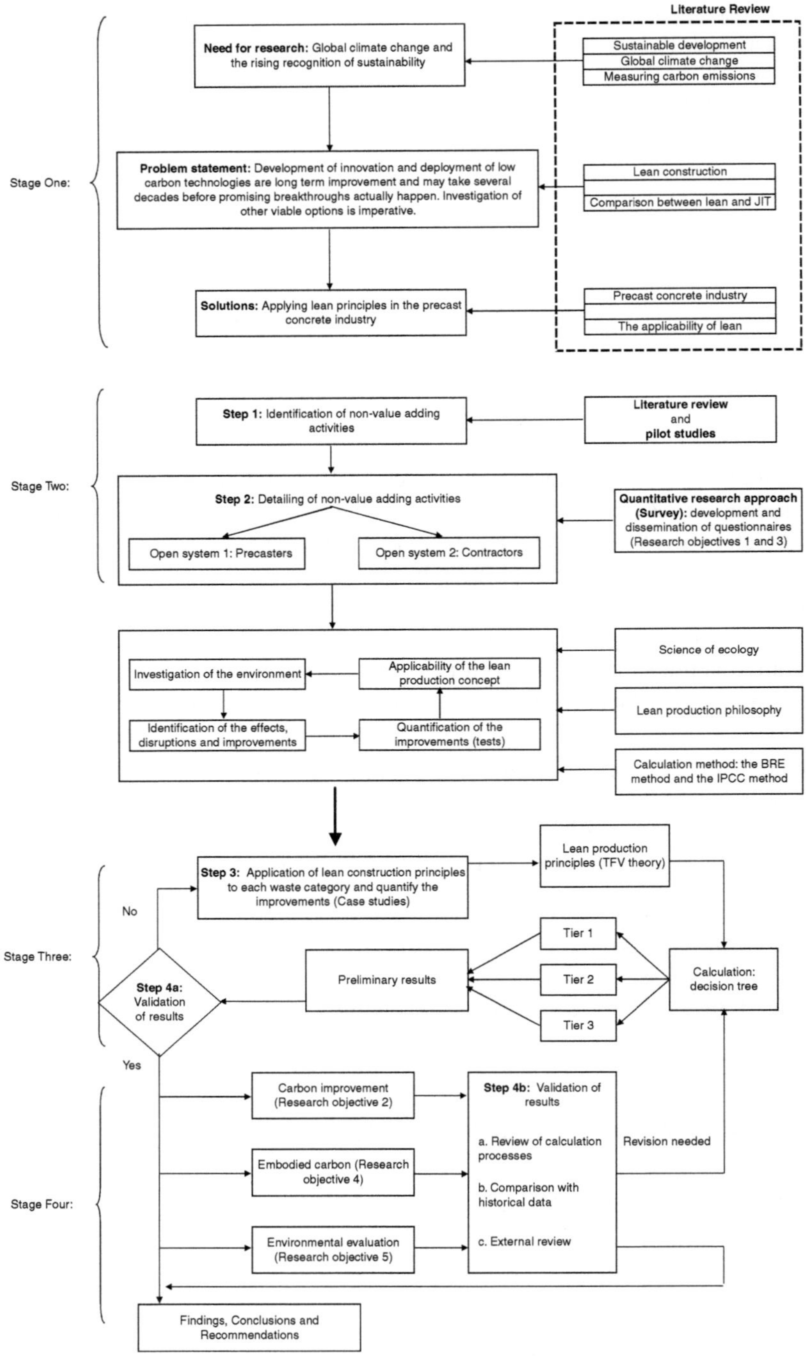

Fig. 6.2 Research methodology

this study. Review of these studies was necessary so that an overview of the sources of waste in the construction industry could be obtained. In addition, in order to identify the areas where the lean production concept could be applied for further improvements, secondary data, such as journal articles, reports and conference papers relating to the application of lean production, JIT and lean construction would be useful.

Questionnaire and Interview

Despite the fact that identifying non-value adding activities in the construction industry has been carried out by previous studies, a questionnaire was designed for this study in accordance with the research aim and objectives. The quantitative research approach required the development and dissemination of questionnaires. The questionnaire included questions relating to the probability (or frequency) of the non-value adding activities (P) and the level of impacts (I) that these activities had on energy consumption and carbon emissions. The frequency of the non-value adding activities was rated using a Likert scale of 1–5: (1) Never; (2) Rarely; (3) Occasionally; (4) Often; and (5) Always. In addition, the level of impacts on energy consumption and carbon emissions was rated on a scale from 1 (insignificant detrimental effect), 2 (minor detrimental effect), 3 (moderate detrimental effect), 4 (significant detrimental effect) to 5 (catastrophic effect). More discussions relating to the probability and impact are discussed in Sects. 7.3.3 (for precasters) and 9.3.3 (for contractors). Open-ended questions would also be organized to ask for factors that the respondents might seem to be important. It should be noted that the delivery stage would be investigated in precast factories and in construction sites where the operators for the delivery would be interviewed. The general descriptions of these factors were provided in Chaps. 7 and 9. The questionnaire designed for precasters is attached as Appendix 2, while the questionnaire designed for contractors is attached as Appendix 3.

6.2.1.2 Population, Samples and Sampling Method

Precasters

To identify current non-value adding activities for evaluation by the lean concept in precast concrete production, a fieldwork, which aims to include all precasters located in Singapore, would be conducted. According to BCA (2010), there were 20 registered precasters in the Directory of BCA Registered Contractors and Licensed Builders. These precasters were classified from L1 to L6 by their financial grade (L1 being the lowest financial grade and L6 being the highest financial grade).

These 20 precasters would be approached through email and telephone. Due to the complexity of the questionnaire, semi-structured interviews would be requested with either the production managers or site managers of the factories.

Contractors

To identify current non-value adding activities for evaluation by the lean concept in the construction sites using precast concrete products, a fieldwork, which aims to include all contractors who were frequently awarded with public housing projects in Singapore, was conducted. According to BCA (2010), the registered contractors in Singapore were classified into seven levels, which are A1, A2, B1, B2, C1, C2 and C3, by their financial grade (A1 being the highest financial grade and C3 being the lowest financial grade). According to the Directory of BCA Registered Contractors and Licensed Builders (BCA 2010), there were 87 contractors within the higher financial grade (A1 and A2), who would frequently be awarded with public housing projects in Singapore. These 87 contractors would therefore form the population of this study.

In accordance with the data analysis method (see Sect. 6.2.1.3), a minimum of 30 contractors was required in order to conduct the parametric tests to test the significance between different non-value adding activities. Convenience sampling method would be used to choose the samples from the population. The precasters who were interviewed would be requested to provide one or two contractors who had been or would be involved in precast concrete projects with them. These contractors' general practices on installing precast concrete products would reflect the general construction practices when adopting precast concrete components.

6.2.1.3 Data Analysis Method

In order to identify the key non-value adding activities from both the precasters' and contractors' point of view, a weighted scoring model was adopted in this study. The weighting factors were assigned to the probability of the non-value adding activities and the level of impacts of these activities on energy consumption and carbon emissions. According to Williams (1993), the risk concept could be broken into the two factors described above. The degree and severity of the risk (S) could then be calculated by: $P \times I$. This methodology has been adopted in previous research (e.g. Alwi et al. 2002). The most important source of non-value adding activities would be assigned with the highest weighting factor (which is S in this study). Paired-Samples T Test would be adopted to rank the most important sources of non-value adding activities for contractors. Meanwhile, as the population of precasters in this study was very small (<30), non-parametric test would be used to rank the most important non-value adding activities for precasters. If the mean of different non-value adding activities were not truly different from each other, they were ranked in the same group with identical importance. The details of the non-parametric tests are presented in Sect. 7.3.4. Similarly, the statistical significance between the non-value adding activities from the contractors' point of view should be identified. As the sample size was large ($=30$), parametric tests were adopted. The general procedure to conduct the parametric tests for contractors is presented in Sect. 9.3.4.

6.2.1.4 Pilot Studies

The aim of the pilot studies was to investigate the production process of precast concrete components to identify the potential sources of waste, as well as to validate the applicability of lean production principles in the precast concrete industry. The details of the pilot studies, including the background, results and discussions, have been discussed in Sect. 4.6. The questionnaire which was adopted for the pilot study is attached in Appendix 1.

6.2.2 Assessing Carbon Emissions

The overall assessment method is adapted from the 2006 IPCC guidelines for national greenhouse gas inventories. There were four categories which seemed to be necessary for the assessment. These were embodied emissions, energy emissions, mobile emissions and personnel travel emissions. However, as the personnel travel emissions were not related to either the production or construction process in this case, this category of carbon emissions was not considered in this study.

Embodied emissions referred to the carbon emissions generated during the product's life cycle. According to Inventory of Carbon and Energy (ICE) (Hammond and Jones 2008, p. 1), embodied energy (carbon) can be defined as:

> The embodied energy (carbon) of a building material can be taken as the total primary energy consumed (carbon released) over its life cycle. This would normally include (at least) extraction, manufacturing and transportation. Ideally the boundaries would be set from the extraction of raw materials (including fuels) until the end of the products lifetime (including energy from manufacturing, transport, energy to manufacture capital equipment, heating and lighting of factory, maintenance, disposal, etc.), known as 'Cradle-to-Grave'. It has become common practice to specify the embodied energy as 'Cradle-to-Gate', which includes all energy (in primary form) until the product leaves the factory gate. The final boundary condition is 'Cradle-to-Site', which includes all of the energy consumed until the product has reached the point of use (i.e. building site).

Energy emissions referred to the emissions that are generated by equipment and facilities using electricity as power instead of directly consuming fuel. Mobile emissions referred to the emissions which are generated by fuel consumption from the equipment that were used in the construction project. Mobile emissions would be mainly calculated using the 2006 IPCC guidelines for national greenhouse gas inventories. Personnel travel emissions is one kind of Energy emissions and mobile combustion would be calculated in this study as the main sources of carbon emissions. Fugitive emissions which were generated by intentional or unintentional release of greenhouse gases during the extraction, processing and delivery of fossil fuels to the point of final use would not be calculated in this study due to its irrelevance to the research aim and objectives.

6.2.2.1 Data Collection Method

Where Singapore-specific emission factors were available, embodied emissions would be calculated using the specific emission factors. Otherwise, emission factors from world-wide average and other countries, such as from the U.S. Greenhouse Gas Inventory Reports and the UK Inventory of Carbon and Energy (ICE) will be adopted. It should be noted that although world-wide average emission factors or other country-specific emission factors could be adopted in this study for most construction materials without the input from the Singapore-specific emissions factors, this study would take the first step to detail the Singapore-specific emission factors by calculating the embodied emission factors of precast concrete columns in Singapore. Data collection would be conducted through the 20 precasters listed in the directory of BCA registered companies.

Data collection relating to energy emissions and mobile emissions would be conducted at three levels. World-wide average data would be obtained from the IPCC Emission Factor Database (EFDB), which served as the first level data of emission factors. Singapore-specific fuel emission factors would be collected through sources including National Climate Change Committee (NCCC), the National Environment Agency, etc. The third level of data would be conducted with the support of more detailed information and effort for the Tier 3 calculation. This kind of data would be collected from precasters and contractors. Equipment suppliers, such as Caterpillar Singapore, JCB Singapore and Komatsu Singapore, might be contacted if necessary.

6.2.2.2 Data Analysis Method

The following equation adapted from Eq. 2.1 would be used to calculate the embodied emissions from construction materials.

GHG emissions from construction products and materials:

$$\text{Emissions}_{\text{materials, products}} = \text{Embodied Carbon}^{*}\text{Quantity} \tag{7.1}$$

Where:

$\text{Emissions}_{\text{materials,products}}$ = carbon emissions by type of materials/products (kg CO_2)
Embodied Carbon = embodied emission factors by type of materials/products (kg CO_2/kg)
Quantity = amount of the materials/products consumed (kg)

Due to the limited time and resources to carry out the overall carbon inventory report for the Singapore construction industry, most embodied emissions of construction materials and products would be calculated using Eq. 7.1. Calculation of the carbon emissions generated during the production process is similar to the computation method identified in the 2006 IPCC guidelines for national greenhouse gas inventories. The method would be discussed with the case study in Chap. 8.

6.2.3 *Case Study*

The aim of the case studies was to examine how the lean production philosophy could be applied in precast concrete factories and in the construction sites and how much carbon emissions could be reduced. In accordance with the research objective 2, i.e. to quantify the carbon emissions that can be achieved by applying the lean philosophy, the case study approach would be appropriate. In addition, according to Yin (1994), case study can deal with a full variety of evidence, including documents, artifacts, interviews and observations. Yin (1994, p. 9) proposed that case study approach being taken in such circumstance:

A "how" or "why" question is being asked about a contemporary set of events over which the investigator has little or no control.

One of the greatest concerns in using case study was the lack of rigor of case study research (Yin 1994). Biased views may affect the results of the case study. However, it should be noted that bias can also affect experiments, questionnaires for surveys and historical research (Rosenthal 1966). In this study, in order to overcome biased view, the data that were related to the calculation of carbon emissions are strictly recorded with the recording devices. In addition, the case study was divided into two sections. One section included all the processes the carbon emissions of which could be quantified. The other section included all the processes the carbon emissions of which could not be quantified. Qualitative descriptions are provided.

A second common concern about the case study approach is its scientific generalization. Critics argued that one single case study could not be used for scientific generalization. The answer was that the case studies, like experiments, were generalizable to theoretical propositions and not to populations or universes (Yin 1994). It should therefore be noted that the strategy of "analytic generalization" rather than "statistical generalization" was adopted. The case studies used the previously developed results from the survey as a template with which the empirical results of the case studies would be compared. In addition, in order to overcome the scientific generalization problem, multi-case studies were conducted for both precasters and contractors (as two pilot case studies have been conducted for precasters, only one case study was presented in details).

The design of the case studies in this study followed a series of pre-specified procedures, which are discussed in the following context.

6.2.3.1 Overview of the Case Study Project

Case Study – Precasters

The aim of the case study (precasters) was to quantify the lean improvements by examining how much carbon emissions could be reduced (see research objective 2 in Sect. 1.3) and calculate the embodied carbon of a specific precast concrete column (see research objective 4 in Sect. 1.3). Based on this research question, the use of case studies might be appropriate (Yin 2009). Two pilot studies have been conducted (see Sect. 4.6). The results showed that the lean production philosophy could be used to reduce carbon emissions by eliminating many non-value adding activities, from site layout management, supply chain management, production management and stock management. A single-case study would be appropriate, because:

1. The evaluation of carbon emissions followed the Life Cycle Assessment (LCA) rules. These rules were applicable to every type of precast concrete product that was targeted for a LCA study;
2. The production process of precast concrete columns might vary. However, these variations were very minor as suggested from the discussions in Sects. 4.2 and 4.3; and
3. The production process of precast concrete columns was very alike in precasters in Singapore as suggested by the two pilot studies (see Sect. 4.6).

However, it should be noted that in order for a single-case study to generalize results, this case should represent a typical or critical case (Yin 2003). The precaster chosen for the case study should therefore ideally occupy a large market proportion. The introduction of the case study is presented in Sect. 8.1.

Case Study – Contractors

As there were no pilot studies conducted for contractors, multiple-case studies were necessary following the "replication" logic. Ideally, the three contractors should use the precast concrete columns produced by the precaster who was chosen as the case study for precasters. The introduction of the three cases is presented in Sects. 10.1 and 10.2.

6.2.3.2 Field Procedures – Data Collection

Data collection in the case studies would rely on many sources of evidence, including documentation, interviews and direct observations.

- Documentation. Documentation information relating to the production/installation procedure, quality control procedure and waste records might be referred to in the case studies.
- Interviews. Interviews with the either the production managers or site managers were necessary in order to identify the frequency of the non-value adding activities.
- Direct observations. Direct observations should also be conducted. Information related to waste of raw materials, waste of finished products and energy consumption caused by the non-value adding activities should be recorded by direct observations.

6.2.3.3 Data Analysis

The data analysis should follow a general procedure. These procedures are discussed in Sect. 8.2 (for the case study of precasters) and Sect. 10.3. In addition, as the precast concrete column chosen for the case study was targeted for a LCA study, many LCA rules should be followed when analyzing the data. These rules are discussed in Sect. 8.3 along with the case study.

6.2.3.4 Case Study Report

The unit of analysis of the case study was a specific type of precast concrete column. It was therefore necessary that the case study results should be reported on a per column basis.

- Case study report (precasters). The information in this report should include: the non-value adding activities and their respective frequency, their impact on the level of carbon emissions and the respective amount of carbon emissions adjusted on a per column per production cycle basis.
- Case study report (contractors). The information in this report should include: the non-value adding activities and their respective frequency, their impact on the level of carbon emissions and the respective amount of carbon emissions adjusted on a per column per installation cycle basis.

6.3 Justification

The methods chosen to conduct this study were literature review, a survey and four case studies (one precaster and three contractors). The aim of the literature review was to identify a comprehensive list of non-value adding activities in the production and installation process of a specific type of precast concrete column. In addition, a survey was necessary to examine the probability, impact and severity of these non-value adding activities. Edwards and Talbot (1999) stated that the case study can be used in practitioner research to illustrate a set of principles, to provide some detailed descriptions of a topic of interest, or to explore a field of study and gather information on it. The four case studies presented in this study would therefore add an important dimension to the survey results.

The case studies provided in this study aimed to illustrate how the lean concept could be applied in the production and installation cycles of precast concrete columns to reduce carbon emissions. It should therefore be noted that the strategy of "analytic generalization" rather than "statistical generalization" was adopted. The case studies used the previously developed results from the survey as a template with which the empirical results of the case studies would be compared.

In addition, when addressing how much carbon emissions could be reduced from the case studies, the procedure, not the results, could be replicated as not all precasters and contractors were facing the same non-value adding activities that happened in the case studies.

6.4 Summary

The methodology adopted in this study included desk research, questionnaire and interviews. Desk research was useful to help identify the non-value adding activities that were related to generate unnecessary carbon emissions (research objective 1). In addition, methods to calculate these carbon emissions, as well as the world-wide emission factors were also obtained from desk research. More importantly, the current application of the lean production concept in the construction industry was useful to provide appropriate recommendations. Questionnaire was designed to address the special link between the lean production concept and carbon emissions. All precasters and 30 contractors who had the experiences in precast concrete projects were targeted in this study. It should be noted that as contractors might not be familiar with the lean production concept, the questionnaire should better be completed within the interview. The results would be validated through three processes, which included internal review of emission factors, comparisons between the results and historical data and external review carried out by experts who were familiar with the source category. The checklist will be useful for both precasters and contractors to improve (research objective 3).

Case studies are conducted in accordance with research objective 2, i.e. to quantify the carbon improvements. The embodied carbon of precast concrete columns (research objective 4) and the impact of the lean philosophy on environmental evaluation (research objective 5) are both investigated from the case study.

It should be especially noted that although the questionnaire was designed for precasters and contractors, the delivery stage was not overlooked. Due to the link between the three stages in the whole construction process, the investigation of the delivery stage could be conducted either in precast factories or in construction sites from the delivery operators.

References

Alwi S, Hampson K, Mohamed S (2002) Waste in the Indonesian construction project. In: Proceedings of the 1st international conference of CIB W107 – creating a sustainable construction industry in developing countries, Spier, South Africa

Baptiste I (2001) Qualitative data analysis: common phases, strategic differences. Forum Qual Soc Res 2(3): Art. 22

Building Construction Authority (2010) BCA directory of registered contractors and licensed builders. Available from: http://www.bcadirectory.sg/. Cited 29 Nov 2013

Burrell G, Morgan G (1979) Sociological paradigms and organisational analysis: elements of the sociology of corporate life. Gower, London

Edwards A, Talbot R (1999) The hard-pressed researcher: a research handbook for the caring professors. Pearson Education Limited, Essex

Hammond G, Jones C (2008) Inventory of carbon and energy (ICE). [Online] Available at: http://www.bath.ac.uk/mech-eng/sert/embodied. Accessed 16 Mar 2010

Laudan L (1984) Science and values. University of California Press, Berkeley

Lee SH, Diekmann JE, Songer AD, Brown H (1999) Identifying waste application of construction process analysis. In: Tommelein ID (ed) Proceedings of IGLC-7. U.C. Berkeley, Berkeley, CA

Patterson M, Williams DR (1998) Paradigms and problems: the practice of social science in natural resource management. Soc Nat Res Int J 11(3):279–295

Rosenthal R (1966) Experimental effects in behavioural research. Appleton-Century-Crofts, New York

Whyte WF (1991) Participatory action research. Sage Publications, Thousand Oaks

Williams TM (1993) Risk-management infrastructures. Int Proj Manag 11(1):5–10

Yin RK (1994) Case study research: design and methods, 2nd edn. SAGE Publication, Thousand Oaks

Yin RK (2003) Case study research: design and methods. Sage Publications, Thousand Oaks

Yin RK (2009) Case study research: design and methods. Sage Publications, Los Angeles

Chapter 7
Applications in Precast Concrete Factories

7.1 Introduction

The applicability of the lean production philosophy in precast concrete factories to reduce carbon emissions was analyzed in previous chapters. In this chapter, detailed evaluation of the applications is presented. The evaluation of the lean applications follows the value stream in precast concrete factories, which includes site layout management, supply chain management, production management and stock management.

In this chapter, the research sample is first introduced along with some background information about the precast concrete industry in Singapore. The empirical study on the non-value adding activities of current precast concrete production from the lean perspective is then presented. The analysis of these non-value adding activities follows the value stream of precast concrete production, as well as a general analytical procedure, which includes the following steps:

1. Descriptive analysis. In descriptive analysis, a few general questions related to current precast concrete production processes are discussed with the precasters. Non-value adding activities along with the improvements are identified.
2. Factors description. In this step, the non-value adding activities in each value stage are introduced. A section number is provided for each non-value adding activity for referencing purposes.
3. Ranking procedure. The non-value adding activities are ranked based on two important criteria: the probability of occurrence (P) and the subsequent impact (I). It should be noted that the subsequent impact are evaluated based on a single criterion: the magnitude of the impact on the carbon emission levels of precast concrete factories. The rank of the non-value adding activities (also referred to as severity in the following context) can be obtained by multiplying their probabilities with the impact, which fall within the range of [1, 25].

P. Wu and S.P. Low, *Lean and Cleaner Production: Applications in Prefabrication to Reduce Carbon Emissions*, DOI 10.1007/978-3-642-42062-7_7,

4. Non-parametric tests. This is to identify the non-value adding activities within the same group categorized by the severity. It should be noted that although the scores obtained by the ranking procedure, in which the probabilities are multiplied with the impact, can be used for ranking purposes, the statistical significance between different groups should be identified. As the sample size is small (<30) with unknown distribution (normal or non-normal), non-parametric tests (paired sample) are adopted in this step.
5. Specific analysis. This is to examine the non-value adding activities with large standard deviation of the probability. Attention should be paid to these non-value adding activities, as improvements may be identified for precasters with higher probabilities of occurrence from precasters with lower ones.

7.2 Response Rate and Representativeness of Data

To identify current non-value adding activities for evaluation by the lean concept in precast concrete production, a fieldwork, which aims to include all precasters located in Singapore, was conducted. According to BCA (2010), there were 20 registered precasters in the *Directory of BCA Registered Contractors and Licensed Builders*. These precasters were classified from L1 to L6 by their financial grade (L1 being the lowest financial grade and L6 being the highest financial grade).

In the fieldwork, the non-value adding activities identified from the literature review and pilot studies were rated by the precasters, according to the suggestion made by Williams (1993), who stated that the risk concept could be broken down into two factors: probability (P) and impact (I). The degree and severity of the risk (S) could therefore be described by multiplying the probability with its corresponding impact, as shown in Eq. 7.1.

$$S = P \times I \tag{7.1}$$

From November 2009 to April 2010, the 20 precasters listed in the directory were approached through email and telephone. Due to the complexity of the questionnaire, semi-structured interviews were requested with either the production managers or site managers of the factories. A total of 17 responses were received. The response rate is 85 %. The 17 precasters had been involved in the precast concrete industry for decades, producing a wide range of precast concrete products, such as hollow cores, staircases and window frames. Two precasters with the lowest financial grade (L1) were not included in this study because they did not want to be interviewed. As can be seen from Table 7.1, the fieldwork includes nearly all precasters in the large financial grade. So the information obtained through interviews can therefore provide a fair representation of the Singapore precast concrete industry.

Table 7.1 Profile of respondents

Financial grade	Total number listed in the directory	Number of precasters interviewed	Response rate
L6	9	8	88.89 %
L5	4	4	100 %
L4	2	2	100 %
L3	3	3	100 %
L2	0	Not applicable	Not applicable
L1	2	0	0

7.3 Lean Site Layout Management in Precast Concrete Factories

Along the value stream of precast concrete production, the first value stage is in site layout management. In this stage, the precast concrete factory is designed for production activities. The site layout of the factory may affect the overall efficiency and energy consumption of the production process. It is therefore an important stage to get work done right at the very start.

7.3.1 Descriptive Analysis

A few general questions related to current site layout management were discussed with the precasters before the non-value adding activities in this value stage were presented for ranking. These questions were listed in the questionnaire for the precasters in Sect. 1.8 of the questionnaire. The lean concept required that fundamental problems to be identified, evaluated and addressed. These questions were therefore designed to examine the performance of the current precast concrete production activities towards sustainable development in terms of environmental performance.

Activities that might cause increases in energy consumption and carbon emissions were identified in many areas, including regulations, human resources and site layout design, as shown in Table 7.2. A few implications can be inferred from Table 7.2, including:

1. Training programmes for supervisors and project managers seem to be insufficient to support a smooth production flow. Only seven precasters (41.2 %) claimed to provide enough training programmes on the use of materials, plants and equipment, which were all important areas where carbon emissions can be generated. According to the supervisors and project managers who had negative responses, the responsibilities for mastering the use of materials, plants and equipment were delegated to operators or persons who were in charge of these areas. These operators or persons in charge were entrusted to obtain in-depth

Table 7.2 General questions in the section of site layout management

Question no.	Description	Positive response	Positive response rate		Category
1.8.1	Are supervisors appropriately trained relating to the use of materials, plants and equipment?	7	41.2 %		Human Resources
1.8.2	Is environmental performance considered when designing the site layout?	2	11.8 %		Legislation
1.8.3	Is there any statutory regulation relating to carbon emissions when designing the site layout?	0	0 %		Legislation
1.8.4	Is there any statutory regulation relating to low energy consumption when designing the site layout?	0	0 %		Legislation
1.8.5	Has reconstruction happened before because of failing to meet statutory requirements?	0	0 %		Legislation
1.8.6	Is the site layout plan sent to contractors and subcontractors and placed on the notice board for information?	2	11.8 %		Site layout design
1.8.7	Is the site layout designed in such a manner that would allow the materials to be used in "first in first out"?	12	70.5 %		Site layout design
1.8.8	Has the site changed since it was first built?	0	0 %		Site layout design
1.8.9	What is the percentage of the storage area to the total area?	Largest 75 %	Smallest 30 %	Average 43 %	Site layout design

understanding about the flow of materials and the operation of equipment. In normal cases, the delegation of responsibilities would be reasonable due to the heavy production schedule in the precast concrete factories, as long as such delegation was supported by appropriate employee performance rating systems. However, such systems were not appropriately designed for precast concrete production which may impede the delegation of responsibilities.

2. Environmental performance of the precast concrete factories was rarely considered. Only two precasters (11.8 %) claimed that environmental performance of the precast concrete factories were considered when designing the site layout. According to these two precasters, consideration of the environmental performance of the factory was purely voluntary. All precasters stated that there were no statutory requirements about low carbon emissions and low energy consumptions when submitting the design details for approval. The statutory requirements that must be complied with included health and safety (e.g. Workplace Safety and Health Act 2006), food hygiene (e.g. Environmental Public Health Regulations 2000), occupiers' liability (e.g. Law of Occupiers' Liability in Singapore 1985), etc. Regulations related to low energy consumption and low carbon emissions

(e.g. Singapore's National Climate Change Strategy 2008) were purely voluntary for the precasters to comply with. Reconstruction had never happened before because of a failure to meet these mandatory requirements. This was due to the strict regulations in Singapore. Mandatory regulations had to be complied with before construction licences were issued.

3. Changes to the site layout of precast concrete factories rarely happened. Once the factory was designed and constructed, the site layout plan remained the same for quite a long time. All precasters (100 %) claimed that the site layout plans of their factories had never changed since they were built. However, it seems that there remained a communication gap between the contractors and the subcontractors involved. Only two precasters (11.8 %) placed the details of the site layout plan on the notice board for information. This might cause problems especially when the subcontractors, operators or managers were replaced due to renewal of contracts. New subcontractors, operators and managers who had little working experience in a particular precast concrete factory might have to refer to the site layout plan frequently. By putting the information on the notice board, a smooth flow of information could be obtained.
4. The large storage area compared to the total area might affect the flow of production in the precast concrete factories. The storage area (percentage) fell into the range between 30 and 75 %, with an average of 43 %. It seems that nearly half the site layout area was occupied by storage. This was in contradiction with the fact that twelve precasters (70.5 %) were adopting the "first in first out" practice when using raw materials. When the pull production method was used, the storage area should be smaller than when the push production method was adopted. However, due to supply disruptions and unstable production schedules, the precasters preferred the push production method instead of the pull production method. This contradiction will be explained along with question 3.6.12, in which the production type would be discussed.

7.3.2 Factors Description

The lean production philosophy advocates that work is to be done right at the very start. If a lean factory with low carbon emissions and energy consumption is to be built, it should be designed right at the very start. The questionnaire relating to site layout management was divided into seven categories, each of which represented an area where non-value adding activities could be identified. The seven categories are shown in Table 7.3. It should be noted that although the section relating to the raw materials of precast concrete products seemed to be irrelevant to site layout management, it was analyzed in this section due to its higher priority than those other activities listed in the value stages later (delivery management, production management, etc.). The design and specifications of the raw materials for precast concrete products should be completed before deliveries and production activities were arranged. Due to its relatively small contents and high priority, the section relating to building materials was discussed in site layout management.

Table 7.3 Seven major categories of non-value adding activities in site layout management

Category no.	Description	Category
1.1	Building materials	Materials
1.2	Site facilities	Facilities
1.3	Statutory requirements	Legislation
1.4	Contractors' and subcontractors' requirements	Communication
1.5	Temporary works and services	Production requirements
1.6	Storage area	Storage
1.7	Managing the factory	Operation

The non-value adding activities were classified in each of the seven categories. These classified non-value adding activities formed the operationalized measureables of this study. General description for each non-value adding activity is provided as follows:

- *Category 1.1 Building materials*

  1.1.1 Improper specifications of building materials
  1.1.2 Inaccurate estimation of quantities required
  1.1.3 Does not think of alternative designs that minimize the use of materials
  1.1.4 Does not think of green building materials

The production of precast concrete products should follow the design specifications. In order for the production processes to be conducted smoothly without errors, these specifications should be made correctly and clearly. In addition, if the designers frequently make adjustments to the specifications, production activities in precast concrete factories may be changed, causing disruptions to existing production.

In the pull production system, the quantities of raw materials should be accurately estimated. The production activities are interrupted by ordering either more or less materials required.

Although the design of precast concrete products can follow routine standards, it does not preclude the use of alternative designs which minimize the use of materials. In fact, the effort that precasters devote into developing alternative designs and the ability of doing so are both reflections of the precasters' plan on continuous improvement. Similar to alternative designs, research in the use of green building materials is another reflection of the precasters' plan on continuous improvement.

- *Category 1.2 No overall consideration of building site facilities*

  1.2.1 Equipment
  1.2.2 Infrastructure

When designing an efficient site layout, equipment and infrastructure that will be used in future production activities should be considered. According to Calvert (1995), the plant items utilized on site have a larger than expected impact on most sites and should be selected in accordance with the method statement. Many alternatives, such as central batching versus ready mix, cranes versus hoists, etc.

should be compared before these are located (Xin 2010). Changes to the site layout due to replacements of equipment and infrastructure may cause disruptions to production activities.

- *Category 1.3 Statutory requirements*

Many legal authorizations should be obtained before the construction works for precast concrete factories begin. These legal authorizations may relate to health and safety, the use of site offices, food hygiene, pollution control, occupiers' liability, etc.

- *Category 1.4 Does not pay full attention to contractors' and subcontractors' requirements*

1.4.1 Duration
1.4.2 Office space
1.4.3 Maximum number of men on site
1.4.4 Services required

Full attention should be paid to contractors' and subcontractors' requirements in order to provide the right amount of supporting facilities to them, such as site offices and employee dormitories. Although the energy consumed in these supporting facilities are usually not considered when calculating the embodied energy for precast concrete products, it is still of vital importance for the precaster to include this to achieve lean production.

- *Category 1.5 Inappropriate design of temporary works and services*

1.5.1 Space for access
1.5.2 Tower crane's fully blocked area
1.5.3 Clearance of the blocked area
1.5.4 Siting of static plant
1.5.5 Parking of mobile plant

Neil (1980) proposed 11 basic guidelines to design a construction site layout, which are equally useful as well when designing the site layout for precast concrete production. In the 11 guidelines, Neil (1980) proposed that designers should master the planning of support facilities to avoid unnecessary relocation, remodeling or expansion as the project progresses. In this study, these guidelines apply as well. Many aspects related to the design of works and services should be considered, including the spaces for access, locating and operating of both the static and mobile plants.

- *Category 1.6 Over provide material storage*

1.6.1 Secure store
1.6.2 Weatherproof store
1.6.3 Open store

Storage is a category of waste in the lean production philosophy. Materials can be damaged during inventory when inappropriately stacked. Singling out processes which involve the use of mobile plants may damage the materials as well. Although the energy consumed during inventory is usually not considered as embodied energy of precast concrete products (inventory is regarded as capital infrastructure and is not considered in LCA study unless it is significant under the BRE methodology for the environmental profiles of construction products), it represents a source of waste that should be considered.

- *Category 1.7 Inappropriate management of the site layout*

  1.7.1 Site layout plan is not tested for economic and efficient production
  1.7.2 Site layout plan is not sent to contractors, subcontractors and general foreman
  1.7.3 Site layout plan is not placed on the notice board for information
  1.7.4 Changes to the site layout plan are not notified immediately

The long term contracts are not used between precasters and suppliers in Singapore (please refer to factor 2.1.2 for more explanations). This may cause communication problems when changing suppliers. The transparency concept in the lean production philosophy aims to make the main flow of operations from start to finish visible and comprehensible to all employees (Stalk and Hout 1990). Failing to do so may expose the production flow to interruptions.

7.3.3 Ranking Procedure

The ranking procedure of the non-value adding activities followed a standard risk analysis protocol, which involveds the use of the P-I table. In the P-I table, both qualitative assessment of the probability (P) and the impact (I) were conducted to determine the effect of the non-value adding activities on carbon emissions (S). In this study, the term "*probability*" was defined as:

> the probability or frequency that this category of non-value adding activity would happen in precast concrete production processes.

Accordingly, the term "impact" was defined as:

> The effect of this category of non-value adding activity on the amount of carbon emissions that is generated from precast concrete production processes.

Before analyzing the P-I table, the value range associated with the qualitative assessment should be introduced. In this study, a five-scale value range was adopted in order to conduct qualitative descriptions of the probabilities of non-value adding activities. The five-scale value range of probabilities is described in Table 7.4

Accordingly, the impact associated with the non-value adding activities was assessed by a five-scale value range, which is described in Table 7.5. The severity of every non-value adding activity could therefore be expressed by timing the probability with impact correspondingly. Attention should be paid to the largest rate and smallest rating of severity for every non-value adding activity. For example, for factor 2.1.1 (Large quantity supply base), the largest rating of probability was 5 by one precaster and the largest rating of impact was also 5 by another precaster. However, the largest rating of severity was 20 based on the rating because for precaster who rated probability by 5, the impact of this non-value adding activity was only rated by 4. This explains why the largest rating of severity

Table 7.4 Five-scale value range to assess the probabilities of non-value adding activities

Rating no.	Category	Description
1	Very low	This category of non-value adding activity *never* happen in current precast concrete production processes
2	Low	This category of non-value adding activity *rarely* happen in current precast concrete production processes
3	Medium	This category of non-value adding activity *occasionally* happen in current precast concrete production processes
4	High	This category of non-value adding activity *often* happen in current precast concrete production processes
5	Very high	This category of non-value adding activity *always* happen in precast concrete production processes

Table 7.5 Five scale value range to assess the impact of non-value adding activities

Rating no.	Category	Description
1	Insignificant	This category of non-value adding activity has *insignificant* impact on the carbon emissions level of the precast concrete products (production cycle)
2	Minor	This category of non-value adding activity has *minor* negative impact on the carbon emissions level of the precast concrete products (production cycle)
3	Moderate	This category of non-value adding activity has *moderate* negative impact on the carbon emissions level of the precast concrete products (production cycle)
4	Major	This category of non-value adding activity has *significant* negative impact on the carbon emissions level of the precast concrete products (production cycle)
5	Catastrophic	This category of non-value adding activity has *catastrophic* negative impact on the carbon emissions level of the precast concrete products (production cycle)

was sometimes not equal to the results by timing the largest rating of probability with the largest rating of impact. Same implications can be obtained for the smallest rating as well.

Rated by the five scale value range, the probability, impact and severity of non-value adding activities in site layout management can be seen in Table 7.6. A few implications can be made from Table 7.6:

1. Over providing storage area was very common in the precast concrete industry of Singapore. As can be seen from Fig. 7.1, over providing storage area in terms of open store (1.6.3), secure store (1.6.1) and weatherproof store (1.6.2) were the top three non-value adding activities rated by probability. Other non-value adding activities with high probability of occurrence included:

   - 1.1.3 Does not think of alternative design that minimize the use of materials (4.0)
   - 1.7.3 Site layout plan is not placed on the notice board for information (3.29)
   - 1.1.4 Does not think of green building materials (2.94).

Table 7.6 Probability, impact and severity of the non-value adding activities in site layout management

		Probability (P)				Impact (I)				Severity (S)			
Factor no.	Description	AR	LR	SR	SD	AR	LR	SR	SD	AR	LR	SR	SD
1.1.1	Improper specifications of building materials	1.63	3	1	0.62	3.31	4	2	0.70	5.63	12	2	2.90
1.1.2	Inaccurate estimation of quantities required	1.75	3	1	0.68	3.50	4	2	0.63	6.13	12	2	2.60
1.1.3	Does not think of alternative designs that minimize the use of materials	4.00	5	1	1.03	2.94	4	2	0.77	11.44	16	4	3.69
1.1.4	Does not think of green building materials	2.94	5	1	1.39	3.81	5	3	0.66	10.38	16	4	4.44
1.2.1	No overall consideration of equipment used in future precast concrete production processes	1.53	3	1	0.62	3.00	4	2	0.87	4.59	12	2	2.40
1.2.2	No overall consideration of infrastructure used in future precast concrete production processes	1.47	2	1	0.51	2.88	4	2	0.86	4.00	6	2	1.17
1.3	Does not comply with mandatory statutory requirements	1.00	1	1	0.00	3.53	5	1	1.01	3.53	5	1	1.01
1.4.1	Does not pay full attention to contractors' and subcontractors' requirements – Duration	1.41	2	1	0.51	2.35	4	1	1.00	3.35	8	1	1.93
1.4.2	Does not pay full attention to contractors' and subcontractors' requirements – Office space	1.47	2	1	0.51	2.71	4	2	0.69	3.94	8	1	2.14
1.4.3	Does not pay full attention to contractors' and subcontractors' requirements – Maximum number of men on site	1.59	2	1	0.51	2.41	4	1	0.87	4.00	8	1	2.18
1.4.4	Does not pay full attention to contractors' and subcontractors' requirements – Services required	1.76	3	1	0.56	2.88	4	2	0.60	5.24	9	2	2.19

1.5.1	Inappropriate design of temporary works and services – Space for access	1.59	2	1	0.51	3.41	4	2	0.62	5.41	8	3	2.03
1.5.2	Inappropriate design of temporary works and services – Tower cranes' fully blocked area	1.71	3	1	0.59	3.18	4	2	0.53	5.47	12	3	2.40
1.5.3	Inappropriate design of temporary works and services – Clearance of the blocked area	1.59	2	1	0.51	2.94	4	2	0.66	4.65	8	2	1.84
1.5.4	Inappropriate design of temporary works and services – Sitting of static plants (e.g. hoists, tower cranes, concrete mixing plants)	1.41	2	1	0.51	3.35	5	2	0.86	4.65	8	2	1.90
1.5.5	Inappropriate design of temporary works and services – Parking of mobile plants	1.53	2	1	0.51	2.82	5	1	1.01	4.29	8	1	1.99
1.6.1	Over provide material storage – Secure store	4.18	5	3	0.53	3.18	4	2	0.81	13.29	20	8	3.95
1.6.2	Over provide material storage – Weatherproof store	4.18	5	3	0.53	3.06	4	2	0.75	12.82	20	8	3.83
1.6.3	Over provide material storage – Open store	4.29	5	4	0.47	3.41	5	2	0.94	14.53	25	8	4.19
1.7.1	Site layout plan is not tested for economic and efficient production	1.53	3	1	0.62	4.00	5	2	0.71	6.00	10	4	2.42
1.7.2	Site layout plan is not sent to contractors, subcontractors and general foreman	2.06	3	1	0.75	1.88	4	1	0.99	3.88	12	1	2.64
1.7.3	Site layout plan is not placed on the notice board for information	3.29	5	1	1.21	1.71	4	1	0.77	5.53	20	2	4.06
1.7.4	Changes to the site layout plan are not notified immediately	2.53	4	1	1.18	2.24	4	1	1.15	5.35	16	1	3.57

Notes: *AR* average rating, *LR* largest rating, *SR* smallest rating, *SD* standard deviation

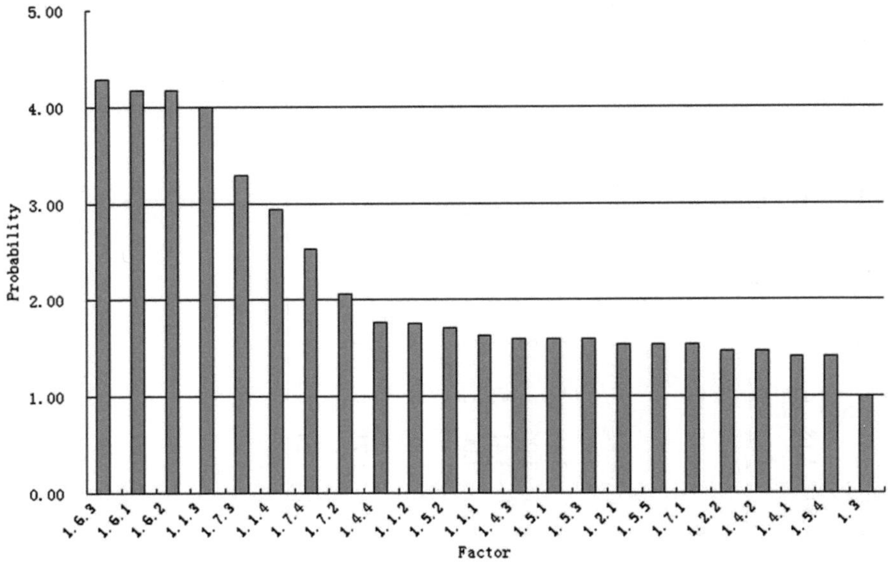

1.6.3 Over provide material storage – Open store
1.6.1 Over provide material storage – Secure store
1.6.2 Over provide material storage – Weatherproof store
1.1.3 Does not think of alternative designs that minimize the use of materials
1.7.3 Site layout plan is not paced on the notice board for information
1.1.4 Does not think of green building materials
1.7.4 Changes to the site layout plan are not notified immediately
1.7.2 Site layout plan is not sent to contractors, subcontractors and general foreman
1.4.4 Does not pay full attention to contractors' and subcontractors' requirements – Services required
1.1.2 Inaccurate estimation of quantities required
1.5.2 Inappropriate design of temporary works and services – Tower crane's fully blocked area
1.1.1 Improper specifications of building materials
1.4.3 Does not pay full attention to contractors' and subcontracts' requirements – Maximum number of men on site
1.5.1 Inappropriate design of temporary works and services – Space for access
1.5.3 Inappropriate design of temporary works and services- Clearance of the blocked area
1.2.1 No overall consideration of equipment used in the future precast concrete production processes
1.5.5 Inappropriate design of temporary works and services – Parking of mobile plant
1.7.1 Site layout plan is not tested for economic and efficient production
1.2.2 No overall consideration of infrastructure used in future precast concrete production processes
1.4.2 Does not pay full attention to contractors' and subcontractors' requirements – Office space
1.4.1 Does not pay full attention to contractors' and subcontractors' requirements – Duration
1.5.4 Inappropriate design of temporary works and services – Siting of static plants
1.3 Does not comply with mandatory statutory requirements

Fig. 7.1 The probability of non-value adding activities in site layout management

It seems that current designs of precast concrete products often followed standardized design protocol (1.1.3). Without the pressure for government and other authorities, using green building materials was not implemented (1.1.4), especially when the costs of such materials were very high.

Although it seemed acceptable that the precasters did not place the site layout plan on the notice board for information when the managers and supervisors had known the site quite well, this might cause communications problems when changing managers, contractors and subcontractors due to renewal of contracts (1.7.3).

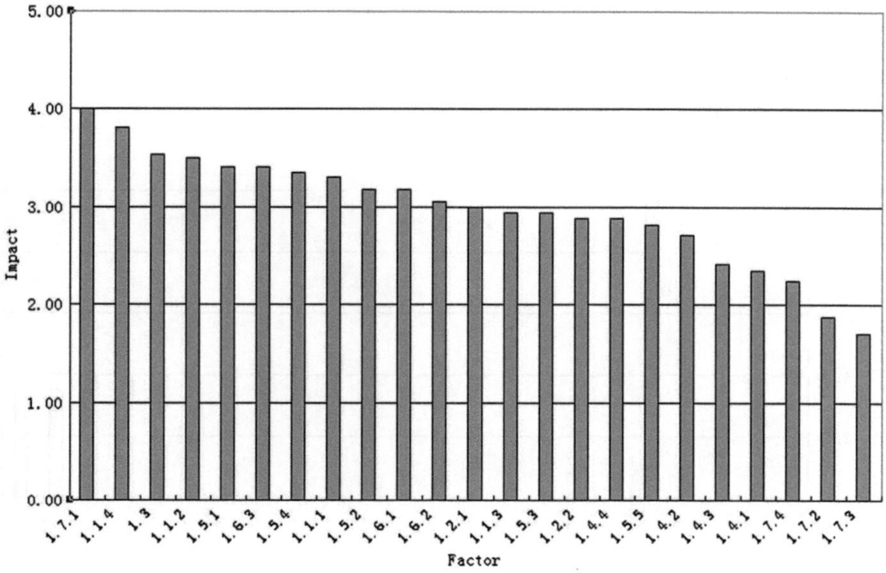

1.7.1 Site layout plan is not tested for economic and efficient production
1.1.4 Does not think of green building materials
1.3 Does not comply with mandatory statutory requirements
1.1.2 Inaccurate estimation of quantities required
1.5.1 Inappropriate design of temporary works and services – Space for access
1.6.3 Over provide material storage – Open store
1.5.4 Inappropriate design of temporary works and services – Siting of static plants
1.1.1 Improper specifications of building materials
1.5.2 Inappropriate design of temporary works and services – Tower crane's fully blocked area
1.6.1 Over provide material storage – Secure store
1.6.2 Over provide material storage – Weatherproof store
1.2.1 No overall consideration of equipment used in the future precast concrete production processes
1.1.3 Does not think of alternative designs that minimize the use of materials
1.5.3 Inappropriate design of temporary works and services- Clearance of the blocked area
1.2.2 No overall consideration of infrastructure used in future precast concrete production processes
1.4.4 Does not pay full attention to contractors' and subcontractors' requirements – Services required
1.5.5 Inappropriate design of temporary works and services – Parking of mobile plant
1.4.2 Does not pay full attention to contractors' and subcontractors' requirements – Office space
1.4.3 Does not pay full attention to contractors' and subcontracts' requirements – Maximum number of men on site
1.4.1 Does not pay full attention to contractors' and subcontractors' requirements – Duration
1.7.4 Changes to the site layout plan are not notified immediately
1.7.2 Site layout plan is not sent to contractors, subcontractors and general foreman
1.7.3 Site layout plan is not paced on the notice board for information

Fig. 7.2 The impact of non-value adding activities in site layout management

2. The three most important factors that would increase the level of carbon emissions were:

   - 1.7.1 Site layout plan is not tested for economic and efficient production (4.0)
   - 1.1.4 Does not think of green building materials (3.81)
   - 1.3 Does not comply with mandatory statutory requirements (3.53)

   The ranking of the impact can be seen in Fig. 7.2. Fifteen precasters (88.24 %) agreed that the site layout should be carefully planned to achieve economic and efficient production in order to reduce energy consumption and carbon emissions (1.7.1).

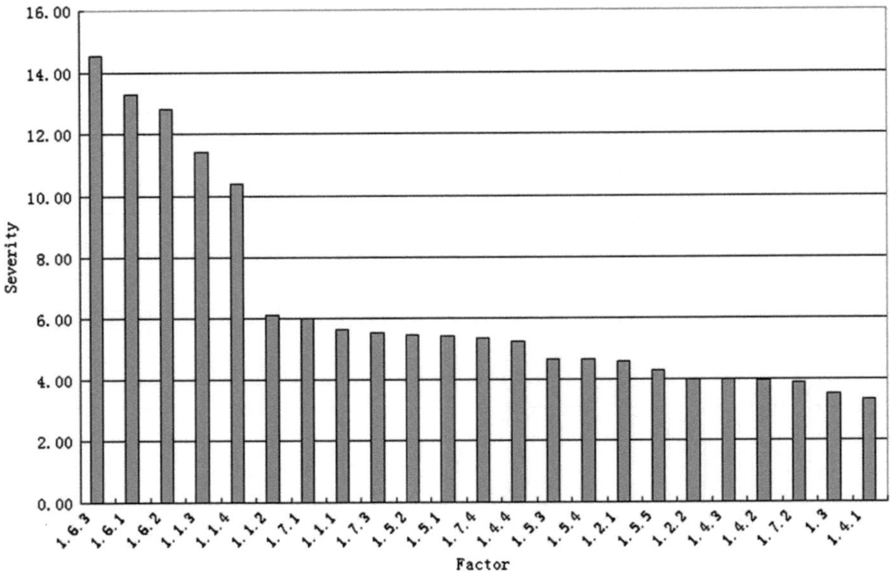

1.6.3 Over provide material storage – Open store
1.6.1 Over provide material storage – Secure store
1.6.2 Over provide material storage – Weatherproof store
1.1.3 Does not think of alternative designs that minimize the use of materials
1.1.4 Does not think of green building materials
1.1.2 Inaccurate estimation of quantities required
1.7.1 Site layout plan is not tested for economic and efficient production
1.1.1 Improper specifications of building materials
1.7.3 Site layout plan is not paced on the notice board for information
1.5.2 Inappropriate design of temporary works and services – Tower crane's fully blocked area
1.5.1 Inappropriate design of temporary works and services – Space for access
1.7.4 Changes to the site layout plan are not notified immediately
1.4.4 Does not pay full attention to contractors' and subcontractors' requirements – Services required
1.5.3 Inappropriate design of temporary works and services- Clearance of the blocked area
1.5.4 Inappropriate design of temporary works and services – Siting of static plants
1.2.1 No overall consideration of equipment used in future precast concrete production processes
1.5.5 Inappropriate design of temporary works and services – Parking of mobile plant
1.2.2 No overall consideration of infrastructure used in future precast concrete production processes
1.4.3 Does not pay full attention to contractors' and subcontracts' requirements – Maximum number of men on site
1.4.2 Does not pay full attention to contractors' and subcontractors' requirements – Office space
1.7.2 Site layout plan is not sent to contractors, subcontractors and general foreman
1.3 Does not comply with mandatory statutory requirements
1.4.1 Does not pay full attention to contractors' and subcontractors' requirements – Duration

Fig. 7.3 The severity of non-value adding activities in site layout management

By replacing conventional building materials with green materials, the embodied carbon of precast concrete products could be reduced as well (1.1.4). The precasters stated that it was very important to comply with mandatory requirements because rework arising from failure to do so might significantly increase the carbon emission levels (1.3).

3. When considering both the probability and impact, the severity of these non-value adding activities is shown in Fig. 7.3. The most severe factors were:

   - 1.6.3 Over provide material storage – Open store (14.53)
   - 1.6.1 Over provide material storage – Secure store (13.29)

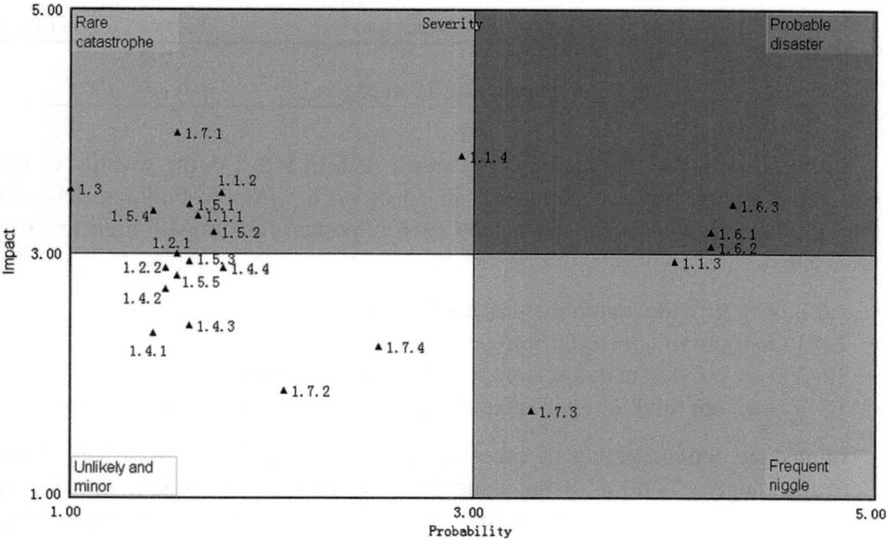

Fig. 7.4 P-I table for non-value adding activities in site layout management

- 1.6.2 Over provide material storage – Weatherproof store (12.82)
- 1.1.3 Does not think of alternative designs that minimize the use of materials (11.44)
- 1.1.4 Does not think of green building materials (10.38).

The precasters agreed that by building up large storage against supply and production disruptions, carbon emission levels might be increased in terms of damaged materials, classifying activities and singling activities (Sect. 1.6). In addition, failing to take alternative designs and green materials into consideration were two important non-value adding activities (1.1.3 and 1.1.4).

On the other hand, factors with high probability of occurrence but low impact, such as 1.7.3 (Site layout plan is not placed on the notice board for information), and factors with high impact but low probability of happening, such as 1.7.1 (Site layout plan is not tested for economic and efficient production) and 1.3 (Does not comply with mandatory statutory requirements), were rated with low severity. For example, although a carefully planned site layout which enhances economic and efficient production (1.7.1) was very important as rated by impact, the severity was very low because all precasters seemed to perform very well in this area (P = 1.53).

For illustration purposes, the probabilities and impact of the non-value adding activities in site layout management were plotted in a P-I table (sometimes referred to as the P-I grid). As shown in Fig. 7.4, there were four categories of non-value adding activities, which were: unlikely and minor, rare catastrophe, frequent niggle and probable disaster.

Table 7.7 Test statistics for factor 1.6.3 and 1.6.1

Test statistics (Wilcoxon)	Factors 1.6.3 and 1.6.1
Z	−1.442
Asymptotic Sig. (2 tailed)	0.149

It should be noted that as the color turned dark in Fig. 7.4, the severity of the non-value adding activity increased. In accordance with the findings obtained through the weighted factor model, the most important factors with high severity included:

- 1.6.3 Over provide material storage – Open store
- 1.6.1 Over provide material storage – Secure store
- 1.6.2 Over provide material storage – Weatherproof store
- 1.1.3 Does not think of alternative designs that minimize the use of materials

Immediate actions should be taken to "avoid" such activities from happening. Correspondingly, "control" actions should apply to high-probability, low-impact activities (frequent niggle); "transfer" actions should apply to low-probability, high-impact activities (rare catastrophe); and "accept" actions should apply to low-probability, low-impact activities (unlikely and minor). Details of the mitigation strategies will be explained in Sect. 6.7.

7.3.4 Non-parametric Tests

This section aims to identify the non-value adding activities within the same group ranked by severity. Attention should be paid to those non-value adding activities with close degree of severities. The statistical significance between different groups should be identified. As the sample size was small (<30) with unknown distribution (normal or non-normal), non-parametric tests were used in this step.

In this study, non-parametric tests developed for paired samples were adopted to test the significance between two different non-value adding activities at 95 % confidence interval, because the sequence of the sample data could not be changed.

For example, factors 1.6.3 (Over provide material storage- Open store) and 1.6.1 (Over provide material storage – Secure store) were rated as the two most important non-value adding activities in site layout management in the precast concrete factories. As can be seen from Table 7.7, p value was larger than 0.05, which indicated that factors 1.6.3 and 1.6.1 were within the same group of severity. Non-parametric tests (paired sample) between the following factor and 1.6.3 continued until p was less than 0.05. The first group of non-value adding activities with the highest severity could therefore be identified.

In this study, factor 1.1.3 was significantly different from factor 1.6.3 (A.$\rho = 0.044$) and was therefore listed as the top factor of ranking group 2. Following the procedure, all non-value adding activities were grouped into four groups, as can be seen in Table 7.8.

Table 7.8 Ranking and grouping of non-value adding activities in site layout management

Ranking	Factor no.	Description	Severity		
			AR	A.ρ	SD
1	1.6.3	Over provide material storage – Open store	14.53	N/A	4.19
1	1.6.1	Over provide material storage – Secure store	13.29	0.149	3.95
1	1.6.2	Over provide material storage – Weatherproof store	12.82	0.073	3.83
2	1.1.3	Does not think of alternative designs that minimize the use of materials	11.44	0.044 N/A	3.69
2	1.1.4	Does not think of green building materials	10.38	0.253	4.44
3	1.1.2	Inaccurate estimation of quantities required	6.13	0.001 N/A	2.60
3	1.7.1	Site layout plan is not tested for economic and efficient production	6.00	0.774	2.42
3	1.1.1	Improper specifications of building materials	5.63	0.474	2.90
3	1.7.3	Site layout plan is not placed on the notice board for information	5.53	0.380	4.06
3	1.5.2	Inappropriate design of temporary works and services – Tower cranes' fully blocked area	5.47	0.721	2.40
3	1.5.1	Inappropriate design of temporary works and services – Space for access	5.41	0.723	2.03
3	1.7.4	Changes to the site layout plan are not notified immediately	5.35	0.396	3.57
3	1.4.4	Does not pay full attention to contractors' and subcontractors' requirement – Services required	5.24	0.412	2.19
3	1.5.3	Inappropriate design of temporary works and services – Clearance of the blocked area	4.65	0.180	1.84
3	1.5.4	Inappropriate design of temporary works and services – Siting of static plants	4.65	0.195	1.90
3	1.2.1	No overall consideration of equipment used in future precast concrete production processes	4.59	0.067	2.40
3	1.5.5	Inappropriate design of temporary works and services – Parking of mobile plants	4.29	0.073	1.99
4	1.2.2	No overall consideration of infrastructure used in future precast concrete production processes	4.00	0.017 N/A	1.17
4	1.4.3	Does not pay full attention to contractors' and subcontractors' requirement – Maximum men on site	4.00	0.916	2.18
4	1.4.2	Does not pay full attention to contractors' and subcontractors' requirement – Office space	3.94	0.861	2.14
4	1.7.2	Site layout plan is not sent to contractors, subcontractors and general foreman	3.88	0.621	2.64
4	1.3	Does not comply with mandatory statutory requirements	3.53	0.254	1.01
4	1.4.1	Does not pay full attention to contractors' and subcontractors' requirement – Duration	3.35	0.186	1.93

Notes: *A.ρ* Asymptotic Significance (2-tailed)

As can be seen in Table 7.8, factors 1.6.3, 1.6.1 and 1.6.2 belonged to ranking group 1 with the highest severity. Much attention should therefore be made to reduce the storage area in precast concrete factories. Factor 1.1.3 (Does not think of alternative design) and 1.1.4 (Does not think of green building materials) were categorized into ranking group 2 with high severity, which indicated the importance of alternative design and green building materials for this industry to achieve low-carbon production. On the other hand, factors in ranking group 4 did not seem urgent to be eliminated and it might not be efficient to focus on these activities too much with limited resources.

7.3.5 *Specific Analysis*

Specific qualitative analysis was conducted when large difference exists between the largest rating (LR) and smallest rating (SR) in the probability section. Large difference might imply that the precasters with high probability of occurrence were not performing very well against those with low probability. In this study, when LR equaled 5 and SR equaled 1, the factor was chosen for specific qualitative analysis.

As can be seen from Table 7.6, three factors which were chosen for specific qualitative analysis:

- 1.1.3 Does not think of alternative designs that minimize the use of materials

  Most precasters agreed that it was very common that the design of precast concrete production followed routine standards. It was not the precasters' responsibility to develop alternative designs to reduce the use of materials, neither did they benefit from doing so. However, one precaster (5.88 %) claimed that it was not uncommon that alternative designs were submitted for evaluation. This was probably because that this precaster was a government-sponsored entity and research was part of the obligations it assumes. The large difference in the probability rating was therefore an outcome of government intervention.
- 1.1.4 Does not think of green building materials

  The use of green building materials (e.g. low-carbon concrete, new additives to reduce the use of concrete) in precast concrete production varied among the precasters in Singapore. Many precasters claimed that by following the research of green building materials, they could seize the newly emerging opportunities to expand their market share. However, three precasters (17.64 %) stated that due to large amount of orders, the priority was to arrange production so that the finished products could be handed to customers on time. Considering the use of green building materials would cause distractions ($P = 5$). These three precasters all belonged to the L6 financial grade and might be exposed to large amount of orders.
- 1.7.3 Site layout plan is not placed on the notice board for information

  Whether or not the site layout plan was placed on the notice board for information was subject to the precaster's own willingness. Only one precaster (5.88 %) claimed that the site layout plan was always placed on the notice board when changing contractors, subcontractors and managers ($P = 1$). Three precasters (17.65 %) believed that this was not necessary because by hiring contractors, subcontractors and managers, they were entrusted with the work delegated ($P = 5$).

7.4 Lean Supply Chain Management in Precast Concrete Factories

In this section, the supply chain management practices of the 17 precast concrete factories were investigated. Non-value adding activities in this section were derived from the literature review relating to JIT sourcing, TQM as well as JIT management practices.

7.4.1 *Descriptive Analysis*

Similar to the section on site layout management, a few general questions related to current supply chain management were discussed with the precasters before the non-value adding activities were provided for ranking. Details of the responses can be seen in Table 7.9.

A few implications can be identified from Table 7.9, including:

1. Three important aspects of JIT sourcing, namely single sourcing, small lot size and long-term relationship, were not adopted at all in the Singapore precast concrete industry. All precasters claimed that single sourcing may upset the production if the supplier failed to deliver the materials in time.

   Small lot size was unnecessary due to the large storage area designed at the beginning, which was 43 % of the total area in average. In addition, the award of

Table 7.9 General questions in the section of supply chain management

Question no.	Description	Positive response	Positive response rate	Category
2.3.1	Is single sourcing adopted in the supply chain?	0	0 %	JIT sourcing
2.3.2	Is small lot size adopted in the supply chain?	0	0 %	JIT sourcing
2.3.3	Is long-term contract awarded to achieve loyalty?	0	0 %	JIT sourcing
2.3.4	Is the Singapore precast concrete industry vulnerable to supply disruptions?	8	47.06 %	Industry information
2.3.5	Is the company operating under a stable production schedule?	16	94.12 %	Industry information
2.3.6	Between the expected costs of small lot delivery and the savings of reduced inventory, has the company conducted any trade-off investigation?	0	0	Continuous improvement
2.3.7	What is the current delivery method adopted in the company?	Refer to implication 3		Delivery
2.3.8	How does the company evaluate suppliers?	Refer to implication 3		Selecting suppliers

contracts to suppliers were made annually or once every 2 years. When such long-term relationships with suppliers were established, precasters might incur cost increase by failing to identify suppliers with lower prices.

2. Eight precasters (47.06 %) stated that the Singapore precast concrete industry was vulnerable to supply disruptions. Delays that were caused by suppliers occurred from time to time. On the other hand, nine precasters (52.94 %) claimed that suppliers in Singapore operated well enough to deliver raw materials on time. Sixteen precasters (94.12 %) had a stable production schedule based on which an estimation of the raw materials required could be conducted. There seemed to be a contradiction between the industry background and the willingness to apply JIT. More than half of the precasters (52.94 %) believed that the suppliers performed very well and nearly all precasters (94.12 %) had a stable production schedule. However, none of them were applying JIT sourcing at the time of this study or intended to do so in the foreseeable future.
3. The evaluation of suppliers was based on several criteria: a quantitative measure of quality, certification programmes, delivery performance and most importantly, the price structure. These criteria were applied by all precasters when selecting suppliers. However, two other important aspects were missing, which were: a qualitative evaluation conducted at the supplier's plant (applied in two precasters) and the geographical location of the supplier (applied in three precasters). According to the precasters interviewed, a qualitative evaluation at the supplier's plant was not necessary because the quality could be guaranteed by different certification programmes. In addition, as the cost of transportation was integrated into the price of products, the geographical location of the suppliers seemed to be unimportant when selecting these suppliers. However, two aspects were overlooked by doing so. A qualitative evaluation at the supplier's plant was not only to guarantee the quality of the products, but also the ability of the suppliers. Whether or not the supplier would be able to deal with sudden increase in demand could be identified by such qualitative evaluation. Using the "all-in-one" price instead of taking the transportation into consideration might lead to delay caused by traffic conditions, which could not be represented by comparing the prices. Of all the delivery methods provided (direct delivery, with an interposed warehouse, milk round collection, suppliers supply a number of components), all precasters (100 %) were using the direct delivery method. It seemed that if direct delivery was used and transportation cost was integrated into the product price, precasters in Singapore had less control to guarantee a dependable delivery.
4. It seemed that precasters currently did not perform well on continuous improvement in terms of research. None of the precasters were conducting any research related to small lot delivery and reduced inventory. On the one hand, the precasters were suffering from problems caused by large inventory. On the other hand, they were satisfied with current production in case improving actions might cause too many changes to the structure of the company, as well as indirectly increase the production costs.

Table 7.10 Two major categories of non-value adding activities in supply chain management

Category no.	Description	Category
2.1	Selecting suppliers	Suppliers
2.2	JIT management process	Delivery management

7.4.2 *Factors Description*

The lean management philosophy adopts the pull production system. By adopting the pull production system, JIT sourcing is necessary. The questionnaire for supply chain management was therefore designed into two major categories: selecting suppliers and JIT management process, which can be seen in Table 7.10.

In each of these two categories, a few classified non-value adding activities were identified. A general description for each non-value adding activity is provided as follows:

- *Category 2.1 Selecting suppliers*

2.1.1 Large quantity supply base which leads to inflexibility

Small lot-sizes were considered as the hallmark of JIT sourcing and were believed flexible enough to overcome the obstacles of higher delivery costs and loss of discount rates (Banerjee and Kim 1995).

2.1.2 No long-term contract to achieve loyalty between suppliers and precasters

Achieving long-term relationship was believed to have several benefits, such as easing the communications problems between precasters and suppliers, obtaining discount and building trust.

2.1.3 Transportation is not taken into consideration

As stated earlier, transportation cost was currently integrated into the overall product price. The geographical location of the suppliers was hardly a factor to consider when selecting suppliers. However, many problems could be caused by sourcing on the basis of price alone, such as losing control of the delivery time.

2.1.4 No quality audits of the supplier

Although quality could be assured by many certification programmes, there were still many benefits by conducting quality audits, especially when JIT delivery was scheduled. However, it seemed that price was still the major consideration when selecting suppliers.

- *Category 2.2 JIT management process*

2.2.1 Demand fluctuations

If a stable production schedule could be anticipated, JIT deliveries could be more readily applied. However, this does not mean that JIT deliveries could not be applied under demand fluctuations. More prompt communication was needed under demand fluctuations.

2.2.2 Not fully prepared for the arrival of raw materials

Under extremely perfect conditions, materials should be used immediately when delivered. However, due to various reasons, such as waiting for inspection and equipment, this was not always the case in the precast concrete factories.

2.2.3 Lack of both advance order and confirmation order

Tommelein and Li (1999) provided JIT delivery mapping alternatives for vertical supply chain integration. In this mapping strategy, both advance order and confirmation order were used to make sure that deliveries arrive on time. Cancellation of orders could be executed without harming suppliers' benefits.

2.2.4 Unsatisfied data exchange with suppliers

As stated earlier, under demand fluctuations, prompt communication between precasters and suppliers is of vital importance when applying the JIT delivery system. The data exchange methods include fax, email, mobile and some other electronic softwares.

7.4.3 Ranking Procedure

Similar to the section on site layout management, the non-value adding activities in supply chain management were rated by both probability and impact under a five-point scale value range. The results are shown in Table 7.11. A few implications can be made from Table 7.11, including:

1. The most frequently occurring non-value adding activities in the supply chain management were:
   - 2.1.2 No long-term contract to achieve loyalty between suppliers and precasters (4.29)
   - 2.1.1 Large quantity supply base which leads to inflexibility (4.12)
   - 2.1.3 Transportation is not taken into consideration (3.94)
   - 2.2.3 Lack of both advance order and confirmation order (3.88).

   It seemed that the precast concrete industry in Singapore still followed the routines from the mass production system. There were no long-term contracts between the suppliers and the precasters (2.1.2). Most contracts were based on an annual or biennial agreement where the suppliers would be re-evaluated at the end of the agreement. Large quantity supply often happened and was rated over 4 in the five-point scale value range (2.1.1).

   As the transportation cost was integrated in the product price, transportation was usually not the prioritized consideration of the precasters (2.1.3). In addition, a JIT management process which involves both advance order and confirmation order was hardly adopted by the precasters (2.3.3). Most precasters preferred placing one order, which was believed to have the same function but

Table 7.11 Probability, impact and severity of the non-value adding activities in supply chain management

Factor no.	Description	Probability (P)				Impact (I)				Severity (S)			
		AR	LR	SR	SD	AR	LR	SR	SD	AR	LR	SR	SD
2.1.1	Large quantity supply base which leads to inflexibility	4.12	5	3	0.60	3.47	5	2	0.80	14.29	20	8	3.77
2.1.2	No long-term contract to achieve loyalty between suppliers and precasters	4.29	5	3	0.59	3.24	4	2	0.66	13.94	20	8	3.58
2.1.3	Transportation is not taken into consideration	3.94	5	1	1.25	3.71	4	2	0.59	10.71	20	3	6.11
2.1.4	No quality audits of the supplier prior to the award of contracts	2.94	5	1	1.39	3.35	5	2	0.93	12.88	20	3	5.16
2.2.1	Demand fluctuations	3.00	4	2	0.79	3.00	4	1	0.94	9.00	16	3	4.02
2.2.2	Not fully prepared for the arrival of raw materials	2.35	3	1	0.79	3.12	4	2	0.78	7.18	12	3	2.77
2.2.3	Lack of both advance order and confirmation order	3.88	5	1	0.86	3.35	4	2	0.70	13.18	20	3	4.29
2.2.4	Unsatisfied data exchange with suppliers	1.47	2	1	0.51	3.06	4	3	0.24	4.53	8	3	1.74

Notes: *AR* average rating, *LR* largest rating, *SR* smallest rating, *SD* standard deviation

less complications. Data exchange between the precasters and suppliers seems good enough when unsatisfied data exchange (2.2.4) was only rated by 1.47. The ranking of the probability of non-value adding activities in supply chain management is shown in Fig. 7.5.

2. The impact of the non-value adding activities in supply chain management lay in the range between moderate and major, as can be seen in Fig. 7.6. One precaster stated that a long-term contract has benefits in simplifying communications with the suppliers and obtaining discounts when purchasing materials (2.1.2). However, this might lead to price inflexibility, causing the precasters to lose the benefits from choosing suppliers with lower prices. The reason that a high rating was assigned to this long-term relationship was the effort that can be saved on choosing suppliers when such relationship was built, as long as suppliers could provide reasonably low prices.
3. The most severe non-value adding activity was 2.1.1 (Large quantity supply base which leads to inflexibility), with a rating of 14.29, as shown in Fig. 7.7.

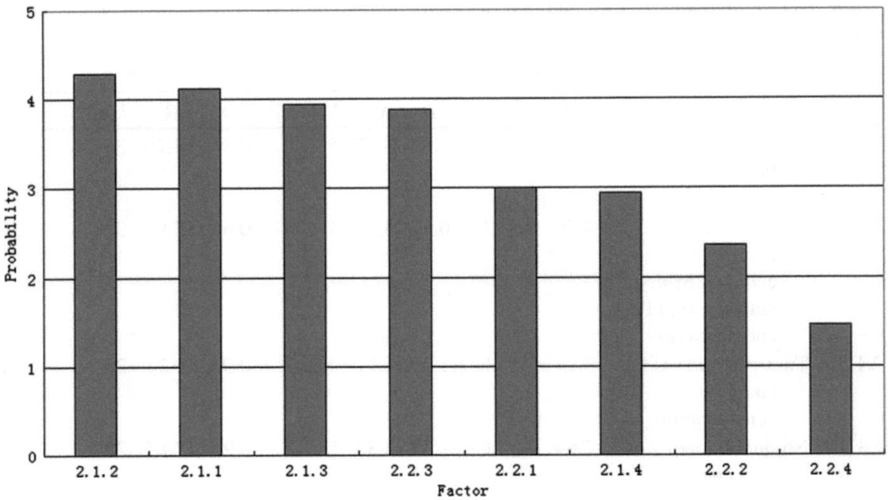

2.1.2 No long-term contract to achieve loyalty between suppliers and precasters
2.1.1 Large quantity supply base which leads to inflexibility
2.1.3 Transportation is not taken into consideration
2.2.3 Lack of both advance order and confirmation order
2.2.1 Demand fluctuations
2.1.4 No quality audits of the supplier prior to the award of contracts
2.2.2 Not fully prepared for the arrival of raw materials
2.2.4 Unsatisfied data exchange with suppliers

Fig. 7.5 The probability of non-value adding activities in supply chain management

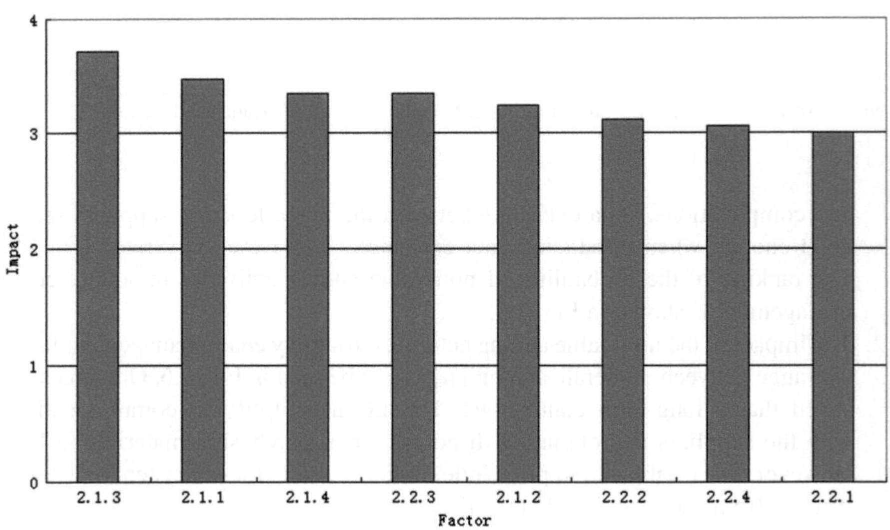

2.1.3 Transportation is not taken into consideration
2.1.1 Large quantity supply base which leads to inflexibility
2.1.4 No quality audits of the supplier prior to the award of contracts
2.2.3 Lack of both advance order and confirmation order
2.1.2 No long-term contract to achieve loyalty between suppliers and precasters
2.2.2 Not fully prepared for the arrival of raw materials
2.2.4 Unsatisfied data exchange with suppliers
2.2.1 Demand fluctuations

Fig. 7.6 The impact of non-value adding activities in supply chain management

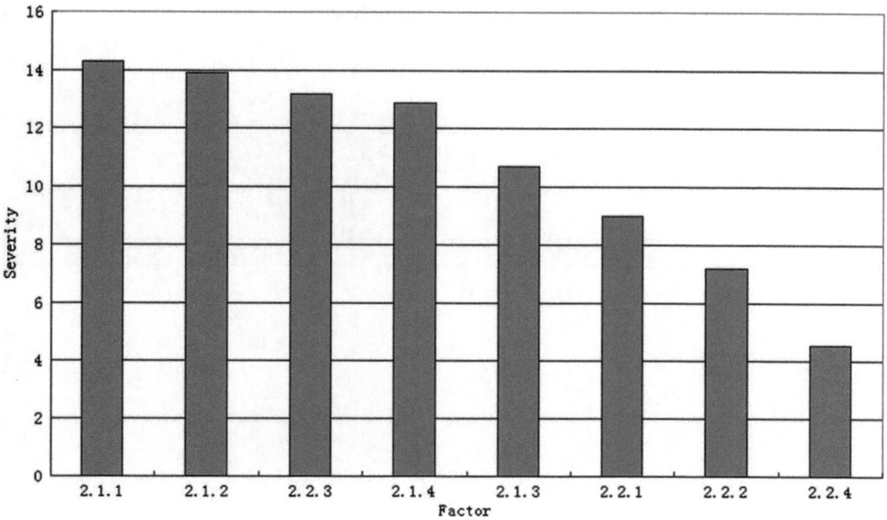

2.1.1 Large quantity supply base which leads to inflexibility
2.1.2 No long-term contract to achieve loyalty between suppliers and precasters
2.2.3 Lack of both advance order and confirmation order
2.1.4 No quality audits of the supplier prior to the award of contracts
2.1.3 Transportation is not taken into consideration
2.2.1 Demand fluctuations
2.2.2 Not fully prepared for the arrival of raw materials
2.2.4 Unsatisfied data exchange with suppliers

Fig. 7.7 The severity of non-value adding activities in supply chain management

It seemed that precasters agreed that such large quantity supply can cause disruptions to production by building large stockpile. However, precasters have not conducted any research to reduce storage, probably due to the potential costs associated with the research. On the other hand, factors 2.2.4 (Unsatisfied data exchange with suppliers) and 2.2.1 (Demand fluctuations) had very low severity due to the low probability, 3.06 and 3.00 respectively. Precasters performed well when handling materials that arrived (2.2.2) and communicating with suppliers (2.2.4). The severity of these non-value adding activities showed that precasters performed badly on JIT sourcing aspects, such as 2.1.1 (Large quantity supply base which leads to inflexibility) and 2.2.3 (Lack of both advance order and confirmation order).

For illustration purposes, the probabilities and impact of the non-value adding activities in supply chain management were plotted in a P-I table, as shown in Fig. 7.8.

It seemed that most non-value adding activities fell into the category of probable disaster. This was probably because these non-value adding activities were the drawbacks of the mass production systems examined by the lean production philosophy. Since most precasters used the mass production system, it was not surprising that the severity of these non-value adding activities was highly rated.

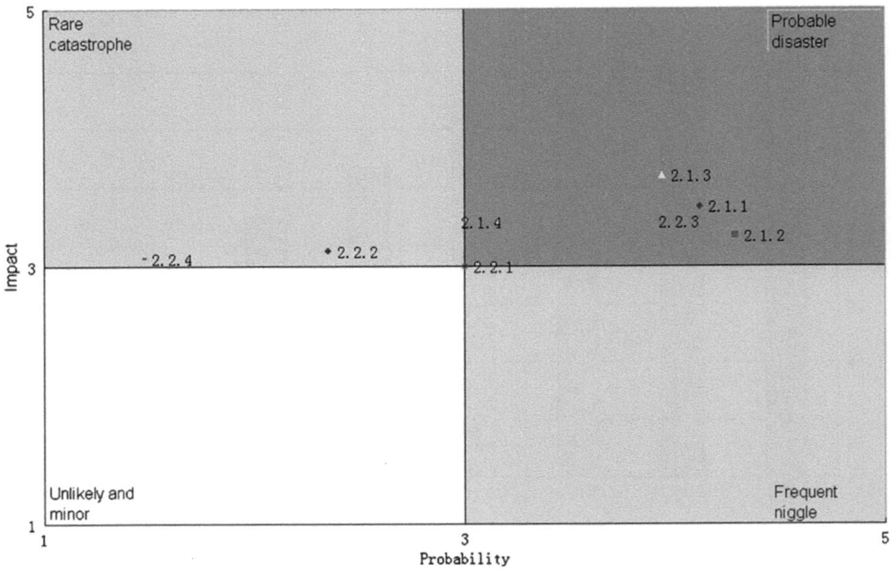

Fig. 7.8 P-I table for non-value adding activities in supply chain management

7.4.4 Non-parametric Tests

Similar to Sect. 7.3.4, non-parametric tests (paired sample) were conducted to rank the non-value adding activities into groups. The non-parametric tests followed the general procedure which included the following steps:

1. Conduct non-parametric tests for the top two factors with the highest severity. If p value is calculated larger than 0.05, the non-parametric tests which compares the third factor with the top factor continues until p value is less than 0.05.
2. Factors with p value less than 0.05 is chosen as the top factor in the second group. All the following factor will be compared to this one unless p value is less than 0.05.
3. Step 1 and step 2 continue unless non-parametric tests are conducted for all factors.
4. Double rows are used for factors that are on top of each ranking group, as can be seen in Table 7.12.

The results of the non-parametric tests for the non-value adding activities in the supply chain management can be seen in Table 7.12. Three groups were identified by the non-parametric tests. It seemed that large quantity supply base (2.1.1), no long term-contracts (2.1.2), no JIT management process (2.2.3) and no quality audits (2.1.4) were the factors with high negative impact on the carbon emission levels in precast concrete factories. On the other hand, data exchange (2.2.4) should not be focused on overly because the precasters were performing very well in these areas.

Table 7.12 Ranking and grouping of non-value adding activities in supply chain management

Ranking	Factor no.	Description	Severity		
			AR	A.ρ	SD
1	2.1.1	Large quantity supply base which leads to inflexibility	14.29	N/A	3.77
1	2.1.2	No long-term contract to achieve loyalty between suppliers and precasters	13.94	0.833	3.58
1	2.2.3	Lack of both advance order and confirmation order	13.18	0.434	4.29
1	2.1.4	No quality audits of the supplier prior to the award of contracts	12.88	0.409	5.16
2	2.1.3	Transportation is not taken into consideration	10.71	0.035 N/A	6.11
2	2.2.1	Demand fluctuations	9.00	0.477	4.02
2	2.2.2	Not fully prepared for the arrival of raw materials	7.18	0.092	2.77
3	2.2.4	Unsatisfied data exchange with suppliers	4.53	0.002	1.74

Notes: $A.\rho$ Asymptotic Significance (2-tailed)

7.4.5 Specific Analysis

Examined by the criterion that LR equals 5 and SR equals 1, three non-value adding activities were chosen for specific analysis. These non-value adding activities were:

- 2.2.3 Lack of both advance order and confirmation order

Fifteen precasters (88.24 %) admitted that only one order was used when purchasing materials (P = 4 and P = 5). However, one precaster stated that this JIT management procedure with two orders was currently adopted (P = 1). The company was small and only produced limited amount of products for public housing projects. According to the project manager, as the design specifications often changed, using both advance order and confirmation order would help the company deal with such changes, especially when the company was trying to adopt the pull system to reduce storage. By using the two-order system, the company was less vulnerable to changes.

- 2.1.4 No quality audits of the supplier prior to the award of contracts

Two precasters (11.76 %) claimed that quality audits are conducted before contracts are awarded to the suppliers chosen (P = 1). Despite the certification programmes, precasters would still obtain an in-depth understanding the suppliers they are cooperating with. However, this was currently not widely adopted in the precast concrete industry where 14 precasters were evaluating suppliers based on the price structure (P = 5 and P = 4), and assumed that quality could be guaranteed by certification programmes.

- 2.1.3 Transportation is not taken into consideration

The decisions related to whether or not transportation should be taken into consideration varied among the precasters in Singapore. Seven precasters (41.18 %) believed that with the all-in-one price, transportation was not a prioritized consideration as long as the suppliers managed to deliver on time (P = 5 and P = 4). Among the seven precasters, one project manager stated that:

> In the Singapore precast concrete industry, suppliers are trusted to make on-time delivery. In fact, they are performing very well, probably because Singapore is such a small country and logistics in Singapore are very good.

However, eight precasters (47.06 %) stated that when selecting suppliers, transportation should be considered (P = 1 and P = 2). As one project manager stated:

> We produce based on orders. Since the production schedule is determined and consumers prefer newly produced products, it is better that the suppliers can deliver on time. Transportation should be considered.

7.5 Lean Production Management in Precast Concrete Factories

In this section, the precast concrete production processes were examined by the lean production philosophy. The non-value adding activities in this section were derived from a literature review (e.g. Ohno 1988; Low and Mok 1999; Alwi et al. 2002) and pilot studies.

7.5.1 Descriptive Analysis

A few general questions related to the current precast concrete production were discussed with the precasters. These general questions covered various areas in the production processes, including regulations, environment, equipment, human resources, continuous improvement and production type and are shown in Table 7.13.

A few implications can be inferred from Table 7.13, including:

1. Seven precasters (41.48 %) stated that they were aware of some regulations that were related to environmental performance, in which the most important one related to the Environmental Protection and Management Regulations. This regulation covered different categories, including energy conservation, boundary noise limits for factory premises, hazardous substances and so on. However, according to the precasters, there were currently no regulations that were designed to control the carbon emissions in factories. Recommendations were otherwise provided for the precasters to reduce the energy consumption and carbon emissions, based on such as the Singapore's National Climate Change Strategy.
2. Only one precaster (5.88 %) was taking actions to reduce the energy consumption and carbon emissions in the factory. Most precasters admitted that until the pressure of reducing energy consumption and carbon emissions was explicitly expressed by law and regulations, the factories would still operate as usual due to the potential costs that might be incurred.

Table 7.13 General questions in the section of production management

Question no.	Description	Positive response	Positive response rate	Category
3.6.1	Is the company aware of the regulations related to environmental performance?	7	41.48 %	Regulations
3.6.2	Is the company currently taking any action to reduce the non-value adding activities listed?	1	5.88 %	Environment
3.6.3	Is the company currently taking an action to reduce energy consumption and carbon emissions in the factory?	1	5.88 %	Environment
3.6.4	Are the customer requirements taken into consideration when designing and manufacturing precast concrete products?	8	47.06 %	Production
3.6.5	Are the manufacturing equipment carefully maintained during the production life cycle?	16	94.12 %	Equipment
3.6.6	When selecting the operator for a specific kind of equipment, is he trained on sustainable operations?	1	5.88 %	Human resources
3.6.7	Are there any training programmes for the operators who work in the manufacturing process?	16	94.12 %	Human resources
3.6.8	Is there an employee performance rating system that takes environment-friendly operations as one consideration?	0	0	Human resources
3.6.9	Is the company conducting any research on how to improve future production?	2	11.76 %	Continuous improvement
3.6.10	Are there any internal periodic meetings to discuss improvement?	17	100 %	Continuous improvement
3.6.11	Do the internal periodic meetings involve external professionals of specific improvement issues?	2	11.76 %	Continuous improvement
3.6.12	What is the production type of this factory?	Push 14	Pull 3	Production type

3. Three precasters (17.65 %) stated that the pull production system was adopted as the production method, while 14 precasters (82.35 %) reported that the push production method was used. However, when observed in these factories, it was discovered that this pull production system was in fact a Horizontally Integrated Hybrid System (HIHS). Raw materials were ordered and stored in factories followed by the push system. When there was an order of precast concrete products, raw materials were then used. This explained why there was such a large storage area even when the pull production system was adopted.

Eight precasters (47.06 %) stated that they could take the customer requirements into consideration when there was a special order. On the other hand, nine precasters (52.94 %) were producing what were largely required in the market through the push system.

4. Sixteen precasters (94.12 %) provided training programmes for operators in the manufacturing process. These training programmes covered various aspects, mainly relating to safety issues. However, only one precaster (5.88 %) was providing training programmes on sustainable operations. As observed earlier, there seems to be a time lag between the urgency of being sustainable and the actual actions that were taken to improve and progress towards sustainability. Until such urgency was explicitly expressed by law and regulations instead of recommendations, it would appear that training programmes would remain the same with few or no contents of sustainable operations.

   Equipment used in precast concrete were carefully maintained by most precasters (94.12 %). Due to the large costs of such equipment, such as tower cranes, mobile cranes, gantry systems and other lifting and production equipment, a detailed maintenance programme was necessary. Sixteen precasters (94.12 %) stated that the maintenance programmes matched the rate of use. Additional maintenance programmes were combined with regular programmes to prevent breakdown if the equipment were operating with a heavy schedule. However, one precaster (5.88 %) was not doing so because of small quantity of demand. Only a monthly maintenance programme was conducted in this factory.
5. Few research studies were conducted by precasters for further improvement. Only two precasters (11.76 %) were conducting such research, e.g. reducing energy consumption and the use of newly invented additives. However, all precasters (100 %) were organizing periodic internal meetings to discuss issues that could help to improve future production. These periodic internal meetings were normally organized once a month. Two precasters (11.76 %) indicated that external professionals may be included in such meetings to help the company, while most precasters (88.24 %) were not doing so because of the need to preserve confidentiality.

7.5.2 Factors Description

Five categories of non-value adding activities were identified through a literature review (e.g. Ohno 1988; Low and Mok 1999; Alwi et al. 2002). These categories can be seen in Table 7.14.

In each of these five categories, detailed non-value adding activities were listed for ranking, which included:

- *Category 3.1 Materials*

  3.1.1 Waste of raw materials in the production process
  3.1.2 Raw materials do not meet specifications
  3.1.3 Materials damaged during handling
  3.1.4 Unnecessary materials handling
  3.1.5 Too much inventory in factory
  3.1.6 Loss of materials in factory

Table 7.14 Five major categories of non-value adding activities in production management

Category no.	Description
3.1	Materials
3.2	Cycle time
3.3	Finished products
3.4	Operations
3.5	Human resources

The use of materials in precast concrete production should be carefully examined. The non-value adding activities listed above could lead to re-order and re-delivery of raw materials. Energy consumed and carbon emissions generated from such activities were the targets to be eliminated in this study.

- *Category 3.2 Cycle time*

  3.2.1 Wait time for inspection
  3.2.2 Wait time for the delivery of materials
  3.2.3 Wait time for labor
  3.2.4 Wait time for equipment

The cause of wait time in precast concrete production was investigated and listed for ranking. For example, as observed in the precast concrete factories, the transportation vehicle was left idling when the quality control step was being carried out. This belonged to the category of wait time for inspection. Similarly, non-value adding activities could happen in terms of waiting time for the delivery, labor and equipment.

- *Category 3.3 Finished products*

  3.3.1 Repair due to damaged products during inventory
  3.3.2 Repair due to damaged products when handling
  3.3.3 Double-handling or delivery due to unsatisfied quality or specifications

According to the precasters, there was a 2 % waste of the finished products, either in inventory or due to handling activities. Delivery of finished products to wrong construction sites did happen for some precasters, despite the low probability of occurrence. The wrong deliveries made were mainly caused by unclear delivery notes and insufficient care of the drivers.

- *Category 3.4 Operations*

  3.4.1 Inappropriate site layout
  3.4.2 Equipment breakdown
  3.4.3 Inappropriate selection of equipment

Inefficiencies in precast concrete production could be caused by the site layout plan. If the design of the site layout was not tested on economic and efficient issues, it might fail to support a smooth production flow. Equipment breakdown and inappropriate selection of equipment might cause interruption to the precast concrete production processes as well.

- *Category 3.5 Human resources*

  3.5.1 Inadequate work crews
  3.5.2 Inexperienced employees
  3.5.3 Lack of supervision

It goes without saying that improving energy efficiency and reducing carbon emissions should involve everyone who is directly or indirectly related to the use of equipment. Supervisors, facilities managers and operators should have the greatest potential to reduce energy consumption, which highlights the importance of human resources in achieving sustainability in precast concrete factories. The term "inexperienced" used here refers to the knowledge of lean rather than the general production process.

7.5.3 Ranking Procedure

The probability, impact and severity of the non-value adding activities in production management are listed in Table 7.15. A few implications can be drawn from Table 7.15, including:

1. The non-value adding activities with high probability of occurrence included (as shown in Fig. 7.9):

   - 3.1.5 Too much inventory in factory (4.47)
   - 3.1.1 Waste of raw materials in the production process (3.94).

   When the precast concrete factories were built, the storage area remained large, causing possible interruptions to production processes (3.1.5). Waste of materials during production was very common in precast concrete production, especially in the aspects of concrete and steel (3.1.1). According to the precasters in the study, the average rate of waste during production was 3.04 %, with the highest level at 5 % and the lowest at 1 %. Although 14 precasters (82.35 %) claimed to have provided recycling plans for the wastes (mainly for steel) and such plans could significantly reduce the energy consumption and carbon emissions, it should be noted that energy were consumed through recycling activities that should not have happened in the first place.

   On the other hand, factors with the lease probability of occurrence included:

   - 3.4.3 Inappropriate equipment choice (1.06)
   - 3.1.6 Loss of materials in factory (1.18)
   - 3.4.1 Inappropriate site layout (1.24)

   It seems that precasters were performing well in selecting equipment (3.4.3), securing products (3.1.6) and designing site layout (3.4.1).

Table 7.15 Probability, impact and severity of the non-value adding activities in production management

Factor		Probability (P)				Impact (I)				Severity (S)			
no.	Description	AR	LR	SR	SD	AR	LR	SR	SD	AR	LR	SR	SD
3.1.1	Waste of raw materials in the production process	3.94	5	2	0.90	3.24	4	2	0.83	12.35	20	6	3.71
3.1.2	Raw materials do not meet specifications	1.88	3	1	0.60	3.47	5	2	0.87	6.53	12	2	2.58
3.1.3	Materials damaged during handling	3.24	5	2	0.66	3.47	5	2	0.80	11.00	16	6	2.55
3.1.4	Unnecessary materials handling	2.88	5	2	0.70	3.12	4	2	0.78	9.12	15	4	3.31
3.1.5	Too much inventory in factory	4.47	5	3	0.62	3.06	4	1	0.90	13.65	20	5	4.50
3.1.6	Loss of materials in factory	1.18	2	1	0.39	3.35	5	2	0.79	3.94	8	2	1.56
3.2.1	Wait time for inspection	2.47	4	2	0.62	3.00	4	1	0.87	7.47	16	2	3.04
3.2.2	Wait time for the delivery of materials	2.65	4	2	0.61	3.06	4	1	0.83	8.18	16	2	3.17
3.2.3	Wait time for labor	2.24	4	1	0.75	3.06	4	2	0.75	6.94	16	2	3.21
3.2.4	Wait time for equipment	2.81	4	1	0.81	2.94	4	2	0.75	6.59	16	2	3.41
3.3.1	Repair due to damaged products during inventory	3.12	5	2	0.86	3.06	4	1	0.83	9.12	12	5	2.37
3.3.2	Repair due to damaged products when handling	3.12	5	2	0.86	3.00	4	1	0.79	8.94	12	5	2.25
3.3.3	Double-handling or delivery due to unsatisfied quality or specifications	2.29	3	1	0.59	3.71	5	2	0.92	8.35	12	4	2.42
3.4.1	Inappropriate site layout	1.24	2	1	0.44	3.59	5	3	0.62	4.47	8	3	1.87
3.4.2	Equipment breakdown	2.71	3	2	0.47	3.76	5	3	0.56	10.35	15	6	2.87
3.4.3	Inappropriate equipment choice	1.06	2	1	0.24	3.12	5	1	1.11	3.35	8	1	1.62
3.5.1	Inadequate work crews	2.53	4	1	0.80	4.18	5	3	0.53	10.47	20	5	3.56
3.5.2	Inexperienced employees	2.59	4	1	0.87	4.06	5	3	0.56	10.76	20	3	4.45
3.5.3	Lack of supervision	2.35	5	1	1.06	4.12	5	3	0.60	9.94	25	3	5.33

Notes: *AR* average rating, *LR* largest rating, *SR* smallest rating, *SD* standard deviation

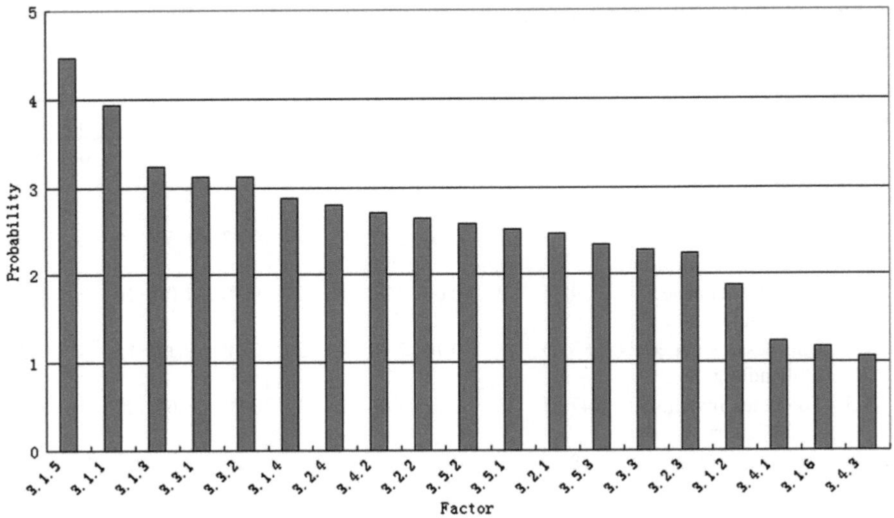

3.1.5 Too much inventory in factory
3.1.1 Waste of raw materials in the production process
3.1.3 Materials damaged during handling
3.3.1 Repair due to damaged products during inventory
3.3.2 Repair due to damaged products when handling
3.1.4 Unnecessary materials handling
3.2.4 Wait time for equipment
3.4.2 Equipment breakdown
3.2.2 Wait time for the delivery of materials
3.5.2 Incompetent employees
3.5.1 Inadequate work crews
3.2.1 Wait time for inspection
3.5.3 Lack of supervision
3.3.3 Double-handling or delivery due to unsatisfied quality or specifications
3.2.3 Wait time for labor
3.1.2 Raw materials do not meet specifications
3.4.1 Inappropriate site layout
3.1.6 Loss of materials in factory
3.4.3 Inappropriate equipment choice

Fig. 7.9 The probability of non-value adding activities in production management

2. The three most important factors with high negative impact on carbon emissions were (as shown in Fig. 7.10):

   - 3.5.1 Inadequate work crews (4.18)
   - 3.5.3 Lack of supervision (4.12)
   - 3.5.2 Inexperienced employees (4.06)

   Precasters agreed that human resources were the most valuable resources in reducing carbon emissions. With appropriately trained employees on sustainable operations, as well as sufficient supervision, the goal of reducing carbon could be achieved. Other important factors included:

   - 3.4.2 Equipment breakdown (3.76)
   - 3.3.3 Double-handling or delivery due to unsatisfied quality or specifications (3.71)
   - 3.4.1 Inappropriate site layout (3.59)

3. The severity of the non-value adding activities in production management can be seen in Fig. 7.11. As can be seen from Fig. 7.11, the four most severe factors in production management were:

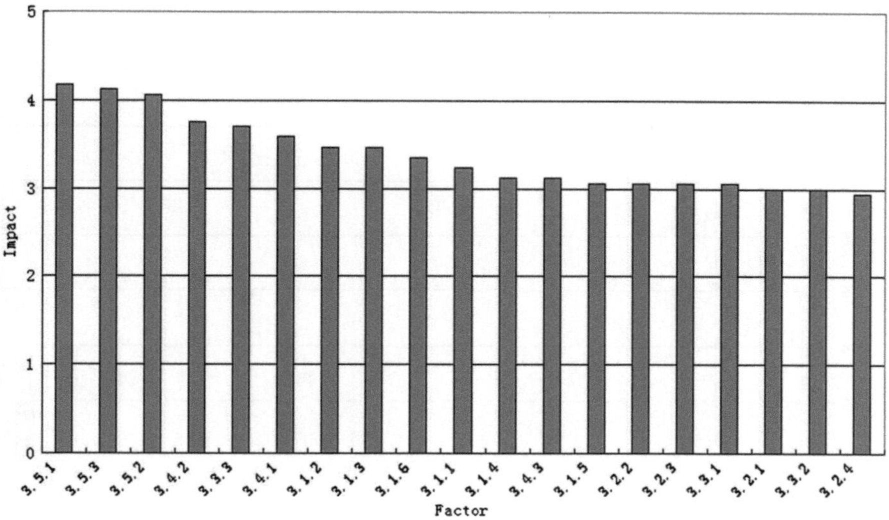

3.5.1 Inadequate work crews
3.5.3 Lack of supervision
3.5.2 Inexperienced employees
3.4.2 Equipment breakdown
3.3.3 Double-handling or delivery due to unsatisfied quality or specifications
3.4.1 Inappropriate site layout
3.1.2 Raw materials do not meet specifications
3.1.3 Materials damaged during handling
3.1.6 Loss of materials in factory
3.1.1 Waste of raw materials in the production process
3.1.4 Unnecessary materials handling
3.4.3 Inappropriate equipment choice
3.1.5 Too much inventory in factory
3.2.2 Wait time for the delivery of materials
3.2.3 Wait time for labor
3.3.1 Repair due to damaged products during inventory
3.2.1 Wait time for inspection
3.3.2 Repair due to damaged products when handling
3.2.4 Wait time for equipment

Fig. 7.10 The impacts of non-value adding activities in production management

- 3.1.5 Too much inventory in factory (13.65)
- 3.1.1 Waste of raw materials in the production process (12.35)
- 3.1.3 Materials damaged during handling (11.00)
- 3.5.2 Inexperienced employees (10.76)

These factors indicated that controlling inventory, managing wastes and promoting employees' ability are of vital importance to reduce carbon emissions in precast concrete factories. It seems that the precasters agreed that by placing too much inventory in the precast concrete factories, the production activities might be interrupted, leading to high energy consumption and high wastage of raw materials. More importantly, materials might be damaged during the transferring and singling out activities. Employees with knowledge of lean were very important. Without these employees, it was very difficult for the precasters to achieve a smooth work flow.

On the other hand, some factors seemed to have little impact on the level of carbon emissions, due to their relatively low probability of occurrence. These factors were:

- 3.4.3 Inappropriate equipment choice (3.35)
- 3.1.6 Loss of materials in factory (3.94)
- 3.4.1 Inappropriate site layout (3.59)

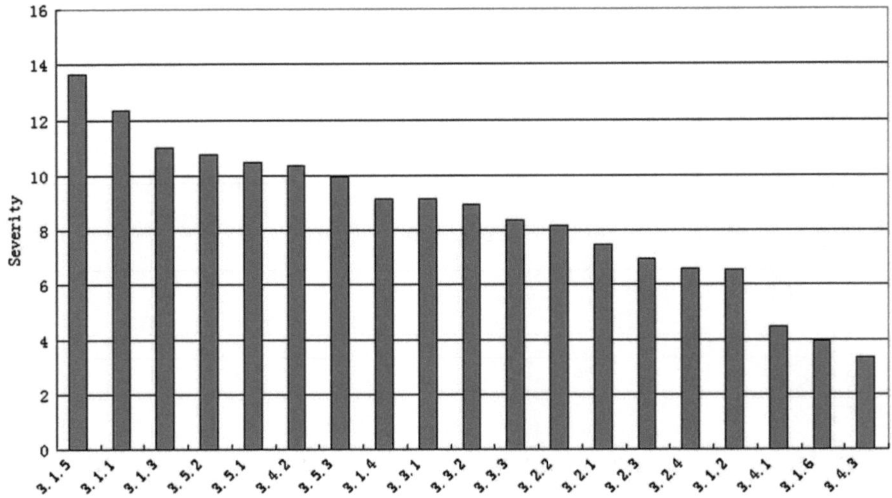

3.1.5 Too much inventory in factory
3.1.1 Waste of raw materials in the production process
3.1.3 Materials damaged during handling
3.5.2 Incompetent employees
3.5.1 Inadequate work crews
3.4.2 Equipment breakdown
3.5.3 Lack of supervision
3.1.4 Unnecessary materials handling
3.3.1 Repair due to damaged products during inventory
3.3.2 Repair due to damaged products when handling
3.3.3 Double-handling or delivery due to unsatisfied quality or specifications
3.2.2 Wait time for the delivery of materials
3.2.1 Wait time for inspection
3.2.3 Wait time for labor
3.2.4 Wait time for equipment
3.1.2 Raw materials do not meet specifications
3.4.1 Inappropriate site layout
3.1.6 Loss of materials in factory
3.4.3 Inappropriate equipment choice

Fig. 7.11 The severity of non-value adding activities in production management

Appropriate equipment was carefully selected with the involvement of equipment suppliers. Loss of materials happened once in only one of the precast concrete factories. Other than the large inventory in the precast concrete factories, the site layout was designed good enough to achieve low-carbon production.

Similarly, these non-value adding activities could be plotted in a P-I table to identify the category to which each factor belongs. The P-I table is shown in Fig. 7.12. As can be seen from Fig. 7.12, most non-value adding activities fell into the category of raw catastrophe and probable disaster due to high impact. Special attention should be paid to activities that are near the boundary of each category. Mixed strategies should be taken when eliminating such activities. Detailed mitigations strategies are explained in Sect. 7.7.

7.5.4 Non-parametric Tests

Four ranking groups were identified by the non-parametric tests, as can be seen in Table 7.16. Similar procedures adopted in Sects. 7.3.4 and 7.4.4 to conduct the

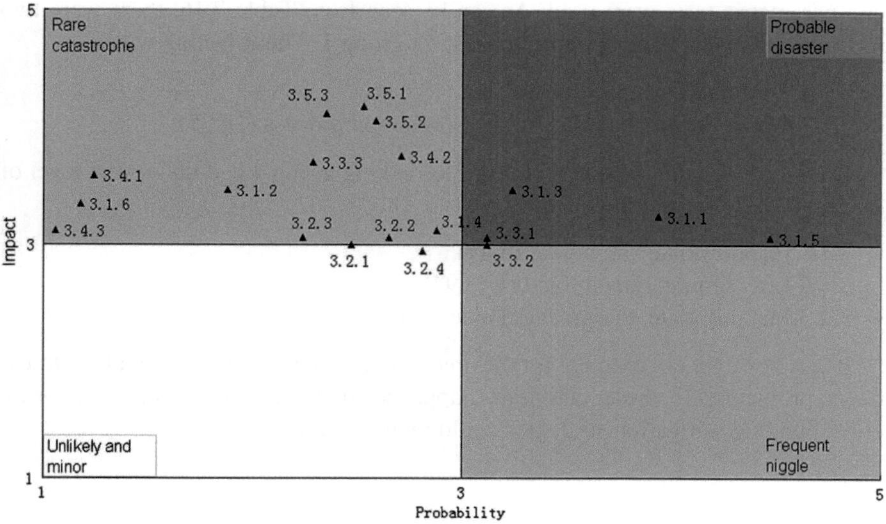

Fig. 7.12 P-I table for non-value adding activities in production management

Table 7.16 Ranking and grouping of non-value adding activities in production management

Ranking	Factor no.	Description	Severity AR	A.ρ	SD
1	3.1.5	Too much inventory in factory	13.65	N/A	4.5
1	3.1.1	Waste of raw materials in the production process	12.35	0.222	3.71
2	3.1.3	Materials damaged during handling	11.00	0.043	2.55
				N/A	
2	3.5.2	Inexperienced employees	10.76	0.834	4.45
2	3.5.1	Inadequate work crews	10.47	0.381	3.56
2	3.4.2	Equipment breakdown	10.35	0.429	2.87
2	3.5.3	Lack of supervision	9.94	0.279	5.33
2	3.1.4	Unnecessary materials handling	9.12	0.065	3.31
3	3.3.1	Repair due to damaged products during inventory	9.12	0.032	2.37
				N/A	
3	3.3.2	Repair due to damaged products when handling	8.94	0.317	2.25
3	3.3.3	Double-handling or delivery due to unsatisfied quality or specifications	8.35	0.152	2.42
3	3.2.2	Wait time for the delivery of materials	8.18	0.251	3.17
4	3.2.1	Wait time for inspection	7.47	0.020	3.04
				N/A	
4	3.2.3	Wait time for labor	6.94	0.336	3.21
4	3.2.4	Wait time for equipment	6.59	0.084	3.41
4	3.1.2	Raw materials do not meet specifications	6.53	0.081	2.58
5	3.4.1	Inappropriate site layout	4.47	0.003	1.87
				N/A	
5	3.1.6	Loss of materials in factory	3.94	0.303	1.56
5	3.4.3	Inappropriate equipment choice	3.35	0.098	1.62

Notes: *A.ρ* Asymptotic Significance (2-tailed)

non-parametric tests were used. As can be seen from Table 7.16, there were two factors which were categorized into ranking group 1. These factors were:

- 3.1.5 Too much inventory in factory (13.64)
- 3.1.1 Waste of raw materials in the production process (12.35)

On the other hand, factors that fell into ranking group 4 had the lowest level of severity. These factors included:

- 3.4.1 Inappropriate site layout (4.47)
- 3.1.6 Loss of materials in factory (3.94)
- 3.4.3 Inappropriate equipment choice (3.35)

It was however not efficient for the precast concrete industry to specially address these problems. If these activities happened frequently in specific factories, corresponding mitigation strategies could then be adopted.

7.5.5 Specific Analysis

The only factor that needed to be addressed here is factor 3.5.3, i.e. lack of supervision. While 11 precasters stated that supervision was sufficient in the current precast concrete production ($P = 1$ and $P = 2$), two precasters were facing a lack of supervision ($P = 4$ and $P = 5$). The reasons were from many levels. Under heavy production schedule, the delegation of responsibilities was reasonable and often applied in the factory. However, if the persons delegated were not appropriately trained as supervisors, interruptions to production processes may happen due to lack of supervision. In addition, lack of supervision could be caused by the ability and number of supervisors. The amount of supervision did not seem to be related to the competency of employees. Precasters with highly competent employees could still face a lack of supervision.

7.6 Lean Stock Management in Precast Concrete Factories

Stock management was the last value stage in the overall value chain of precast concrete production. As energy was consumed in terms of building up the inventory, as well as singling out the products for delivery, an inefficient stock management system would cause interruptions to the production process.

7.6.1 Descriptive Analysis

A few general questions related to stock management were discussed with the precasters, as shown in Table 7.17. Three main sections were identified, including delivery, inventory and human resources.

Table 7.17 General questions relating to stock management

Question no.	Description	Positive response	Positive response rate	Category
4.6.1	Is identification mark clearly provided to avoid wrong delivery?	17	100 %	Delivery
4.6.2	Has wrong delivery happened due to defective identification marks?	11	64.71 %	Delivery
4.6.3	Is the delivery note orally made or provided by written documents?	Oral 2 (11.76 %)	Written 15 (88.24 %)	Delivery
4.6.4	Are routine inspections conducted before the release of products?	17	100 %	Delivery
4.6.5	Have products been damaged due to inappropriate stacking?	17	100 %	Inventory
4.6.6	Are working instructions provided to employees?	17	100 %	Inventory
4.6.7	Are the working instructions orally provided or provided by written documents?	Oral 15 (88.24 %)	Written 2 (11.76 %)	Inventory
4.6.8	Are periodic stock checks conducted?	17	100 %	Inventory
4.6.9	Is computer stock control system adopted for stock management?	16	94.12 %	Inventory
4.6.10	Is the driver appropriately trained to provide sufficient care to the precast concrete products?	17	100 %	Human resources

Several implications could be derived from Table 7.17 relating to current stock management operations, including:

1. All precasters (100 %) provided identification marks to avoid wrong deliveries. However, 11 precasters (64.71 %) stated that wrong deliveries have happened in the past by delivering wrong products to construction sites, although infrequently at once or twice a year. Delivery of right materials to wrong construction sites have happened infrequently as well, despite the fact that delivery notes were provided by all precasters (100 %). Fifteen precasters (88.24 %) provided written delivery notes while two precasters (11.76 %) orally provided the delivery details. However, it could not be concluded that written delivery notes performed better because wrong deliveries do happened when written delivery notes were provided. This might be caused by insufficient care of the vehicle drivers when handling deliveries. All precasters (100 %) stated that routine inspections were conducted before the release of the products. The routine inspections were conducted by both parties, precasters and contractors, to make sure that the products were in good quality and correct quantities upon arrival.
2. When managing inventory, 17 precasters (100 %) provided working instructions related to stacking requirements to avoid damaging the products. Fifteen precasters (88.24 %) adopted on-the-job training by providing working instructions orally. On the other hand, two precasters (11.76 %) offered written documents as working instructions. However, all precasters (100 %) were encountering

damaged products caused by inappropriate stacking. The percentage of waste was in the range of between 3 and 5 %. When observed in the precast concrete factories, many products were stacked in the large storage area. Due to the stacking, newly manufactured products were arranged for delivery, leaving the old ones exposed to the harsh environment, including sunshine and rain. The longer the waiting time, the higher the occurrence of damage would be.

Periodic stock checks were conducted by all precasters (100 %), although with different frequencies. Monthly, quarterly, and randomly conducted stock checks were the three most commonly used types of stock checks. A mixed use of the three types of checks was adopted by some precasters as well. Sixteen precasters (94.12 %) stated that computer stock control was adopted for stock checks. However, only one precaster (5.88 %) used a specialized stock control software while others were mostly using Windows Excel for this purpose. According to the precasters in this study specialized stock control softwares have been tested and adopted in the past. Compared to Windows Excel, these have no extraordinary advantages.

3. Seventeen precasters (100 %) stated that training programmes were provided to vehicle drivers to make sure that sufficient care was provided during transportation. However, as the transportation work was subcontracted out, this was in fact a delegation of responsibilities. As stated earlier, delegation of responsibilities could only be made along with an appropriate employee rating system. However, this rating system was currently missing in managing the deliveries.

7.6.2 Factors Description

Five categories of non-value adding activities were identified in stock management. It should be noted that the delivery of the finished products was integrated into this stage. The delivery of the finished products could be the last value stage in precast concrete production. It could also be the first value stage of precast concrete erection. This study therefore discussed delivery in both value stages. The five categories of non-value adding activities in this context are shown in Table 7.18.

General description for each non-value adding activity is provided as follows:

- *Category 4.1 Layout and equipment*

  4.1.1 Lack of well maintained stockyard
  4.1.2 Insecure stockyard that leads to the loss of products
  4.1.3 Inappropriate selection of equipment

In order to prevent the waste of products from happening, a well maintained stockyard is required. Bumps on the delivery path can cause damages, especially in the contact area between the delivery vehicle and the precast concrete products. In addition, loss of products happened once according to one of the precasters interviewed. Reinforcement bars were more exposed to theft due to their relatively higher value compared to other products. Inappropriate selection of equipment might also cause damages to the products as well.

Table 7.18 Five major categories of non-value adding activities in stock management

Category no.	Description
4.1	Layout and equipment
4.2	Staffing arrangement
4.3	Stacking arrangement
4.4	Stock records
4.5	Loading, unloading and delivery

- *Category 4.2 Staffing arrangement*

4.2.1 Inappropriate staffing arrangement

In practice, three persons, including a crane driver, a banksman and one charge hand, should be allocated in order to make sure that the loading and unloading activities were correctly and efficiently conducted.

- *Category 4.3 Stacking arrangement*

4.3.1 Unclear identification marks
4.3.2 Inappropriate battens
4.3.3 Unclear working instructions

Damages due to inappropriate stacking were very common in precast concrete factories. Battens should be correctly inserted in order to prevent such damages. In addition, unclear identification marks and work instructions might lead to double-handling, which was also a category of waste examined by the lean concept.

- *Category 4.4 Stock records*

4.4.1 Lack of periodic stock checks
4.4.2 Lack of computer stock control

The main purpose of stock records was to make sure that the right products were delivered to the construction sites at the right time. When the main purpose was fulfilled, the precasters were able to start to organize deliveries in a "first produced, first delivered" manner. By organizing deliveries in such way, damages to the products due to inventory could be reduced.

- *Category 4.5 Loading, unloading and delivery*

4.5.1 Unclear delivery notes which lead to wrong delivery
4.5.2 Lack of sufficient care which lead to damage
4.5.3 Lack of routine inspection prior to release of products

It goes without saying that loading and unloading activities were vitally important to precast concrete products. The probability of damages to finished products increased with the frequency of such activities. Along with sufficient care when conducting loading and unloading activities, clear delivery notes should be provided and routine inspections prior to the release of products should be conducted to prevent double-handling, or double-delivering in the worst case scenarios.

Table 7.19 Probability, impact and severity of the non-value adding activities in production management

Factor no.	Description	Probability (P)				Impact (I)				Severity (S)			
		AR	LR	SR	SD	AR	LR	SR	SD	AR	LR	SR	SD
4.1.1	Lack of well maintained stockyard	1.59	3	1	0.71	2.35	4	1	0.86	3.82	9	1	2.30
4.1.2	Insecure stockyard that leads to the loss of products	1.06	2	1	0.24	2.24	4	1	0.97	2.29	4	1	0.92
4.1.3	Inappropriate selection of equipment	1.06	2	1	0.24	2.88	4	2	0.70	3.00	4	2	0.71
4.2.1	Inappropriate staffing arrangement	2.12	3	1	0.60	3.94	5	3	0.56	8.35	12	3	2.60
4.3.1	Unclear identification marks	2.24	3	1	0.66	3.82	5	3	0.53	8.29	12	3	2.47
4.3.2	Inappropriate battens	2.35	5	1	0.86	3.59	4	2	0.62	8.41	20	4	3.62
4.3.3	Unclear working instructions	1.76	3	1	0.66	3.53	4	2	0.62	6.24	12	2	2.56
4.4.1	Lack of periodic stock checks	1.35	2	1	0.49	3.35	5	2	0.70	4.47	8	2	1.62
4.4.2	Lack of computer stock control	1.71	5	1	1.05	3.12	5	1	1.11	5.41	25	2	5.29
4.5.1	Unclear delivery notes which lead to wrong delivery	1.94	3	1	0.83	3.65	4	3	0.49	7.06	12	3	3.27
4.5.2	Lack of sufficient care which lead to damage	2.41	3	1	0.62	3.71	4	2	0.59	9.00	12	4	2.92
4.5.3	Lack of routine inspection prior to release of products	1.82	4	1	0.81	3.65	4	3	0.49	6.65	16	3	3.16

Notes: *AR* average rating, *LR* largest rating, *SR* smallest rating, *SD* standard deviation

7.6.3 Ranking Procedure

The probability, impact and severity of non-value adding activities in stock management are presented in Table 7.19. Based on Table 7.19, a few implications can be drawn, including:

1. The probability of non-value adding activities in stock management was relatively low, as shown in Fig. 7.13. The factors with the high probability included:

   - 4.5.2 Lack of sufficient care which lead to damage (2.41)
   - 4.3.2 Inappropriate battens (2.35)

   Interpreted using the five-scale value range, this category of non-value adding activity rarely happened. It seemed that precasters were performing very well in

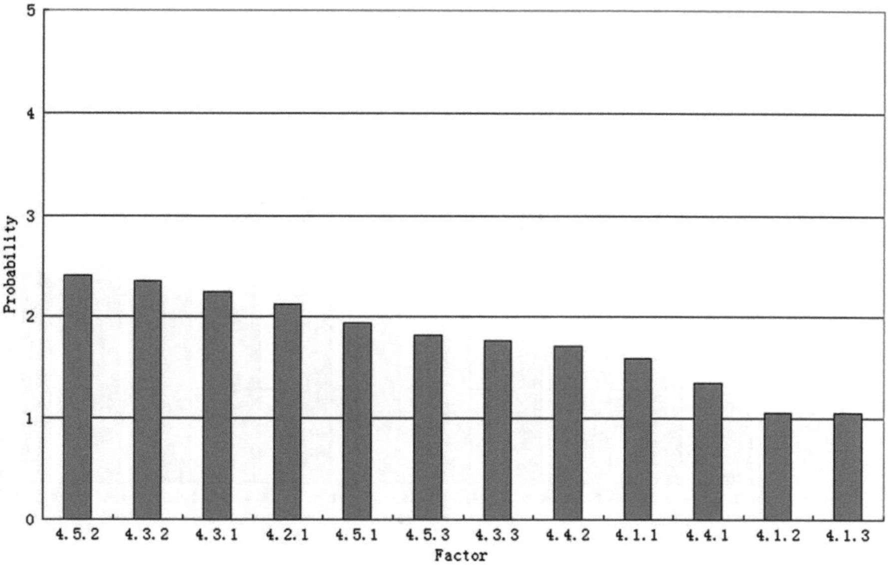

4.5.2 Lack of sufficient care which lead to damage
4.3.2 Inappropriate battens
4.3.1 Unclear identification marks
4.2.1 Inappropriate staffing arrangement
4.5.1 Unclear delivery notes which lead to wrong delivery
4.5.3 Lack of routine inspection prior to release of products
4.3.3 Unclear working instructions
4.4.2 Lack of computer stock control
4.1.1 Lack of well maintained stockyard
4.4.1 Lack of periodic stock checks
4.1.2 Insecure stockyard that leads to the loss of products
4.1.3 Inappropriate selection of equipment

Fig. 7.13 The probability of non-value adding activities in stock management

stock management, especially in selecting equipment (factor 4.1.3) and securing stockyard (factor 4.1.2). Areas that needed to be improved included sufficient care when conducting loading and unloading activities (4.5.2), inappropriate battens (4.3.2) and identification marks (4.3.1).

2. Factors with high impact on carbon emission levels included (as shown in Fig. 7.14):

   - 4.2.1 Inappropriate staffing arrangement (3.94)
   - 4.3.1 Unclear identification marks (3.82)
   - 4.5.2 Lack of sufficient care which lead to damage (3.71)

   It seemed that staffing arrangement played an important role in stock management when precasters aimed to reduce carbon emissions (4.2.1). When selecting the products ordered for delivery, three persons should be provided, including a crane driver, a banksman and one charge-hand.

   In addition, clear identification marks should be provided to avoid wrong delivery (4.3.1). Water-proof identification marks should be advocated due to the high frequency of rain in Singapore. Sufficient care should be provided by employees during both stock management when selecting the products and delivery management when delivering the components (4.5.2). The impact of these non-value adding activities is shown in Fig. 7.14.

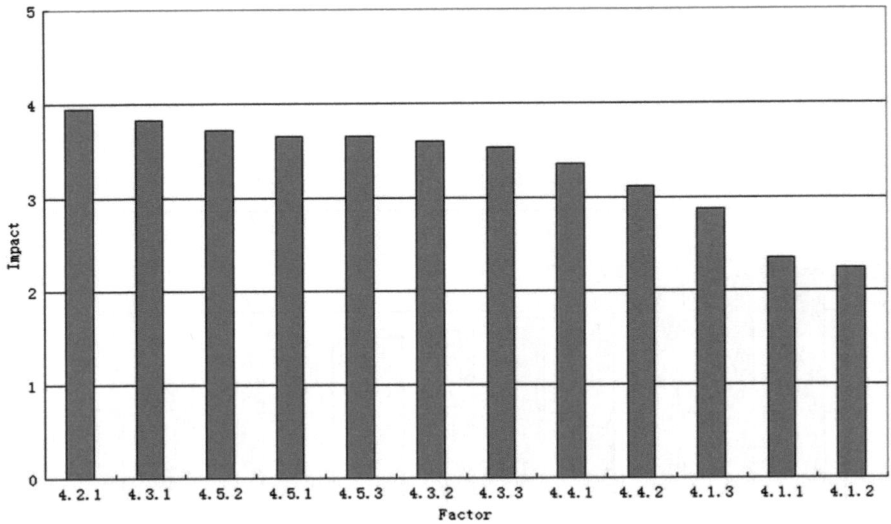

4.2.1 Inappropriate staffing arrangement
4.3.1 Unclear identification marks
4.5.2 Lack of sufficient care which lead to damage
4.5.1 Unclear delivery notes which lead to wrong delivery
4.5.3 Lack of routine inspection prior to release of products
4.3.2 Inappropriate battens
4.3.3 Unclear working instructions
4.4.1 Lack of periodic stock checks
4.4.2 Lack of computer stock control
4.1.3 Inappropriate selection of equipment
4.1.1 Lack of well maintained stockyard
4.1.2 Insecure stockyard that leads to the loss of products

Fig. 7.14 The impact of non-value adding activities in stock management

3. When considering both the probability and the impact, the four most important factors were:

   - 4.5.2 Lack of sufficient care which lead to damage (9.00)
   - 4.3.2 Inappropriate battens (8.41)
   - 4.2.1 Inappropriate staffing arrangements (8.35)
   - 4.3.1 Unclear identification marks (8.29)

   These non-value adding activities highlighted the severity of sufficient care, correct battens, staffing arrangement and identification marks in stock management. On the other hand, factors with lower severity included:

   - 4.1.2 Insecure stockyard that leads to the loss of products (2.29)
   - 4.1.3 Inappropriate selection of equipment (3.00)
   - 4.1.1 Lack of well maintained stockyard (3.82)

   When providing recommendations for precasters to improve their operations, such activities should be highlighted, albeit not as strongly, due to their relatively low severity (Fig. 7.15).

As can be seen from Fig. 7.16, most non-value adding activities fell into the category of rare catastrophe, which seems reasonable based on the high impact of these non-value adding activities. For example, if due to unclear identification

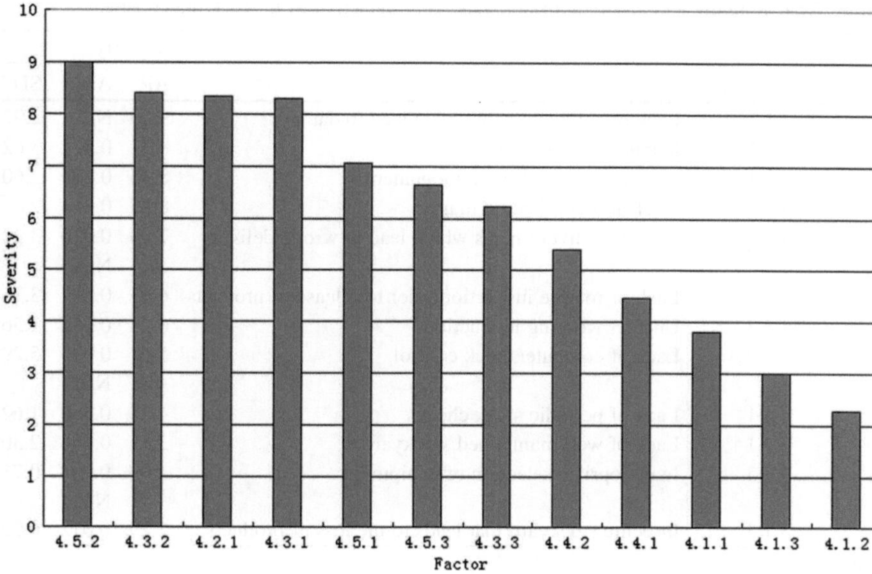

4.5.2 Lack of sufficient care which lead to damage
4.3.2 Inappropriate battens
4.2.1 Inappropriate staffing arrangement
4.3.1 Unclear identification marks
4.5.1 Unclear delivery notes which lead to wrong delivery
4.5.3 Lack of routine inspection prior to release of products
4.3.3 Unclear working instructions
4.4.2 Lack of computer stock control
4.4.1 Lack of periodic stock checks
4.1.1 Lack of well maintained stockyard
4.1.3 Inappropriate selection of equipment
4.1.2 Insecure stockyard that leads to the loss of products

Fig. 7.15 The severity of non-value adding activities in stock management

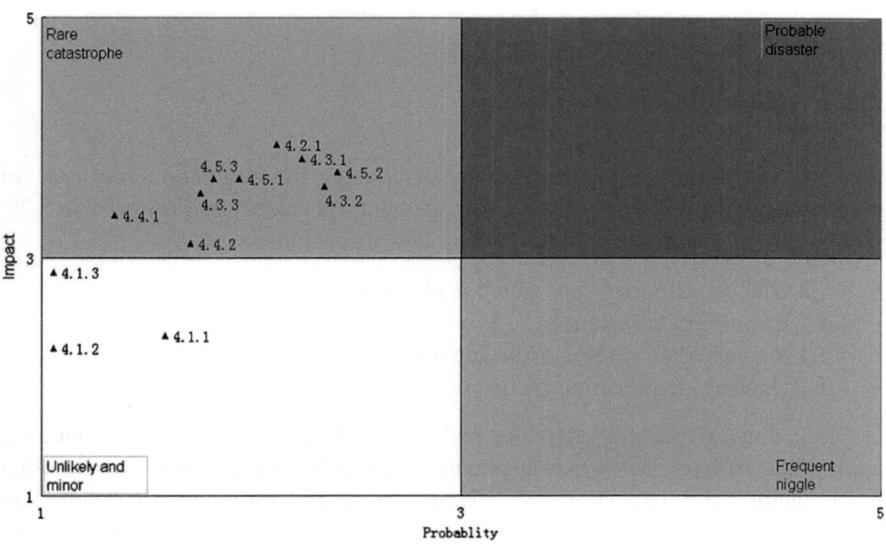

Fig. 7.16 P-I table for non-value adding activities in stock management

Table 7.20 Ranking and grouping of non-value adding activities in stock management

			Severity		
Ranking	Factor no.	Description	AR	A.ρ	SD
1	4.5.2	Lack of sufficient care which lead to damage	9.00	N/A	2.92
1	4.3.2	Inappropriate battens	8.41	0.321	3.62
1	4.2.1	Inappropriate staffing arrangement	8.35	0.475	2.60
1	4.3.1	Unclear identification marks	8.29	0.344	2.47
2	4.5.1	Unclear delivery notes which lead to wrong delivery	7.06	0.020 N/A	3.27
2	4.5.3	Lack of routine inspection prior to release of products	6.65	0.507	3.16
2	4.3.3	Unclear working instructions	6.24	0.152	2.56
3	4.4.2	Lack of computer stock control	5.41	0.030 N/A	5.29
3	4.4.1	Lack of periodic stock checks	4.47	0.888	1.62
3	4.1.1	Lack of well maintained stockyard	3.82	0.158	2.30
4	4.1.3	Inappropriate selection of equipment	3.00	0.006 N/A	0.71
5	4.1.2	Insecure stockyard that leads to the loss of products	2.29	0.012	0.92

Notes: *A.ρ* Asymptotic Significance (2-tailed)

marks (4.3.1), wrong precast concrete products were sent to the construction site, a re-delivery would then be scheduled, resulting in a huge waste of resources. However, due to its low probability, it was better to transfer the responsibility to a supervisor to conduct regular checks of the identification marks. Low probability indicates that it was not economical to provide training programmes to the employees all over again. The details of mitigation strategies are discussed in Sect. 7.7.

7.6.4 Non-parametric Tests

Five ranking groups were identified by conducting the non-parametric tests for non-value adding activities in stock management. As can be seen from Table 7.20, ranking group 1 with the highest level of severity included:

- 4.5.2 Lack of sufficient care which lead to damage
- 4.3.2 Inappropriate battens
- 4.2.1 Inappropriate staffing arrangement
- 4.3.1 Unclear identification marks

These non-value adding activities highlighted the importance of sufficient care, appropriate battens, staffing arrangement and identification marks. On the other hand, factor 4.1.3 (Inappropriate selection of equipment) and 4.1.2 (Insecure stockyard that leads to the loss of products) were classified into ranking group 4 and group 5, respectively. When providing recommendations for precasters to achieve lean stock management, the selection of equipment and secure stockyard appears to be of lower priority than those activities in the higher ranking groups.

7.6.5 *Specific Analysis*

Based on the rating of probability in Table 7.19, factors 4.3.2 (Inappropriate battens) and 4.4.2 (Lack of computer stock control) were selected for specific analysis (LR = 5 and SR = 1). The explanations are given as follows:

- Working instructions related to the location of battens should be improved by one precaster (P = 5). The number of battens, location of battens as well as other important information should be offered by precasters to avoid damages. Twelve precasters (70.59 %) stated that damages to finished products due to battens never or rarely happen (P = 1 and P = 2) in current precast concrete production. However, given the fact that the working instructions were mostly provided orally because of on-the-job training, written instructions should be prepared in the future if such activities were targeted to be eliminated.
- The large difference in adopting computer stock control has been explained in implication 2 in Sect. 6.6.1. Only one precaster (5.88 %) was using a specialized stock control software while others were mostly using Windows Excel. According to the precasters investigated, Windows Excel could offer similar functions at a much lower price.

7.7 Mitigation Strategies and Actions for Precasters

In this section, mitigation strategies and mitigation actions were developed for precasters to manage the non-value adding activities examined in the previous sections. It should be noted that mitigation strategies defined the type of mitigation efforts while mitigation actions formed the detailed mitigation plans. For example, when the precasters aimed to eliminate high-probability and low-impact non-value adding activities, the mitigation strategies would be directed at "control", because such activities were normally caused by repetitive actions. Following this direction, it was decided that training programmes would be provided for all employees and written instructions would be adopted. Both the training programmes and written instructions belonged to the category of mitigation actions, which detailed the mitigation plans.

7.7.1 *The General Procedure to Develop Mitigation Actions*

The development of mitigation strategies for the precasters to eliminate non-value adding activities followed a standard procedure, which is shown in Fig. 7.17. It should be noted that the process flow was a continually improving loop process.

Fig. 7.17 The process flow to define mitigation actions for single non-value adding activity

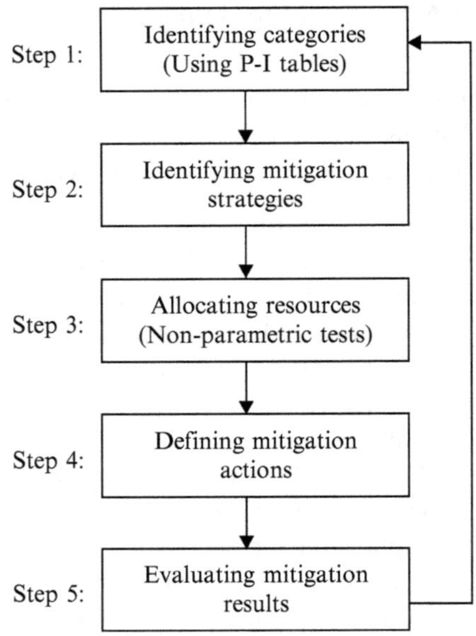

Non-value adding activities might not be fully eliminated by the first application when continual monitoring and improving were necessary. The steps are explained as follows:

- **Step 1**: Identifying categories. In this step, the activity was plotted into the P-I table to obtain the category it belonged to. As stated previously, four categories were identified in the P-I table, which were unlikely and minor, rare catastrophe, frequent niggle and probable disaster. Attention should be paid to activities located near the junction area of every two categories. The zoning of each category was not fixed. Flexibility around the junction areas should therefore be adopted in managing the non-value adding activities.

  For activities that fell into the category of unlikely and minor, it seemed that no actions were necessary due to the low level of severity. Precasters were either performing very well in these areas or the impact was too low to be considered. Focusing too much on these activities could lead to inefficiency due to limited resources.

  Factors in the category of rare catastrophe were caused by low-probability and high- impact operations. Strategies could be taken to transfer the risk to parties who were better positioned to absorb the impact. Since precasters were performing very well most of the time, it was not economical to discard the old training programmes and start all over again.

  For example, factor 2.2.2 (Not fully prepared for the arrival of raw materials) fell into the category of rare catastrophe with low probability and high impact. Supervisors should be assigned to such an activity and training programmes should be offered to supervisors to make sure that the materials were handled over smoothly.

"Control" was applied to high-probability and low-impact non-value adding activities, such as factor 1.7.3 (Site layout plan is not placed on the notice board for information), which could be eliminated by assigning appropriate persons to fix the problem.

"Avoid" was applied to activities in the category of probable disaster, such as factor 1.6 (too much inventory), 2.1.1 (large quantity supply base), 2.2.3 (lack of both advance order and confirmation order), 3.1.3 (materials damaged during handling). In order to eliminate these activities, a review and re-examination of the current precast concrete production practice was needed. These non-value adding activities were probably caused by either the design or the production method that come along with the push production system. A review and re-examination was therefore necessary before mitigation actions could be taken.

- **Step 2**: Identifying mitigation strategies. Detailed mitigation strategies were developed in this stage. For example, when deciding on the right mitigation strategies for factor 3.1.3 (Materials damaged during handling), it was recommended that the handling process is reviewed with the involvement of different participants, such as the operator, the banksman and the charge-hand. The decision to provide either more training programmes to operators or offer more supervision during handling was dependent on the review process. In this stage, the mitigation strategies were detailed within each category of severity.
- **Step 3**: Allocating resources. This step was conducted by comparing the relative priority in the non-parametric tests. Factors within the same category of severity but with higher average rating (AR) in the non-parametric tests should be allocated with more resources, such as more training programmes, more eligible supervisors, etc.
- **Step 4**: Defining mitigation actions. In this step, the mitigation strategies identified above were elaborated into detailed mitigation actions. If training programmes were proposed as mitigation strategies, the number of persons who should be trained, as well as the frequency of such training would form the detailed mitigation actions.
- **Step 5**: Evaluating mitigation results. The non-value adding activities eliminated after the mitigation actions should be plotted on the P-I table to identify the results. A looped application of the process tree should be monitored on a continuous basis.

7.7.2 Developing the Mitigation Actions for Precasters

Four main mitigation strategies have been identified in Sect. 7.7.1, which were "accept", "control", "transfer" and "avoid". Followed by the process tree, the non-value adding activities in precast concrete factories were categorized into eight groups by their relative severity. Mitigation strategies were also provided in Table 7.21. A few implications could be obtained from Table 7.21, including:

1. The most severe activities in the current precast concrete production included:
   - 2.1.1 Large quantity supply base
   - 2.1.2 No long-term contract to achieve loyalty between suppliers and precasters
   - 3.1.5 Too much inventory
   - 1.6.1 Over provide secure storage in design
   - 2.2.3 Lack of both advance order and confirmation order
   - 2.1.4 No quality audits of the suppliers
   - 1.6.2 Over provide weather proof storage in design
   - 3.1.1 Waste of raw materials in the production processes

   In order to eliminate the impact of activities within this group, "avoid" actions should be taken by making an in-depth review of the current operations.

   The delivery performance of the Singapore suppliers was satisfactory. Since 16 precasters (94.12 %) produced on a stable schedule with accurately estimated quantities of demand and nine of them (52.94 %) were experiencing on-time deliveries, the storage area, in terms of open storage, weatherproof storage and secure storage, could be reduced in the design stage. According to Waters-Fuller (1996), one of the most important factors that might impede the adoption of JIT sourcing was related to the huge demand fluctuations. However, as most precasters (94.12 %) were experiencing demand stability, a few JIT sourcing activities could be tested on feasibility, including small quantity supply base, long-term relationship, the two-order system, etc. Quality audits should be conducted by the precasters to make sure that the suppliers were competent to cope with such JIT sourcing activities. Ansari and Modarress (1987) stated that a lack of supplier support was the most significant barrier to success. Government intervention with tax benefits might help to push both the precasters and the suppliers to improve towards lean production, which could be supported with the current production background.

   Waste of raw materials in production processes was very common (P = 3.94). The percentage of waste varied from 2 to 5 %, which would largely increase the amount of carbon emitted. On the other hand, if such waste led to interruptions (e.g. causing re-delivery of raw materials from suppliers), the amount of carbon emitted during transportation might be doubled.
2. Non-value adding activities in ranking group 2 to group 5 have intermediate impact on the carbon emission levels in precast concrete factories. Mitigation actions could vary among these activities, depending on the nature of the activities. Most actions fell into the category of "transfer" and "control". As stated above, "control" mainly applied to high-probability and low-impact activities by improving the overall competence of employees. On the other hand, "transfer" applied to high-impact but low-probability activities by improving the ability of the supervisors or the specific persons in charge.

   Attention should be paid to factors which lie at the junction area between two categories. In this case, mixed strategies should be advocated instead. The problem

Table 7.21 Ranking, grouping and mitigation actions for non-value adding activities in precast concrete factories

		Severity						Severity			
R	Factor	AR	A.ρ	SD	MA	R	Factor	AR	A.ρ	SD	MA
1	1.6.3	14.53	N/A	4.19	AV	2	1.1.3	11.44	0.044	3.69	MS
									N/A		
1	2.1.1	14.29	0.753	3.77	AV	2	3.1.3	11.00	0.916	2.55	AV
1	2.1.2	13.94	0.832	3.58	AV	2	3.5.2	10.76	0.950	4.45	T
1	3.1.5	13.65	0.539	4.50	AV	2	2.1.3	10.71	0.908	6.11	AV
1	1.6.1	13.29	0.149	3.95	AV	2	3.5.1	10.47	0.683	3.56	T
1	2.2.3	13.18	0.382	4.29	AV	2	1.1.4	10.38	0.253	4.44	MS
1	2.1.4	12.88	0.432	5.16	T	2	3.4.2	10.35	0.529	2.87	T
1	1.6.2	12.82	0.073	3.83	AV	2	3.5.3	9.94	0.467	5.33	T
1	3.1.1	12.35	0.098	3.71	AV	2	3.1.4	9.12	0.300	3.31	T
						2	3.3.1	9.12	0.254	2.37	AV
3	3.3.3	8.35	0.009	2.42	T	2	2.2.1	9.00	0.241	4.02	MS
			N/A								
3	4.2.1	8.35	0.975	2.60	T	2	4.5.2	9.00	0.169	2.92	T
3	4.3.1	8.29	0.979	2.47	T	2	3.3.2	8.94	0.161	2.25	MS
3	3.2.2	8.18	0.937	3.17	T	2	4.3.2	8.41	0.064	3.62	T
3	3.2.1	7.47	0.195	3.04	MS						
3	2.2.2	7.06	0.207	2.77	T	4	3.2.3	6.94	0.037	3.21	T
									N/A		
3	4.5.1	6.94	0.232	3.27	T	4	4.5.3	6.65	0.979	3.16	T
						4	3.2.4	6.59	0.276	3.41	MS
5	1.5.2	5.47	0.032	2.40	T	4	3.1.2	6.53	0.399	2.58	T
			N/A								
5	1.5.1	5.41	0.811	2.03	T	4	4.3.3	6.24	0.710	2.56	T
5	4.4.2	5.41	0.211	5.29	T	4	1.1.2	6.13	0.298	2.60	T
5	1.7.4	5.35	0.674	3.57	MS	4	1.7.1	6.00	0.500	2.42	T
5	1.4.4	5.24	0.959	2.19	MS	4	1.1.1	5.63	0.053	2.90	T
5	1.5.3	4.65	0.104	1.84	MS	4	1.7.3	5.53	0.086	4.06	C
5	1.5.4	4.65	0.108	1.90	T						
						7	4.1.3	3.00	0.007	0.71	AC
6	1.2.1	4.59	0.047	2.40	MS				N/A		
			N/A								
6	2.2.4	4.53	0.463	1.74	T						
6	3.4.1	4.47	0.975	1.87	T	8	4.1.2	2.29	0.012	0.92	AC
6	4.4.1	4.47	0.873	1.62	T						
6	1.5.5	4.29	0.936	1.99	AC						
6	1.2.2	4.00	0.102	1.17	AC						
6	1.4.3	4.00	0.588	2.18	AC						
6	1.4.2	3.94	0.688	2.14	AC						
6	3.1.6	3.94	0.447	1.56	T						
6	1.7.2	3.88	0.277	2.64	AC						
6	4.1.1	3.82	0.365	2.30	AC						
6	1.3	3.53	0.210	1.01	AC						
6	1.4.1	3.35	0.071	1.93	AC						
6	3.4.3	3.35	0.114	1.62	T						

Notes:

A.ρ Asymptotic Significance (2-tailed), *R* ranking group, *AR* average rating, *SD* standard deviation, *MA* mitigation actions, *AC* accept, *T* transfer, *AV* avoid, *C* control, *MS* mixed strategies

was caused by introducing the four severity categories as four squares demarcated by the line P = 3 and I = 3. In fact, there were overlapping areas between each two categories and mixed strategies should therefore be adopted to eliminate such non-value adding activities.

3. Non-value adding activities in ranking groups 6, 7 and 8 have little impact on the carbon emission levels in the precast concrete factories. "Accept" actions could be taken because it would not be efficient to focus on these activities too much. Similarly, attention should be paid to factors that lied in the overlapping area. "Accept" actions were taken for the precast concrete industry to focus on the activities that have larger impact on the carbon emission levels. Whether or not a single precaster should take actions to eliminate such activities was subject to the precaster's own rating results.

7.8 Summary

The non-value adding activities in precast concrete production processes were described, evaluated and ranked in this chapter. Followed by a general analysis procedure, non-value adding activities in each value stage, from site layout management, supply chain management, production management to stock management, were classified into several ranking groups categorized by severity. This ranking procedure was also conducted for all non-value adding activities.

The most important ones were related to large quantity supply base, no long-term contract, too much inventory, over provide secure storage in design, lack of both advance order and confirmation order, no quality audits, over provide weather proof storage in design and waste of raw materials in the production processes. Actions could be taken to eliminate these non-value adding activities because the performance of the Singapore precast concrete industry could support lean applications.

Many non-value adding activities with intermediate impact on the carbon emission levels were also identified in this chapter. These included waste of raw materials in the handling procedure and inexperienced employees. It was believed that by reducing the probability of occurrence of such activities, precasters could achieve low-carbon production. A case study is provided in the following chapter for the precasters to identify the potential improvements that could be obtained by applying the lean production concept.

References

Ansari A, Modarress B (1987) Potential benefits of JIT purchasing for US manufacturers. Prod Invent Manag J 28(2):30–33

Alwi S, Hampson K, Mohamed S (2002) Waste in the Indonesian construction project. In: Proceedings of the 1st international conference of CIB W107 – creating a sustainable construction industry in developing countries, Spier, South Africa

Banerjee A, Kim SL (1995) An integrated JIT inventory model. Int J Oper Prod Manage 15(9):237–244

Building Construction Authority (2010) BCA directory of registered contractors and licensed builders. Available from: http://www.bcadirectory.sg/. Cited 29 Nov 2013

Calvert RE (1995) Introduction to building management. Butterworth-Heinemann Ltd., Oxford

Low SP, Mok SH (1999) The application of JIT philosophy to construction: a case study in site layout. Constr Manage Econ 17(5):657–668

Neil JM (1980) Teaching site layout for construction. ASCE meeting, Portland, OR, 14–18 April, pp 1–11

Ohno T (1988) Toyota production system: beyond large-scale production. Diamond, Inc., Tokyo

Stalk GJ, Hout TM (1990) Competing against time. Free Press, New York

Tommelein ID, Li EY (1999) Just-in-time concrete delivery mapping alternatives for vertical supply chain integration. In: Tommelein ID (ed) Proceedings of IGLC-7, Berkeley, CA

Waters-Fuller N (1996) The benefits and costs of JIT sourcing: a study of Scottish suppliers. Int J Phys Distrib Logist Manag 26(4):35–50

Williams TM (1993) Risk-management infrastructures. Int Proj Manage 11(1):5–10

Xin N (2010) A construction site layout planning decision-making system for dynamic, unequal-area and multi-objective optimization problem. VDM Verlag Publishing, Saarbrucken

Chapter 8
Case Study of a Precast Concrete Factory

8.1 Introduction

Reducing the environmental footprint of concrete is of vital importance for the construction industry to achieve sustainability. Some researchers were currently concerned with how a concrete structure might be designed to be green and sustainable. Due to the importance of global climate change and greenhouse gases, the term "CO_2 emissions" was often used as one indicator to highlight whether or not the product was environmentally sustainable. Many environmental labelling programmes have been initiated to include CO_2 emissions as an important consideration, such as the UK Ecopoints and Blue Angle.

The Singapore Construction 21 Committee (1999) encouraged contractors to adopt precast concrete products because of their benefits such as faster construction, fire protection, productivity improvement, etc. Glass (2000) observed that the market for precast concrete was rising due to the large demand for prefabricated housing in the UK, Germany, the Netherlands and other developed countries. Precast concrete products were believed to have some "green" benefits, such as reducing waste, improving total quality, reducing energy consumption and carbon emissions. However, there were few studies which focused on estimating the embodied carbon of precast concrete products. This chapter aims to initiate the estimation of the embodied carbon of precast concrete products, using a specific type of precast concrete column (16HPC1) produced by one precaster (referred to as Precaster A in the following context) in Singapore using the case study approach. The carbon reduction achieved by applying the lean production philosophy was assessed, both quantitatively and qualitatively. The methodology and results of this estimation would be useful when building the life cycle inventory of construction materials and products for the Singapore construction industry. It would also help to build a lean score in eco-labelling systems, especially in the aspects of carbon-labelling for the Singapore construction industry.

P. Wu and S.P. Low, *Lean and Cleaner Production: Applications in Prefabrication to Reduce Carbon Emissions*, DOI 10.1007/978-3-642-42062-7_8,

Precaster A was set up in 1994 to spearhead the adoption of prefabrication technology in Singapore. A total land area of 17,000 m^2 and a built-up area of 21,000 m^2 were occupied. The main activities of Precaster A included:

- Design and produce precast concrete products;
- Manage and deliver precast concrete products; and
- Research on innovative construction materials.

As research played a very important role in Precaster A, several lean applications have already been applied, which included adopting the pull system, focusing on continuous improvement, etc. However, it still suffered from a few problems, such as large storage area and weak employees.

8.2 General Procedure to Quantify the Lean Improvements

A general procedure to calculate the lean improvements was developed for Precaster A. The general procedure included two major subprocesses, which were the screening process and the estimation process. The list of non-value adding activities presented in Chap. 7 was used as a checklist for Precaster A. As seen in Fig. 8.1, the screening process included three steps:

1. Screening step 1: In this step, the relative importance of the factor was identified. The probability of occurrence of non-value adding activities in each value stage were rated by Precaster A. Factors with no probability of occurrence (P = 1) were dropped from the assessment. It should be noted that absolute care should be paid because operational activities may be different from what Precaster A has rated. Screening step 1 should be based mainly on observations made in Precaster A, along with ratings on the probability of occurrence.
2. Screening step 2: Factors that required assessment and screened in step 1 were re-categorized into different emitting groups. Carbon emissions in precasters were generated by several sources, which included equipment, electricity, waste of raw materials, waste of finished products and capital facilities (such as site offices and employee dormitories). Factors under different emitting group were assessed by different equations.
3. Screening step 3: Factors which could not be categorized in any group mentioned in step 2 might not be eligible for a quantitative assessment. Qualitative descriptions of the impact of such activities to the level of carbon emissions were then provided.

When factors could be categorized into different emitting groups and quantitative assessments could be obtained, the estimation process was then conducted. The estimation of the improvements was conducted for each emitting category. The equations used in the estimation process included:

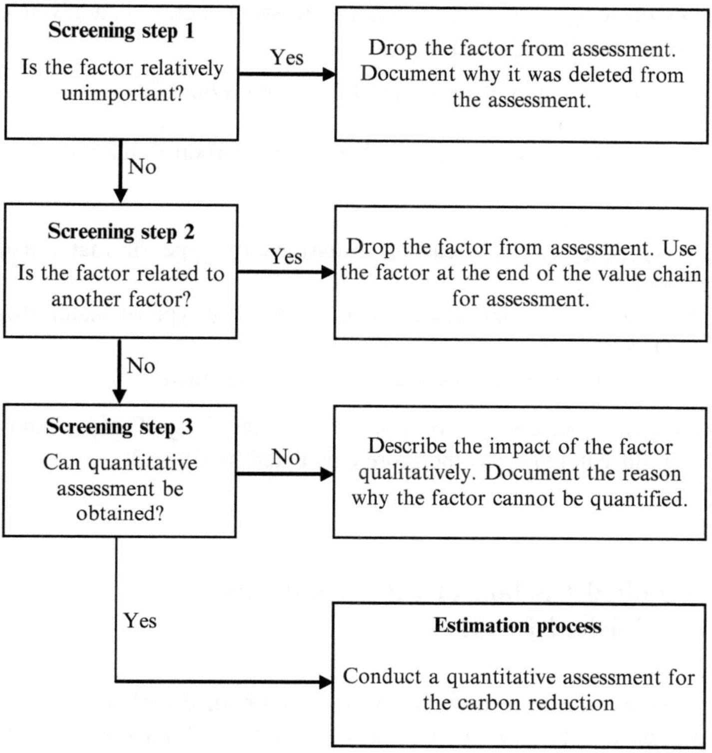

Fig. 8.1 Screening procedure to identify factors that could be estimated

- For equipment:

$$\text{Emissions}_{\text{GHG,fuel}} = \text{Fuel Consumption}_{\text{fuel}} * \text{Emission Factor}_{\text{GHG,fuel}} \tag{8.1}$$

Where:

$\text{Emissions}_{\text{GHG, fuel}}$ = emissions of a given GHG by type of fuel (kg GHG)
$\text{Fuel Consumption}_{\text{fuel}}$ = amount of the fuel combusted (TJ)
$\text{Emission Factor}_{\text{GHG, fuel}}$ = default emission factor of a given GHG by type of fuel (kg gas/TJ)

- For electricity:

$$\text{Emissions}_{\text{GHG,electricity}} = \text{Electricity usage} * \text{Emission Factor}_{\text{GHG,electricity}} \tag{8.2}$$

Where:

$\text{Emissions}_{\text{GHG, electricity}}$ = emissions of a given GHG by electricity (kg GHG)
Electricity usage = amount of the electricity used (KWh)

$\text{Emission Factor}_{\text{GHG, electricity}}$ = default emission factor of a given GHG by electricity (kg GHG/kWh)

- For waste of raw materials or waste of finished products

$$\text{Emissions}_{\text{materials, products}} = \text{Embodied carbon} * \text{Quantity} \tag{8.3}$$

Where:

$\text{Emissions}_{\text{materials, products}}$ = carbon emissions by type of materials/products (kg CO_2)
Embodied carbon = embodied emission factors by type of materials/products (kg CO_2/kg)
Quantity = amount of the materials/products consumed (kg)

Capital facilities were normally supported by electricity. The amount of carbon emissions in this category was therefore calculated by Eq. 8.2.

8.3 Embodied Carbon of Raw Materials and Finished Products

In order to assess the lean improvements, the embodied carbon of raw materials used in precast concrete production, as well as the embodied carbon of finished products should be calculated at the very start. In this study, the raw materials used in precast concrete production include cement, aggregates and reinforcement. The finished product was a specific type of precast concrete column.

8.3.1 Calculation Method

Life Cycle Assessment (LCA) has been widely adopted to evaluate the environmental impact in both the manufacturing and construction sectors (Harris 1999; Petersen and Solberg 2002). LCA assigns elementary flows and potential environmental impact to a specific product system. According to Hammond and Jones (2008), while a LCA study should ideally be set from the extraction of raw materials to the end of the production lifetime (which is usually referred to as cradle to grave), it is now common practice to specify the life cycle as cradle to gate, which includes all inputs until the product leaves the factory gate. A LCA study can be carried out in four steps, as illustrated in Fig. 8.2.

Reliable LCA requires the use of reliable Life Cycle Inventories (LCI), which consists of estimates of the materials and energy inputs and the emissions to air, land and water associated with the manufacture of a product, operation of a process

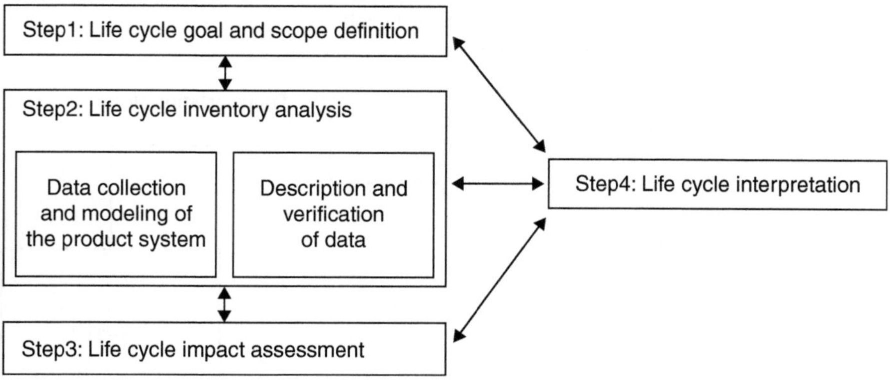

Fig. 8.2 Four steps to conduct a LCA study

or provision of a service (Nisbet et al. 2000). In this study, CO_2 emissions is the only target of the LCA.

The LCI data was investigated in each production process of the precast concrete column. Although Singapore's specific data was preferred, it was acknowledged that this category of data was scarce. The following estimation criteria and assumptions were therefore made in order to conduct the LCA.

8.3.2 Estimation Criteria

A few estimation criteria were established so that the calculation process could follow a standard procedure. The estimation criteria included:

1. Compliance with approved methodologies and standards. According to Hammond and Jones (2008), in the case of modern data, an ideal study would be ISO 14040/44 compliant (the international standard on environmental life cycle assessment). However, due to the differences in production activities, studies that complied with the same standard could provide different results. Further estimation criteria should therefore be provided for data consistency.
2. System boundaries. "Cradle-to-gate" was selected as the boundary condition of this study. Precasters could add up the carbon emissions during inventory and transportation of the finished products when calculating the embodied carbon while using the "cradle-to-site" boundary condition. In the case of the Singapore precast concrete industry, which relies heavily on the import of raw materials, the system boundaries of this study are illustrated in Fig. 8.3. The calculation process involved the following process, which were the manufacturing of raw materials, international transportation of raw materials to local traders, concrete production in the local concrete mixing plant, local transportation to precast concrete factories and the production process of precast concrete products.

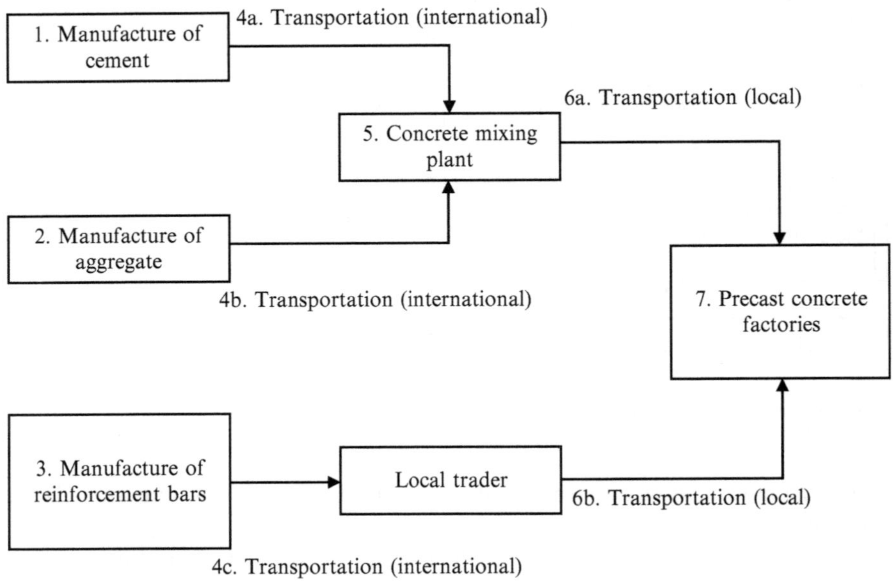

Fig. 8.3 System boundaries of the LCA for precast concrete columns

3. Data selection. Data that originated from the Singapore construction industry would be preferred. However, the embodied carbon and energy of raw materials, such as cement, aggregates and steel were still being investigated at the time of this study. However, local emissions data related to electricity generation could be obtained. This study therefore adopted a combination of local data and foreign data (such as European and world-wide average data), where local data was not available.
4. Emissions factors. This category of data was obtained by referring to a series of other LCI studies. The references for each category data are listed in Table 8.1.

8.3.3 Estimation Assumptions

Three assumptions were made to calculate the embodied carbon for both the raw materials and the finished products. These assumptions were:

1. Functional unit. The precast concrete column, which was typically adopted in public housing projects in Singapore, was adopted as the functional unit of this study, as shown in Fig. 8.4. This type of hollow core column weighs 1.864 t, including 0.178 t reinforcement and 1.686 t concrete. The mix design of the concrete would be explained later in this study in the section on primary materials.
2. It was assumed that waterborne transportation was adopted for the international transportation of raw materials. This assumption was made based on both the

Table 8.1 Information sources for materials and energy consumption data

Materials, embodied carbon and energy consumption	Sources
Mix design of Portland cement concrete	Obtained from precast concrete factory
Embodied carbon emissions of cement	Nisbet et al. (2000), Hammond and Jones (2008), Indexmundi (2006), USGS (2006), and Teo et al. (2010)
Embodied carbon emissions of aggregates	Hammond and Jones (2008)
Embodied carbon emissions of steel	World Steel Association (2008)
Carbon emissions from waterborne transportation	DEFRA (2005) and McKinnon (2008)
Carbon emissions from road transportation	Peyroteo et al. (2007)
Energy inputs from concrete mixing plant	Nisbet et al. (2000)
Energy inputs from precast concrete factory	Observed and recorded in precast concrete factory
Emission factor of electricity generation	NEA (2009)
Emission factor of idling trucks	Stodolsky et al. (2000)
Emission factor of illumination in the precast concrete factory	U.S. EPA (2007)

Fig. 8.4 The precast concrete column that was examined in this study

geographical location of Singapore and the background information of the Singapore precast concrete industry. In addition, it was assumed that the road transportation of finished precast concrete products was adopted. This assumption was more conservative because it is general knowledge that road transportation consumes more energy and generates more carbon emissions than railways or waterborne navigation (Nisbet et al. 2000). Round trip local transportation distances were assumed as follows, based on the geographical location of the suppliers to the precast concrete factory:

- From concrete plant to precast concrete factory: 15 miles (or 24.15 km)
- From steel trader to precast concrete factory: 15 miles (or 24.15 km)

3. Carbon emissions generated from capital facilities (e.g. offices, employee dormitories) were not considered in this study. In addition, carbon emissions due to personal travel emissions were not included in this study. However, this does not mean that capital facilities were not important. Carbon emissions generated by capital facilities was calculated and explained but was not included in the embodied carbon of the precast concrete columns.

8.3.4 Inputs

There were two major inputs of data in the production process of the precast concrete columns, which were raw materials and energy. The inputs of data are explained as follows:

1. Concrete
   Portland cement concrete was adopted for manufacturing the precast concrete columns. The mix information of the concrete and its strength requirement is listed in Table 8.2. According to Consoli et al. (1993), the inputs to a process could not be included in the LCI if the inputs are less than 1 % of the total mass of the processed materials or product, do not contribute significantly to a particular emission, nor have a significant associated energy consumption.
   The specific gravity of admixture 1 and 2 were 1.125 and 1.205 respectively. The mass of admixture 1 and 2 would therefore represent 0.042 and 0.25 % of the total mass of the precast concrete column investigated, which was much less than 1 %. The embodied carbon of both admixtures would be excluded in this study. This calculation method has also been adopted by other researchers (e.g. Nisbet et al. 2000) based on the fact that admixtures within concrete are not likely to be a source of emissions or effluent contamination because they are to a large extent chemically bonded and retained in the concrete product.
2. Mould
   Steel form was chosen in the precast factory investigated. Due to the economic benefits of the steel form, it was usually chosen for precast concrete production. The steel form has practically unlimited service life, its embodied carbon would not therefore be included in the calculation process of the embodied carbon of precast concrete columns. However, it should be noted that if concrete or wood forms are used in the absence of any recycle or reuse plan, the embodied carbon of the form materials should be included in the calculation process.
3. Reinforcement
   Steel was used in this type of precast concrete column because concrete had poor tensile strength. Reinforcement used in the precast concrete column included two core reinforcing bars with diameter of 25 mm and four corner bars with diameter of 13 mm, which helped the column share the compressive loads and resist the tensile stresses. Reinforcing links were also inserted at intervals of 200 mm to prevent the column from buckling under load.

Table 8.2 Mix design of the 16HPC1 precast concrete columns

Concrete mix design								
1. Concrete Grade			C40	Tremie mix	[25 N/mm^2 @ 24 h]			
2. Concrete strength requirement								
Characteristic strength			40 N/mm^2 @ 28 days					
Standard deviation			4.5 N/mm^2					
Target mean strength			50.0 N/mm^2 @ 28 days					
Free water/cement ratio (w/c)			0.356					
Fine aggregate/total aggregate ratio			0.432					
Total aggregate/cement ratio			3.91					
3. Mix information per cubic metre of concrete								
Materials	Cement	Fine aggregate	Coarse aggregate	Water	Admix (1)	Admix (2)	Air	
(units)	(kg)	(kg)	(kg)	(kg)	(ml)	(ml)	(%)	
Mix proportion	450	760	1,000	160	900	4,950	2	**Total volume**
Volume (l/m^3)	142.9	291.2	381.7	160	0.9	4.95	20	1,001.5
Density (kg/m^3)	2,370							

Notes:
Admixture 1 and 2 were both products of BASF Construction Chemicals. As the embodied carbon of admixtures were not considered in this study, the details of the admixtures would not be revealed here

4. Transportation
   The calculation processes of carbon emissions during transportation in this study were divided into two groups: international and local transportation. In international transportation of raw materials, it was assumed that the materials were delivered from the largest sea port of each country. DEFRA (2005) provided emission factors for several types of vessel ranging from a small one at 60 g CO_2 per tonne per km to a large bulk carrier at 7 g CO_2 per tonne per km. A mid-range value of 30 g CO_2 per tonne per km was adopted in this study. This assumption was based on the place of origins where the raw materials were imported, as shown in Table 8.3. The emission factors were averaged by the market share of each country.
   The amount of carbon emissions generated during transportation is highly related to transport distance. The impact of domestic transport is therefore much smaller than international transport. As stated above, it was assumed that distances from both the concrete mixing plant and local steel trader to the precast concrete factory were 15 miles (or 24.15 km).
5. Plant operations
   Energy inputs from the concrete mixing plant and precast concrete factory included electricity and fuel used for the operating equipment. While the data for the concrete mixing plant was adopted from Canadian concrete plant estimates, this could be used in this study because only the energy consumed in the operations was considered. In addition, the electricity consumed in precast concrete factories was observed and recorded in the precast concrete factory from a series of production activities, including gantry operations and vibration.

8.3.5 Embodied Carbon of the Precast Concrete Column

The overall calculation of the embodied carbon of the precast concrete column is shown in Table 8.4. For this type of precast concrete column, 609.59 kg carbon emissions were generated in its life cycle (from cradle to gate). As can be seen from Table 8.4, the manufacture of steel seems to be the most significant source of carbon emissions.

However, there are several points that should be noted:

1. The embodied carbon of raw materials (from cradle to gate) was the largest source of emissions arising from the use of the precast concrete column. Both steel and concrete contributed significantly to the overall embodied carbon of the precast concrete columns. As shown in Sect. 7.5.3, waste of raw materials is the very common in precast concrete factory with a probability of 3.94, it is therefore important to design the column to be both economically and environmentally efficient to reduce the usage of both raw materials. In addition, for the ready mix concrete, cement manufacturing accounts for 70–80 % of the total carbon emissions, which means that the relative importance of the contribution

Table 8.3 Calculation of CO_2 intensity during transportation

Material	Country	Percentage by Qty in recent three years (%)	Delivery port	Transport distance (delivery port to Singapore) (km)	CO_2 intensity during transportation from country i (kg/t)	Overall CO_2 intensity during transportation (kg/t)
Cement	Japan	50.5	Yokohama	5,292	158.76	104.2
	Taiwan	20.5	Kaohsiung	2,952	88.56	
	Malaysia	19.8	Klang	325	9.75	
	Thailand	8.4	Laem Chabang	1,351	40.53	
	China	0.5	Shanghai	3,801	114.03	
Steel	China	70.8	Shanghai	3,801	114.03	121.6
	Korea	17.7	Pusan	4,574	137.22	
	Japan	3.9	Yokohama	5,292	158.76	
Aggregate	Indonesia	44.3	Cigading	893	26.79	35.7
	Malaysia	35.9	Klang	325	9.75	
	China	16.4	Shanghai	3,801	114.03	

Table 8.4 Embodied carbon of 16HPC1 precast concrete column

Raw materials		Emission factors	Carbon emissions	
1. Cement	320.13 kg	0.4970 kg CO_2/kg	159.10 kg CO_2	26.1 %
2. Aggregates	540.66 kg	0.0050 kg CO_2/kg	2.70 kg CO_2	0.4 %
3. Reinforcement	178.00 kg	1.7000 kg CO_2/kg	302.60 kg CO_2	49.6 %
4a. Transportation (cement) (international)	320.13 kg	104.20 kg CO_2/t	33.36 kg CO_2	5.5 %
4b. Transportation (aggregate)(international)	540.66 kg	121.60 kg CO_2/t	65.74 kg CO_2	10.8 %
4c. Transportation (reinforcement) (international)	178.00 kg	35.70 kg CO_2/t	6.35 kg CO_2	1.0 %
5. Concrete plant operation	68.60 kWh/m^3	0.5233 kg CO_2/kWh	25.52 kg CO_2	4.2 %
6a. Transportation (concrete) (local)	24.15 km	0.1200 kg CO_2/km/t	9.77 kg CO_2	1.6 %
6b. Transportation (reinforcement)(local)	24.15 km	0.1200 kg CO_2/km/t	1.03 kg CO_2	0.2 %
7. Precast concrete production	6.50 kWh	0.5233 kg CO_2/kWh	3.40 kg CO_2	0.6 %
		Total	609.59 kg CO_2	100.0 %

of cement to the overall embodied carbon increases as the cement content in the mix design increases. This may require new designs of concrete to adopt fly ash and other admixtures to reduce the use of cement. According to Hammond and Jones (2008), the overall embodied carbon of cement can be 50 % lower, if 50 % of the cement used is replaced by fly ash.

2. Other than countries with local supplies of raw materials (e.g. China, the U.S.), the production of precast concrete columns in Singapore relied heavily on the import of raw materials. This resulted in 16.3 % of the total embodied carbon emissions. On the other hand, domestic transport accounted for 1.8 % of the total carbon emissions during the life cycle of the precast concrete columns.
3. The production processes of precast concrete column seemed to account for only 0.6 % of the total CO_2 emitted. However, it should be noted that this computation was based on the assumption that there were no wastage of materials in precast concrete factories. In fact, if 3–5 % (which is very common in precast concrete factories) wastage of raw materials were included in the production process, the production processes of precast concrete column should account for nearly 5 % of the total CO_2 emitted. In addition, energy consumed in bending, cutting and assembling of steel was not taken into consideration, as well as the energy consumed when managing the inventory for both raw materials and finished products. Otherwise, the whole operational processes in the precast concrete factory could take up to 10 % of the total embodied carbon. Details of such operations are discussed in sect. 8.4.

4. A carbon-labelling programme could be initiated based on this study. This would help the building industry to build its own carbon footprint programme. It seems that there are three possible ways to reduce the embodied carbon of precast concrete columns, either by improving the mix design to reduce the usage of cement and steel, by ordering locally manufactured raw materials, or by reducing the waste and energy consumed during production. Details of the carbon-labelling programme are discussed in Chap. 11.

8.4 Screening and Estimation Process

Non-value adding activities identified in each value stage of precast concrete production were tested and rated by Precaster A. Factors with no probability of occurrence ($P = 1$) were dropped from the assessment due to their relatively low importance, since Precaster A was performing well in these areas.

8.4.1 Site Layout Management

Following the screening procedure, Precaster A appeared to be facing several problems in site layout management, which included:

1. Improper specifications of building materials. According to the project manager interviewed, changes to the specifications of precast concrete products happened occasionally. As explained in Sect. 7.3.3, improper specifications happened occasionally, but usually had relatively high impact on the level of carbon emissions. Due to the changes, the process to design and produce such precast concrete products had to be re-developed, in which case the previously manufactured products would no longer be used. It was a waste of both raw materials and energy consumption in the production processes and should be eliminated. Based on one previous experience of such change, two batches of three precast concrete columns have been produced. The calculation was based on the assumption that improper specifications of building materials happened once during one contract period, which was half a year. As three precast concrete columns were produced daily, the total precast concrete columns produced in half of the year were assumed to be 500, including the consideration of rainy days when production was not possible. Carbon emissions of these six precast concrete columns were evenly distributed to the 500 precast concrete columns.
2. Over provision for material storage. The site layout plans of Precaster A are shown in Figs. 8.5 and 8.6. A 4-storey building and an open yard were included in the site layout plan. The storage area for raw materials was colored in dark grey while the prefabrication yard was colored in grey in the site layout plan. The storage of the finished products was within the production yard. The

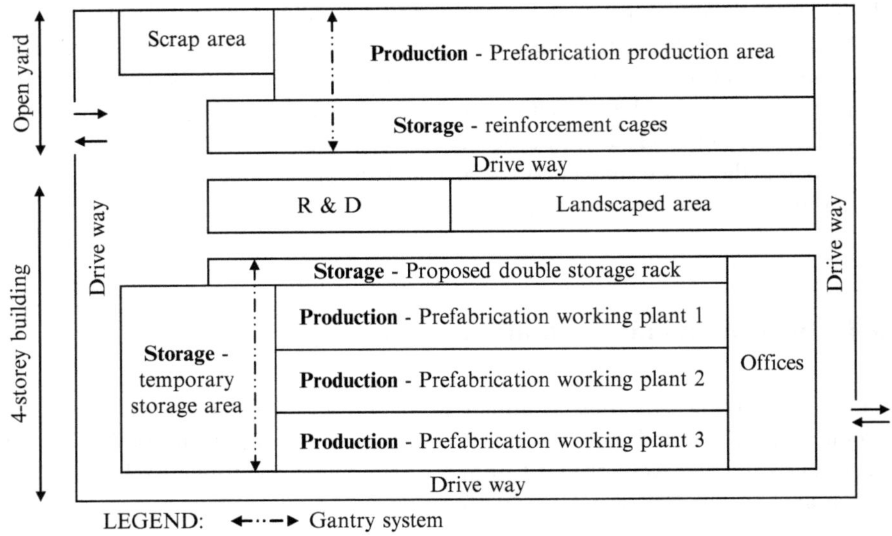

Fig. 8.5 The site layout design of the ground floor (not to scale)

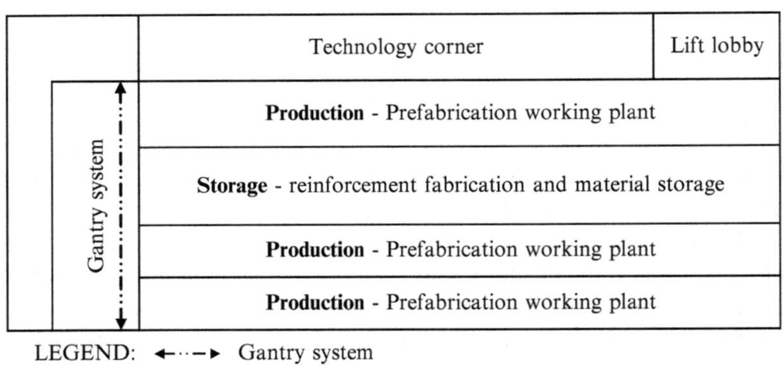

Fig. 8.6 The site layout design of the 2nd, 3rd and 4th floor of the 4-storey building (not to scale)

material storage area occupied 26.4 % of the total area in the site layout design, which was below the average (43 %), as stated in Sect. 7.3.1. It should be noted that this calculation was based on the site layout and only accounted for the material storage. The total storage area was about 40 % of the total area when the storage of the finished products was considered as well.

The direct consequence of such large storage was unnecessary loading and unloading activities caused by singling out and transferring activities. Such singling and transferring activities were carried out by the gantry system and trailers. In Precaster A, precast concrete columns were produced in the prefabrication production yard of the open yard, as shown in Fig. 8.5. Singling out and transferring activities happened in this area were chosen for carbon calculations.

The other source of carbon emissions generated by large storage area was the waste of either raw materials or finished products. According to the project manager interviewed, there was a 2 % waste of raw materials and a 3 % waste of finished products.

3. The site layout is not carefully planned to achieve economic and efficient production. Although Precaster A claimed to have an economic and efficient site layout design (P = 1), the site layout was not designed to be lean. As can be seen in Fig. 8.6, the reinforcements delivered from suppliers were sent to the 2nd and 3rd floor of the 4-storey building for fabrication. When the fabrication was completed, reinforcement cages were then delivered to the storage area in the open yard. Similar to problem 2, the energy consumption by the singling out and transferring activities is the most important direct consequence of such a site layout. On the other hand, damages to raw materials might also happen due to such arrangements. Although most precasters (P = 1.53) stated that the site layout was usually designed to achieve economic and efficient production, as explain in Sect. 7.3.3, it should be examined by the lean principles case by case.
4. Does not think of green building materials. Precaster A stated that research related to new green building materials was followed but rarely applied, in accordance with previous findings in Sect. 7.3.3 that the use of green materials was rarely considered. For example, according to Nielsen (2008), reducing the clinker content by substitution with supplementary cementitious materials such as fly ash really has a dramatic impact on the carbon footprint of the concrete. According to Prusinski et al. (2006), carbon emissions savings for precast concrete ranged from 137 to 222 kg/m^3 if slag cement mixtures were adopted. However, as can been seen in the mix design from Table 8.2, only regular cement (Portland cement) was adopted in current precast concrete production. Unless the willingness to use such green materials was expressed by clients, green building materials were not usually adopted.
5. Site layout plan is not placed on the notice board for information. On the notice board of Precaster A, safety issues were emphasized, as well as the provision of a detailed contact list. The amount of information was not sufficient to support a smooth production flow. Examined by the lean concept, the information on the notice board should be sufficient and transparent. According to Blumenthal (2008), two important aspects should be displayed on the notice board, which are waste streams and logistics of daily production. By doing so, the lead time can be reduced, thus reducing the costs.

The overall assessment of the lean improvements can be seen in Table 8.5. Primary calculation refers to the most significant consequence that leads to the increase in carbon emissions, while secondary calculation refers to the less significant ones that might lead to an increase in carbon emissions.

Table 8.5 Quantification of the lean improvements in site layout management

Factor No.	Description	Primary calculations (kg CO_2/column)	Secondary calculations (kg CO_2/column)	Qualitative assessments
1.1.1	Improper specification of building materials	7.32	N/A	N/A
1.6	Over provide material storage	0.58	22.40[a]	N/A
1.7.1	The site layout is not carefully planned to achieve economic and efficient production	0.96	9.29[a]	N/A
1.1.4	Does not think of green building materials	N/A	N/A	Yes[b]
1.7.3	Site layout is not placed on the notice board for information	N/A	N/A	Yes[c]

Notes:
[a]9.29 kg CO_2 was caused by waste of raw materials and was part of the 22.4 kg CO_2 caused by both waste of raw materials and illumination
[b]Carbon reduction achieved by the use of green building materials is highly dependent on the materials used. According to Hammond and Jones (2008), the overall embodied carbon of cement can be 50 % lower, if 50 % of the cement used is replaced by fly ash. Prusinski et al. (2006) stated that for concrete with 3,000 psi (28 day compressive strength), if 50 % slag cement is used, a carbon reduction about 42.5 % can be achieved. On the other hand, according to the Concrete Centre (2009), if self-compacting concrete is adopted, vibration is no longer needed to ensure compaction. For Precaster A, a carbon reduction of 0.2 % could be achieved. It should be noted that such reduction was obtained without the consideration of vibration process in ready-mix concrete production. Carbon reduction would be much higher if such process is considered
[c]The most important benefit of displaying sufficient information on the notice board is the reduction of lead time and the increase of profit. According to Swain and Mossman (2003), a 20 % reduction in lead time can cause a 5 % increase in profit

8.4.2 Supply Chain Management

Similar to sect. 8.4.1, activities in supply chain management in Precaster A were investigated to identify the improvements in carbon reduction. These activities included:

1. Large quantity supply base. Large quantity supply base could lead to large storage area and cause the same problems identified in category 1.6. Precaster A used large quantity supply base as most precasters did. As stated in Sect. 7.4.3, large quantity supply base happened frequently in precast concrete production, with a probability of 4.12. An amount of 0.58 kg CO_2 was emitted by singling out and transferring activities. In addition, as can be seen in Figs. 8.5 and 8.6, a total of 3,672 m^2 covered storage area was provided. Maintained illuminance (lux) was assumed to be 100 based on the illumination requirements in the factory. Energy consumption was therefore assumed to be 6.82 $kWh/m^2/year$ based on the illuminance. The total energy consumption that could be reduced in the contract period, which is half a year, was 12,521.52 kWh, resulting in a carbon reduction of 6,552.52 kg.

Table 8.6 Quantification of the lean improvements in supply chain management

Factor No.	Description	Primary calculations (kg CO_2/column)	Secondary calculations (kg CO_2/column)	Qualitative assessments
2.1.1	Large quantity supply base	0.58[a]	13.11	N/A
2.1.2	No long-term contract to achieve loyalty between suppliers and precasters	N/A	N/A	Yes[b]
2.1.3	Transportation is not taken into consideration	N/A	0.58	Yes[c]

Notes:
[a]0.58 kg CO_2 was caused by over providing material storage and has been included in Sect. 1.6
[b]Carbon emissions could be reduced in terms of less number of delivery and servicing trips, fewer deliveries in peak hours and less waste of raw materials during transportation
[c]Besides the carbon reduction achieved by fewer delivery and less waste, the lead time could be reduced. As stated earlier, a 20 % reduction in the lead time can result in a 5 % increase in profit

2. No long-term contract to achieve loyalty between suppliers and precasters. Long-term contract between suppliers and precasters may help to reduce the number of delivery and servicing trips, reduce trips in peak hours and reduce waste (Evanson 2008). When such relationship was built, precasters might ask the suppliers to hire operators with high awareness of sustainability in terms of certification programmes (e.g. Freight Operator Recognition Scheme in the UK). These benefits could help reduce carbon emissions but could not be assessed in this case study because it involved a long-term monitoring process which could not be completed given the time constraint.
3. Transportation is not taken into consideration. Price structure was the main consideration when selecting suppliers. By doing so, precasters lost control of the delivery. Of all the delivery methods provided, only direct delivery was adopted in Precaster A. If other lean delivery methods were adopted, such as milk round collection and delivery with an interposed warehouse, precasters would have more control on the arriving time of the delivery, thus reducing the lead time. In fact, if the precaster was using his own transportation trailer and the return trip is fully utilized, a carbon reduction of 0.58 kg can be achieved for steel transportation. Since concrete transportation required the use of in-transit mixers, the return trip for concrete might not be utilized at the moment.

The overall assessment is shown in Table 8.6. It should be noted that benefits other than carbon reduction could be achieved by taking transportation into consideration, as explained in note 3 of Table 8.6.

8.4.3 Production Management

The non-value adding activities in production management were observed in Precaster A to quantify the carbon reduction. It should be noted that a few

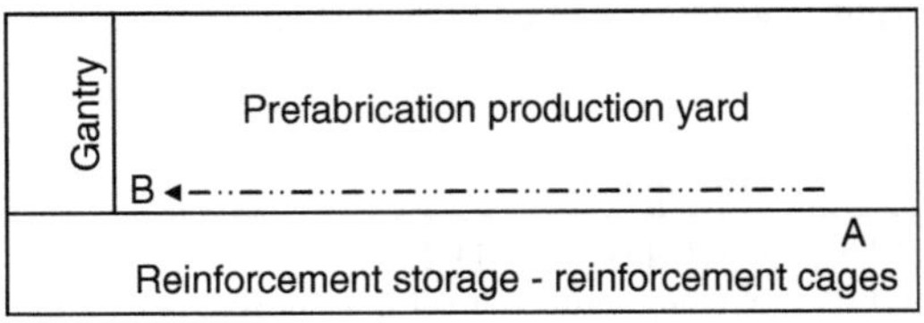

Fig. 8.7 A type of unnecessary movement in precast concrete production

non-value adding activities led to the same source of carbon emissions. For example, factor 3.1.5 (too much inventory) would lead to the same amount of carbon emissions with factor 1.6, because Precaster A was not taking any action to reduce the size of the inventory at the time of this study. The quantifications of these factors are integrated into one calculation.

1. Waste of raw materials and damages of raw materials. Both factors led to the 2 % waste of raw materials, as mentioned in sect. 8.4.1. The primary contribution the carbon emission level was 9.29 kg, which was caused by the 2 % waste of raw materials. In accordance with previous findings in Sect. 7.5.3, it is found that waste of raw materials and damages of raw materials happened frequently in precast concrete production and had a high impact on the level of carbon emissions ($I > 3$).
2. Raw materials do not meet specifications. According to the project manager interviewed, re-order and re-delivery of raw materials happened occasionally due to unsatisfied quality. It was assumed that during the contract period, two re-deliveries occurred. Two arrangements for local transport of concrete and steel should therefore be arranged. Similarly, this amount of carbon emissions was evenly distributed to the 500 precast concrete columns produced in the contract period. It was assumed that every delivery of concrete could support the production of 10 columns and every delivery of reinforcement could support the production of 20 columns. A total waste of 0.47 kg CO_2 was emitted for each precast concrete column.
3. Unnecessary materials handling. When observed in Precaster A, it was found that the employees lacked the awareness about the importance of a smooth work flow. For example, when the gantry operator intended to pick up the reinforcement cage for placing and had moved the gantry to location A, he was asked to carry out the lifting process in location B, as shown in Fig. 8.7. This unnecessary movement was caused by inexperienced employees and the lack of a written production manual. During the five days of observation in Precaster A, this unnecessary movement happened three times and it was assumed that this frequency remained during the contract period. The secondary calculation was related to the potential damages to raw materials.
4. Wait time. The two most important categories of wait time in Precaster A were wait time for inspection and labor. When the finished precast concrete products were placed on the trailer and final quality control checks were conducted, the trailer was left idling. When the trailer driver was handing the paper work to the supervisors, the trailer was left idling as well. A total time of 6 min were

recorded. Six precast concrete columns were delivered on each trailer and carbon emissions of the 6-min idling were evenly distributed, which resulted in an increase of 0.173 kg CO_2 for each column.

5. Double-handling or delivery due to unsatisfied quality or specifications. According to the project manager interviewed, double-handling due to quality problems happened occasionally in the demoulding and lifting process. For precast concrete columns, sockets and anchors were provided. The connection between these sockets and anchors and the concrete was very strong in some cases that double-handling was necessary. This was very common in the production of precast concrete columns, although each release might need different degree of effort. It is assumed that in every six precast concrete columns, such operations were performed for one column. Energy consumption was recorded to be 0.917 kWh.

   Wrong delivery do happened in Precaster A, although in an extremely low frequency about once or twice a year. It was assumed that one wrong delivery happened in the contract period with the delivery of six precast concrete columns. A total of 32.411 kg CO_2 was generated by the wrong delivery.
6. Inadequate work crews, weak employees and lack of supervision. Precaster A stated that the competency of both employees and supervisors could be improved. If such improvements could be achieved, the waste of raw materials and finished products, as well as the non-value adding activities could be eliminated.

The overall assessment is shown in Table 8.7. Special attention should be paid to the frequency of the non-value adding activities. Although the non-value adding activities in Precaster A happened very rarely, inappropriate production arrangements accounted for 38.62 % (1.313/3.4) of the carbon emissions in the production section, which represented a large improvement potential that could be achieved by applying the lean concept.

8.4.4 Stock Management

The quantification of non-value adding activities in stock management is shown in Table 8.8. The quantification process is described as follows:

1. Inappropriate staffing arrangement. A crane driver, a banksman and a chargehand should be provided in the precast concrete factory to carry out lifting process. However, this was not always the case in Precaster A. As observed in the factory, when moving the precast concrete columns to the storage area, the gantry operator was doing all the work. Many sudden accelerating and braking were caused when only one person was involved to conduct such loading and unloading activities. Sudden accelerating and braking would cause an increase in the level of carbon emissions.
2. Unclear identification marks and unclear delivery notes. The amount of 0.07 kg CO_2 was caused by wrong deliveries, as calculated in factor 3.3.3. It was assumed that two wrong deliveries, one under each category, happened in the

Table 8.7 Quantification of the lean improvements in production management

Factor No.	Description	Primary calculations (kg CO_2/column)	Secondary calculations (kg CO_2/column)	Qualitative assessments
3.1.1	Waste of raw materials in the production process	9.29	N/A	N/A
3.1.3	Materials damaged during handling			
3.1.2	Raw materials do not meet specifications	0.47	N/A	N/A
3.1.5	Too much inventory in factory	0.58[a]	13.11[a]	N/A
3.1.4	Unnecessary materials handling	0.003	9.29[b]	N/A
3.2.1	Wait time for inspection	0.17	N/A	N/A
3.2.3	Wait time for labor			
3.3.1	Repair due to damaged products during inventory	18.29[c]	N/A	N/A
3.3.2	Repair due to damaged products when handling			
3.3.3	Double-handling or delivery due to unsatisfied quality or specifications	0.09 0.07[d]	18.29[c]	N/A
3.5.1	Inadequate work crews	N/A	N/A	Yes[e]
3.5.2	Inexperienced employees			
3.5.3	Lack of supervision			

Notes:
[a]Please refer to factor 2.1.1
[b]Unnecessary materials handling was one source that led to waste of raw materials with a total of 9.29 kg CO_2 emissions
[c]3 % waste of finished products was used in this calculation
[d]The amount of 0.07 kg CO_2 was caused by wrong delivery and was not within the scope of "cradle to gate". This amount would not be used in calculating the embodied carbon of precast concrete columns
[e]The incompetence of human resources was believed to be the underlying problems causing unnecessary carbon emissions. With absolute care during operations, such carbon emissions could be eliminated

contract period. It should be noted that the amount of 0.14 kg did not fall within the boundary of "cradle-to-gate" and should not be considered when calculating the embodied carbon of the precast concrete column.

3. Lack of sufficient care. Precaster A stated that with sufficient care, waste of finished products and wrong delivery could be avoided. In fact, operators who were closely related to facilities operations have the greatest potential to reduce energy consumption and carbon emissions. However, the training programmes provided to the operators were currently not sufficient to support sustainable operations and needed to be improved.

The overall assessment for non-value adding activities in stock management is shown in Table 8.8. Similarly, the amount of 0.14 kg CO_2 per column was not considered in the embodied carbon because it was out of the system boundary of "cradle to gate".

Table 8.8 Quantification of the lean improvements in stock management

Factor No.	Description	Primary calculations ($kg\ CO_2$/column)	Secondary calculations ($kg\ CO_2$/column)	Qualitative assessments
4.2	Inappropriate staffing arrangement	N/A	N/A	Yes[a]
4.3.1	Unclear identification marks	0.14[b]	N/A	N/A
4.5.1	Unclear delivery notes			
4.5.2	Lack of sufficient care	N/A	N/A	Yes

Notes:
[a]The carbon emissions were qualitatively assessed according to U.S. EPA (2007)
[b]The amount of 0.14 kg CO_2 emissions was generated from two wrong deliveries (one under each category) in the contract period

8.5 Results

The carbon reduction that could be achieved by applying the lean production philosophy is shown in Table 8.9. A few implications could be obtained, including:

1. The amount of carbon emissions generated when producing one precast concrete column could be expressed by the following formula:

$$\text{The amount of } CO_2 = \underline{609.59}_{\text{Effective carbon}} + \underline{(18.29 + 9.29) + 9.933}_{\text{Waste}} + \underline{13.11}\,(\text{kg})_{\text{Capital Facility}}$$

   Effective carbon was obtained when there was no waste in the production process, as has been calculated in Sect. 8.3. A total amount of 37.513 kg CO_2 was emitted in terms of waste, which represented 6.15 % of the effective carbon. Another 13.11 kg CO_2 (2.15 %) could be reduced if the precast concrete factory was designed with no inventory of raw materials. The total carbon reduction obtained in Precaster A was 8.3 % of the effective carbon. This 8.3 % was in fact the lean score which can help improve the carbon labelling programmes, the details of which would be explained in Chap. 11.
2. It could be seen from the formula that there were several possible strategies to reduce carbon emissions in precast concrete production. The embodied carbon of raw materials (from cradle to gate) was the largest source of emissions in the precast concrete columns. Both steel and concrete contributed significantly to the overall embodied carbon of the precast concrete columns. As can be seen in Table 8.4, the manufacture of raw materials (cement, aggregates and steel) accounted for about 70 % of the total carbon emissions of the precast concrete columns. It was therefore important to design the precast concrete columns to be both economically and environmentally efficient to reduce the usage of concrete and steel.

Table 8.9 Carbon reduction achieved by applying the lean production philosophy

Category	The amount of carbon emissions (kg CO_2/column)
Waste of finished products	18.29
3.1.5 Too much inventory in factory	
3.3.1 Damaged products during inventory	
3.3.2 Damaged products when handling	
3.3.3 Double-handling or delivery due to unsatisfied quality or specifications	
Illumination savings	13.11
1.6 Over provide material storage	
2.1.1 Large quantity supply base	
3.1.5 Too much inventory in factory	
Waste of raw materials	9.29
1.6 Over provide material storage	
1.7.1 The site layout is not carefully planned to achieve economic and efficient production	
3.1.1 Waste of raw materials in the production process	
3.1.3 Materials damaged during handling	
3.1.4 Unnecessary materials handling	
Inappropriate production arrangements	9.933
1.1.1 Improper specification of building materials	7.25
1.6 Over provide material storage	0.58
1.7.1 The site layout is not carefully planned to achieve economic and efficient production	0.96
2.1.3 Transportation is not taken into consideration	0.58
3.1.2 Raw materials do not meet specifications	0.47
3.1.4 Unnecessary materials handling	0.003
3.3.3 Double-handling or delivery due to unsatisfied quality or specifications	0.09
Delivery performance	0.14
4.3.1 Unclear identification marks	0.07
4.5.1 Unclear deliver notes	0.07
Other qualitatively described lean improvements	Please refer to the qualitative assessment for each factor
1.1.4 Does not think of green building materials	
1.7.3 Site layout is not placed on the notice board for information	
2.1.2 No long-term contract to achieve loyalty between suppliers and contractors	
2.1.3 Transportation is not taken into consideration	
3.5 Human resources	
4.2 Inappropriate staffing arrangement	
4.5.2 Lack of sufficient care	

However, the precaster explained that the design of the precast concrete columns merely followed the routine guidance and the precaster was not conducting any research relating to alternative designs that could minimize the use of materials. It should however be noted that the effort and ability, if any, which the precaster devoted to developing alternative designs were reflections of the precaster's plans on continuous improvement, which would help the precaster to deal with increasing pressure on reducing carbon emissions. In addition, cement manufacturing accounts for 70–80 % of the total carbon emissions for the ready mix concrete (Nielsen 2008) and 26.10 % of the total carbon emissions for the precast concrete columns. The importance of the contributions of cement to the overall embodied carbon increased as the cement contents in the mix designs increased. This might require new designs of concrete to adopt fly ash and other admixtures to reduce the use of cement. As explained earlier, if 50 % of the Portland cement is replaced by slag cement, a savings in carbon emissions of 24.11 % can be achieved for the precast concrete columns, compared with 100 % Portland cement mixtures.

Other than countries with local supplies of raw materials (e.g. China, the U.S.), the production of precast concrete columns in Singapore relied heavily on the import of raw materials. As can be seen in Table 8.4, international transport of raw materials accounted for 17.30 % of the total embodied carbon emissions. On the other hand, domestic transport accounted for 1.80 % of the total carbon emissions. If the precast concrete columns were produced in countries with local supplies of raw materials, the carbon emissions value could be significantly reduced.

As shown in Table 8.4, the production processes of the precast concrete columns seemed to account for only 0.60 % of the total carbon emissions emitted. However, it should be noted that this computation was based on the assumption that there were no wastage of either raw materials or finished products in the precast concrete factories. In fact, according to the precaster, there was a 2 % wastage of raw materials and a 3 % wastage of the finished products in the production processes. In addition, there were many non-value adding activities that contributed to increase the level of carbon emissions. Ohno (1988) identified seven categories of waste: overproduction, correction, material movement, processing, inventory, waiting and motion. Low and Mok (1999) discovered that there were several types of waste during the manufacturing process, including waste from overproduction, waste from waiting time, transportation waste and waste of motion, inventory waste, waste of product defects, etc. Many non-value adding activities that contributed to increase the level of carbon emissions have been identified earlier. This might help the precasters to identify the non-value adding activities in their own precast concrete factories to achieve low-carbon production.

Reducing carbon emissions caused by non-value adding activities was especially important when green building materials were used and the embodied carbon of the precast concrete products was reduced. Delivery and operation activities played a more important role when the carbon emissions from raw materials were reduced. For example, when green building materials, as well as alternative designs were adopted and the embodied carbon of the column was reduced to 304.80 kg CO_2 per column (half of the original embodied carbon), the overall amount of carbon emissions that could be reduced by the lean concept

Table 8.10 The breakdown of carbon reduction when the embodied carbon is reduced

Category	The amount of carbon emissions (kg CO_2/column)	Percentage (%)
Embodied carbon	304.80	100
Waste of raw materials	4.645	1.52
Inappropriate production arrangements	9.933	3.26
Illumination savings	13.11	4.30
Waste of finished products	9.145	3.00
Total waste	36.833	12.08

was increased to 12.08 %, as can be seen in Table 8.10. This indicated that for precast concrete products with low inputs of raw materials, the application of the lean production philosophy is more prominent.

3. Wait time for inspection and labor should be especially noted here because delivery is the link between precasters and contractors. The trailer drivers were not appropriately trained on the awareness of sustainable operations, as observed in Precaster A. It is therefore reasonable to assume that equipment idling might happen in delivery and when unloading at construction sites. According to the American Trucking Association (1998), if a truck was left idling for one hour, an amount of 10.397 kg CO_2 would be emitted. North Carolina's Environmental Management Commission has adopted a rule to restrict on-road gasoline and diesel-powered vehicles with a loaded weight greater than 10,000 lb from idling for more than 5 consecutive minutes in any 60-min period (U.S. Department of Energy 2010). According to the U.S. Department of Energy (2010), Rhode Island also restricts the vehicle from idling for no more than 5 min by law. It appears timely for Singapore to take steps to develop similar law to raise operators' awareness on sustainable operations.
4. Different precasters might face different non-value adding activities. Although there were many non-value adding activities identified in Chap. 7, it should be noted that the non-value adding activities in each precast concrete factory should be identified case by case. The improvements in this case study were identified for Precaster A. As stated earlier, Precaster A was a leading government-sponsored entity who had better environmental performance than other precasters in Singapore. The carbon reduction achieved by applying the lean production philosophy might be larger than 8.3 % of the effective embodied carbon.

8.6 Summary

Many non-value adding activities that generated unnecessary carbon emissions have been identified. For example, unnecessary loading and unloading activities in terms of transferring and singling out activities could be eliminated if the site layout was designed to be lean at the very beginning. By producing according to the customer requirements, deviation costs, which might be caused by change of specifications, could be reduced. It was proposed that a value stream mapping process should be conducted to identify the non-value adding activities in the precast concrete production process at the very start. Instantaneous results in terms of low carbon emissions could be obtained by eliminating these non-value adding activities.

In addition, a high quality standard should be maintained in the precast concrete factories, which could be achieved by the support of high transparency (such as clear identification marks and delivery notes) and total quality control. The employees should be appropriately trained to conduct a smooth work flow with a high level of sustainable awareness. This was usually a continuous improvement process that may not produce instantaneous results.

Unlike introducing new technologies and green materials to the construction industry, lean production principles could be used to reduce carbon emissions with low implementation costs and low entry barrier. In addition to the benefit of low carbon emissions, the application of lean led to economic benefits as well, since the carbon emissions, which the lean concept focused on, were caused by unnecessary energy and fuel consumption. The economic benefits would increase if the fuel prices continued to rise.

The lean concept provided useful information on the precast concrete production process that caused unnecessary carbon emissions. This piece of information could be used to improve the carbon labelling practices in many environmental labelling programs, such as the Ecopoints, the BCA Green Mark Scheme (Singapore), etc. Based on the improvements that were brought about by the lean concept, the precast concrete industry could then move closer towards being a low-carbon industry. The details of the contribution would be presented in Chap. 11.

References

American Trucking Association (1998) Green truck website. [Online] Available at: http://www.truckline.com/air_emissions/1420.html. Accessed 9 Aug 2010

Blumenthal A (2008) Lean logistics, lessons learnt from Japan. [Online] Available at: http://www.leanconstruction.org.uk/lci_uk_summit_2008. Accessed 9 Aug 2010

British Precast Concrete Federation (2009) The concrete industry sustainability performance report, 1st report. The Concrete Centre, Surrey

Consoli F et al (1993) Guidelines for life-cycle assessment: a "code of practice". Society for Environmental Toxicology and Chemistry (SETAC), Pensacola

DEFRA (2005) Guidelines for company reporting on greenhouse gas emissions. [Online] Available at: http://www.defra.gov.uk/environment/business/reporting/pdf/envrpgas-annexes.pdf. Accessed 14 Jan 2011

Evanson D (2008) Lean construction logistics requirements within London. [Online] Available at: http://www.leanconstruction.org.uk/lci_uk_summit_2008. Accessed 20 Aug 2010

Glass J (2000) The future for precast concrete in low-rise housing. British Precast Concrete Federation, Leicester

Hammond G, Jones C (2008) Inventory of carbon and energy (ICE). [Online] Available at: http://www.bath.ac.uk/mech-eng/sert/embodied. Accessed 16 Mar 2010
Harris DJ (1999) A quantitative approach to the assessment of environment impact of building materials. Build Environ 34(6):751–758
Indexmundi (2006) Cement: world production, by country. [Online] Available at: http://www.indexmundi.com/en/commodities/minerals/cement/cement_t22.html. Accessed 14 Jan 2010
Low SP, Mok SH (1999) The application of JIT philosophy to construction: a case study in site layout. Constr Manage Econ 17(5):657–668
Mckinnon A (2008) CO_2 emissions from freight transport in the UK. Report prepared for the Climate Change Working Group of the Commission for Integrated Transport. Logistics Research Centre
National Environment Agency (NEA) (2009) Information on emission factors. [Online] Available at: http://www.nccc.gov.sg/cdm. Accessed 19 Jan 2009
Nielsen CV (2008) Carbon footprint of concrete buildings seen in the life cycle perspective. In: Proceedings of NRMCA 2008 Concrete Technology Forum, Denver
Nisbet M, VanGeem MG, Gajda J, Marceau M (2000) Environmental life cycle inventory of Portland cement concrete. Portland Cement Association, Skokie
Ohno T (1988) Toyota production system: beyond large-scale production. Diamond, Inc., Tokyo
Petersen AK, Solberg B (2002) Greenhouse gas emissions, life-cycle inventory and cost-efficiency of using laminated wood instead of steel construction. Case: beams at Gardermoen airport. Environ Sci Policy 5(2):169–182
Peyroteo A, Silva M, Jalali S (2007) Life cycle assessment of steel and reinforced concrete structures: a new analysis tool. In: Portugal SB 2007 sustainable construction, material and practices. IOS Press, Amsterdam, pp 1–6
Prusinski JR, Marceau ML, Van Geem MG (2006) Life cycle inventory of slag cement concrete. In: Naik T (ed) Proceedings of the 8th international conference on fly ash, silica fume, slag and natural pozzolans in concrete. CANMET/ACI, pp 1–26
Singapore Construction 21 Committee (1999) Construction 21: reinventing construction. Ministry of Manpower and Ministry of National Development, Singapore
Stodolsky F, Gaines L, Vyas A (2000) Analysis of technology options to reduce the fuel consumption of idling trucks, ANL/ESD-43. Argonne National Laboratory, Argonne
Swain B, Mossman A (2003) Smooth and lean – hathaway roofing case. Quality World 29(3):27–30
Teo AL, Lin GM, Lim JW (2010) Energy-related CO_2 emissions from the manufacturing and transportation of cement. Int J Sustain Soc 2(4):327–340
U.S. Department of Energy (2010) National idling reduction network news. [Online] Available at: http://www1.eere.energy.gov. Accessed 9 Aug 2010
U.S. Environmental Protection Agency (2007) Emissions factors and AP42, Compilation of air pollutant emission factors. Available from: http://www.epa.gov/ttnchie1/ap42/. Cited 29 Nov 2013
U.S. Geological Survey (USGS) (2006) Minerals yearbook 2006. [Online] Available at: http://minerals.usgs.gov/minerals/pubs/commodity. Accessed 15 Jan 2010
World Steel Association (WSA) (2008) Sustainability report of the world steel industry. [Online] Available at: http://www.worldsteel.org/?action=publicationdetailandid=52. Accessed 14 Jan 2010

Chapter 9
Precast Concrete Products in Construction Sites

9.1 Introduction

The applicability of the lean production philosophy in the construction sites using precast concrete products (hereinafter referred to as precast concrete construction sites) to reduce carbon emissions was analyzed in previous chapters. In this chapter, detailed evaluation of the applications is presented. The evaluation of the lean applications followed the value stream in the precast concrete construction sites, which included site layout management, transportation management, stock management and erection management.

In this chapter, the research sample is first introduced along with some background information about the construction projects that used the precast concrete products. The empirical study on the non-value adding activities after precast concrete production is then presented. Similar to Chap. 7, the analysis of these non-value adding activities in the precast concrete construction sites followed the value stream, as well as a general analytical procedure, which included:

1. Descriptive analysis. In the descriptive analysis, general questions related to site layout management, transportation management, stock management and erection management practices were discussed with the contractors. Non-value adding activities along with the possible improvements were identified for contractors to rate based on a five-point scale.
2. Factors description. In this step, the non-value adding activities in each value stage were introduced. Its impact on the level of carbon emissions of the construction project was briefly described. A factor number was provided for each non-value adding activity for referencing purposes.
3. Ranking procedure. The non-value adding activities were ranked based on two important criteria: the probability of occurrence (P) and the subsequent impact (I). It should be noted that the subsequent impact was evaluated based on a single criterion: the magnitude of the impact on the carbon emissions levels in the precast concrete construction sites. The severity (S) of the non-value adding

P. Wu and S.P. Low, *Lean and Cleaner Production: Applications in Prefabrication to Reduce Carbon Emissions*, DOI 10.1007/978-3-642-42062-7_9,

activities could be obtained by multiplying their probabilities with the impact, which fell within the range of [1, 25].

4. Parametric tests. This was conducted to identify the non-value adding activities within the same group categorized by the severity. It should be noted that although the scores obtained by the ranking procedure, in which the probabilities were multiplied with the impact, could be used for ranking purposes, the statistical significance between different factors should be identified. As the population was large and the sample size was 30, parametric tests were conducted. Paired sample t-tests were conducted because the sequence of the sample data could not be changed.
5. Specific analysis. This is to examine the non-value adding activities with large standard deviation of the probability. Attention was paid to these non-value adding activities, as improvements might be identified for contractors with higher probabilities of occurrence from contractors with lower ones.

9.2 Response Rate and Representativeness of Data

To identify the non-value adding activities for evaluation by the lean production philosophy in the precast concrete construction sites, a fieldwork involving 30 contractors who have had experience in managing construction projects using precast concrete products in the Singapore construction industry was conducted. The number of contractors targeted was based on the hypothesis testing methodology. In order to conduct an equality comparison of the population means of two populations with unknown variance, a large sample size ($n \geq 30$) was needed. Due to the time constraint, 30 contractors were chosen for this study.

From May 2010 to Oct 2010, 32 contractors were approached through email and telephone to invite them to participate in this study. Due to the complexity of the questionnaire, semi-structured interviews were requested with either the project managers or the site managers of the contractors. A site visit along with the semi-structured interview was also requested. A total of 30 responses was received. The response rate was 93.75 %. The first few contractors were recommended by the precasters interviewed earlier. These contractors were requested to provide two more contractors who have had experience with precast concrete projects. The 30 contractors were all in the group with the two highest financial grades (A1 and A2) in the *Directory of BCA Registered Contractors and Licensed Builders*. They were carrying out construction projects at the time of this study. The information obtained through interviews could therefore provide a fair representation of the Singapore construction industry.

In the fieldwork, the non-value adding activities identified from the literature review were rated by the contractors. Similar to the questionnaire for the precasters, the risk concept was broken down into two factors: probability (P) and impact (I). The degree and severity of the risk (S) was described by multiplying the probability with its corresponding impact.

9.3 Lean Site Layout Management in the Precast Concrete Construction Sites

Along the value stream of precast concrete installation, the first value stage was site layout management. In this stage, the construction site was designed for construction and installation activities. An effective site layout would promote a smooth work flow either in the transportation stage or the erection stage.

9.3.1 *Descriptive Analysis*

A few general questions related to existing site layout management practices in the precast concrete construction sites were discussed with the contractors before the non-value adding activities in this value stage were presented for ranking. These questions are listed in section 1.8 in the questionnaire for the contractors. The lean concept requires fundamental problems to be identified, evaluated and addressed, which highlights the importance of a lean site layout at the very start. The questions were therefore designed to examine the site layout to identify the factors which might increase the level of carbon emissions in the precast concrete construction sites. These questions are shown in Table 9.1.

Activities that might cause an increase in energy consumption and carbon emissions were identified in many areas, including regulations, human resources and site layout design. A few implications can be inferred from Table 9.1, including:

Table 9.1 General questions in the section for site layout management

Question no.	Descriptions	Positive responses	Positive response rate	Category
1.8.1	Are the supervisors/project managers appropriately trained relating to the use of materials, plant and equipment?	30	100 %	Human Resources
1.8.2	In the current site layout design, is environmental performance taken into consideration?	2	6.67 %	Legislation
1.8.3	Is there any statutory regulation relating to low energy consumption/low carbon emissions when designing and building the construction site?	0	0 %	Legislation
1.8.4	Are the site offices fully occupied at the moment?	30	100 %	Site layout design
1.8.5	Are the precast concrete columns used in a "first in first out" manner?	8	26.67 %	Site layout design
1.8.6	Is the site layout placed on the notice board for information?	17	56.67 %	Site layout design

1. Training programmes for supervisors and project managers were well provided by the contractors. All contractors (100 %) stated that training programmes relating to the use of materials, plants and equipment were adequately provided to either the project managers or the site supervisors. If the contractors were facing problems relating to the use of materials, plants and equipment, appropriate supervisors in charge of the respective area would be requested to solve the problems.
2. Environmental performance of the construction sites was rarely considered. Only two contractors (6.67 %) claimed that environmental performance of the construction sites was considered when designing the site layout. According to these two contractors, such environmental performance of the construction sites was related to water recycling (for both contractors) and noise control (for one contractor). The noise control actions were taken because there were complaints from the neighborhood. All contractors stated that there were no statutory requirements about low carbon emissions and low energy consumptions when submitting the design details for approval. The statutory requirements that must be complied with included health and safety (e.g. Workplace Safety and Health Act 2006), food hygiene (e.g. Environmental Public Health Regulations 2000), occupiers' liability (e.g. Law of Occupiers' Liability in Singapore 1985), etc. Regulations related to low energy consumption and low carbon emissions (e.g. Singapore's National Climate Change Strategy 2008) are purely voluntary for the contractors to comply with. Although a scenario for total reconstruction had never happened before because of a failure to meet these mandatory requirements, cease of construction activities for a short period of time (normally several days) had happened to five contractors (16.67 %), mainly due to poor health and safety issues.
3. All contractors (100 %) stated that the site offices at the construction sites were fully occupied. Some of them were even facing a shortage of workstations for their employees. One of the project manager interviewed stated that:

   It is a waste of money if the site offices are not fully occupied and the client won't approve it.

   By fully occupying the site offices, contractors were performing well in managing the carbon emissions from capital facilities. No carbon reduction could be achieved by reducing the size of the site offices.
4. Eight contractors (26.67 %) stated that the precast concrete products were used, or at least planned to be used in a "first in first out" manner. On the other hand, 22 contractors (73.33 %) stated that using the precast concrete products in a "first in first out" basis was not necessary because of the small storage area available in the construction sites. Unlike the storage area in the precast concrete factories, the storage area in the construction sites was relatively small, where the amount of precast concrete products required for the next 1 or 2 days was stored. Using the materials in a "first in first out" manner would therefore not be necessary.

   Seventeen contractors (56.67 %) were placing the site layout on the notice board for information. The most commonly exhibited information on the notice board

Table 9.2 Seven major categories of non-value adding activities in site layout management

Category no.	Description	Category
1.1	Construction materials	Materials
1.2	Site facilities	Facilities
1.3	Statutory requirements	Legislation
1.4	Construction requirements	Communication
1.5	Temporary works and services	Construction requirements
1.6	Storage area	Storage
1.7	Managing the construction site	Operation

included health and safety information, organizational structure and detailed contact lists. For some contractors, the maintenance schedule of the equipment and the progress chart were displayed as well. Similar to the problems in the precast concrete factories, failing to put the site layout and related information on the notice board may cause problems when changing subcontractors, operators or managers due to renewal of contracts, which happened in some projects because of unsatisfied performance. These subcontractors, operators and managers might have to refer to the site layout plan frequently. By putting the information on the notice board, a smooth work flow could be facilitated.

9.3.2 Factors Description

The lean production philosophy advocates that work is to be done right at the very start. If the contractors aim to achieve low-carbon erection, the site layout should be designed to support this objective at the very beginning. The questionnaire relating to site layout management was divided into seven categories, each of which represented an area where non-value adding activities could be identified. The seven categories are shown in Table 9.2. It should be noted that although the section relating to the use of green building materials in the precast concrete products seemed to be irrelevant to site layout management, it was analyzed in this section due to its higher priority than those other activities listed in the value stages later (transportation management, erection management, etc.). The design and specifications of the precast concrete products should be completed before transportation and erection were arranged. Because of its relatively small contents and high priority, the section relating to building materials was discussed in this section.

The non-value adding activities were described in each of the seven categories. These non-value adding activities as classified formed the operationalized measurables of this study. General descriptions for each non-value adding activity were provided as follows:

- *Category 1.1 Construction materials*

  1.1.1 Improper specifications of the precast concrete products
  1.1.2 Inaccurate estimation of the quantities required
  1.1.4 Does not think of the green building materials

For the erection processes to be conducted smoothly without errors, the specifications of the precast concrete products should be spelled out correctly and clearly. In addition, if the contractors frequently make adjustments to the specifications, erection activities in the construction sites may be changed, causing disruptions to existing installation works.

The quantities of the precast concrete products should be accurately estimated within a JIT delivery system. The erection activities would be interrupted by ordering either more or less of the precast concrete products required.

The use of the green building materials is a reflection of the contractors' plans relating to continuous improvement to achieve low carbon construction.

- *Category 1.2 No overall consideration of construction site facilities*

When designing an efficient site layout, equipment and infrastructures that will be used in future erection activities should be considered. According to Calvert (1995), the plant items utilized on site have a larger than expected impact on most sites and should be selected in accordance with the method statement. Many alternatives, such as on-site fabrications versus off-site fabrications, crawler cranes versus tower cranes, etc. should be compared before decisions are made for these to be located (Xin 2010). Changes to the site layout due to replacements of the equipment and infrastructures may cause disruptions to erection activities in the construction sites.

- *Category 1.3 Statutory requirements*

Many legal authorizations needed to be obtained before commencing construction in the site. These legal authorizations might include: health and safety (e.g. Workplace Safety and Health Act 2006), food hygiene (e.g. Environmental Public Health Regulations 2000), occupiers' liability (e.g. Law of Occupiers' Liability in Singapore 1985), etc.

- *Category 1.4 Does not pay full attention to the construction requirements*

  1.4.1 Duration
  1.4.2 Office space
  1.4.3 Maximum number of men on site
  1.4.4 The use of other equipment and plants
  1.4.5 Services required

Full attention should be paid to the construction requirements in order to provide the right amount of supporting facilities, such as site offices and employee dormitories. Although the energy consumed in these supporting facilities is not usually considered when calculating the carbon emissions in the erection cycle, it is still of vital importance for the contractor to include this to achieve low carbon construction.

- *Category 1.5 Inappropriate design of temporary works and services*

  1.5.1 Space for access
  1.5.2 Tower crane's fully blocked area

1.5.3 Clearance of the blocked area
1.5.4 Siting of static plant
1.5.5 Parking of mobile plant

Neil (1980) proposed 11 basic guidelines to design a construction site layout. In the 11 guidelines, Neil (1980) proposed that designers should master the planning of support facilities to avoid unnecessary relocation, remodeling or expansion as the project progresses. These guidelines apply to the construction sites in this study as well. Many aspects related to the design of works and services should be considered, including the space for access, locating and operating of both the static and mobile plants.

- *Category 1.6 Over provide storage area*

1.6.1 Secure store
1.6.2 Weatherproof store
1.6.3 Open store

Storage is a category of waste in the lean production philosophy. Materials could be damaged in inventory when inappropriately stacked. Singling out processes which involve the use of mobile plants may damage the materials as well. Although the energy consumed for maintaining the inventory is usually not considered in LCA studies (inventory is regarded as capital infrastructure and is not considered in LCA study unless it is significant under the BRE methodology for the environmental profiles of construction products), it represents a source of waste that should be considered in the construction site.

- *Category 1.7 Inappropriate management of the site layout*

1.7.1 Site layout plan is not tested for economic and efficient construction
1.7.2 Site layout plan is not sent to subcontractors and general foreman
1.7.3 Site layout plan is not placed on the notice board for information
1.7.4 Changes to the site layout plan are not notified immediately

Long term contracts were not usually used between the contractors and the precasters in Singapore. This may cause communication problems when switching between precasters in the subsequent projects. The transparency concept in the lean production philosophy aims to make the main flow of operations from start to finish visible and comprehensible to all employees (Stalk and Hout 1990). Failing to do so may expose the production flow to interruptions.

9.3.3 Ranking Procedure

Similar to the ranking procedure for the precasters, the ranking procedure for the non-value adding activities in the construction sites followed the standard risk analysis protocol, which involved the use of the P-I table. Quantitative assessment of both the probability (P) and the impact (I) were conducted to determine the

Table 9.3 Five-point scale to assess the probabilities of non-value adding activities

Rating no.	Category	Description
1	Very low	This category of non-value adding activity *never* happen in the erection cycle of the precast concrete products
2	Low	This category of non-value adding activity *rarely* happen in the erection cycle of the precast concrete products
3	Medium	This category of non-value adding activity *occasionally* happen in the erection cycle of the precast concrete products
4	High	This category of non-value adding activity *often* happen in the erection cycle of the precast concrete products
5	Very high	This category of non-value adding activity *always* happen in the erection cycle of the precast concrete products

Table 9.4 Five scale to assess the impact of non-value adding activities

Rating no.	Category	Description
1	Insignificant	This category of non-value adding activity has *insignificant* impact on the carbon emissions level of the precast concrete products (erection cycle)
2	Minor	This category of non-value adding activity has *minor* negative impact on the carbon emissions level of the precast concrete products (erection cycle)
3	Moderate	This category of non-value adding activity has *moderate* negative impact on the carbon emissions level of the precast concrete products (erection cycle)
4	Major	This category of non-value adding activity has *significant* negative impact on the carbon emissions level of the precast concrete products (erection cycle)
5	Catastrophic	This category of non-value adding activity has *catastrophic* negative impact on the carbon emissions level of the precast concrete products (erection cycle)

severity (S) of the non-value adding activities on carbon emissions. In this analysis, the term "*probability*" was defined as:

> The probability or frequency that this category of non-value adding activity would happen in the erection cycle in the construction sites.

Accordingly, the term "impact" was defined as:

> The effect of this category of non-value adding activity on the amount of carbon emissions that is generated from the erection cycle in the construction sites.

The erection cycle referred to the value stages after production. This erection cycle included site layout management of the construction sites, transportation management from the precast concrete factories to the construction sites, stock management and erection management of the precast concrete products. A five-point scale was adopted for qualitative descriptions of the probabilities of non-value adding activities. The five-point scale of probabilities is described in Table 9.3.

Accordingly, the impact associated with the non-value adding activities was assessed by a five-point scale, which is described in Table 9.4. The severity of every

Table 9.5 An example to show the difference between the LR of severity and the results (P × I)

Contractor no.	Probability of occurrence	Impact	Severity
A	2	3	6
B	4	2	8
C	5	1	5
D	2	3	6
Largest rating	5	3	8
Smallest rating	2	1	5

non-value adding activity can therefore be expressed by multiplying the probability with the corresponding impact.

Attention should be paid to the largest rating (LR) and smallest rating (SR) of severity for every non-value adding activity. It should be emphasized that the LR of severity was sometimes not equal to the result by multiplying the LR of probability with the LR of impact, as can be seen from Table 9.5. The LR of probability was 5 and the LR of impact was 3. However, the LR of severity (8) was not equal to the result simply by multiplying the LR of probability with the LR of impact (15). The same implications could be obtained for the smallest rating as well.

Rated by the five-point scale, the probability, impact and severity of non-value adding activities in site layout management could be seen in Table 9.6. A few implications could be made from Table 9.6, including:

1. As shown in Fig. 9.1, the most frequent non-value adding activities in the construction sites where the precast concrete products were used included:
   - 1.7.3 Site layout plan is not placed on the notice board for information (2.63)
   - 1.1.3 Does not think of the green building materials (2.53)
   - 1.7.4 Changes to the site layout plan are not notified immediately (2.47)

   As stated previously, the site layout plan was very important when there are changes in project managers, supervisors and subcontractors during the contract. According to the contractors, green building materials were only used in projects that were targeted for green mark certification. The most commonly used green building materials was the green concrete which was specially manufactured using fly ash or other recycled materials to replace Portland cement. Although the price of such green concrete was slightly higher (S$5 higher per cubic metre for the grade G35 concrete) than the Portland cement concrete, it was seldom used in precast concrete projects unless the projects were targeted for green mark certification. In addition, in precast concrete projects where the green concrete was used, only the minimum requirements in the green mark certification programmes were met. It seems that more incentive schemes from the government should be promoted to encourage the use of more green building materials.

   One significant difference between the precast concrete factories and the construction sites were the storage area. Unlike in precast concrete factories where large storage area was provided, the storage area in the construction sites was relatively small. Contractors placed the order based on a quantity estimation in the following 1 or 2 days. However, it should be noted that non-value adding

Table 9.6 Probability, impact and severity of the non-value adding activities in site layout management

Factor no.	Description	Probability (P)				Impact (I)				Severity (S)			
		AR	LR	SR	SD	AR	LR	SR	SD	AR	LR	SR	SD
1.1.1	Improper specifications of the precast concrete products	1.93	3	1	0.78	3.57	4	2	0.75	6.93	12	3	3.19
1.1.2	Inaccurate estimation of quantities required	1.77	3	1	0.63	3.77	4	3	0.43	6.73	12	3	2.68
1.1.3	Does not think of the green building materials	2.53	4	1	1.07	4.37	5	3	0.56	10.83	20	4	4.43
1.2	No overall consideration of construction site facilities	1.97	3	1	0.72	2.20	3	2	0.41	4.27	9	2	1.64
1.3	Does not comply with mandatory statutory requirements	1.57	3	1	0.63	3.07	5	2	0.83	4.77	12	2	2.49
1.4.1	Does not pay full attention to the construction requirements – Duration	1.53	2	1	0.51	2.13	3	1	0.57	3.10	6	2	1.18
1.4.2	Does not pay full attention to the construction requirements – Office space	1.63	2	1	0.49	2.30	4	1	0.60	3.60	6	2	1.28
1.4.3	Does not pay full attention to the construction requirements – Maximum number of men on site	1.50	2	1	0.51	2.30	4	1	0.60	3.37	6	2	1.47
1.4.4	Does not pay full attention to the construction requirements – The use of other equipment and plants	1.77	3	1	0.63	2.53	4	2	0.78	4.60	12	2	2.46
1.4.5	Does not pay full attention to the construction requirements – Services required	1.70	3	1	0.60	2.13	3	1	0.43	3.60	6	2	1.40
1.5.1	Inappropriate design of temporary	1.73	3	1	0.58	2.87	4	2	0.82	4.87	9	2	1.94

(continued)

Table 9.6 (continued)

Factor no.	Description	Probability (P)				Impact (I)				Severity (S)			
		AR	LR	SR	SD	AR	LR	SR	SD	AR	LR	SR	SD
	works and services – Space for access												
1.5.2	Inappropriate design of temporary works and services – Tower cranes' fully blocked area should be large	2.07	3	1	0.69	2.90	4	2	0.84	5.63	9	3	1.65
1.5.3	Inappropriate design of temporary works and services – Sitting of static plants (e.g. hoists, tower cranes, concrete mixing plants)	1.87	3	1	0.51	3.37	4	3	0.49	6.10	9	3	1.56
1.5.4	Inappropriate design of temporary works and services – Parking of mobile plants	1.87	3	1	0.51	3.17	4	2	0.59	5.90	9	2	1.86
1.6.1	Over provide storage area – Secure store	1.57	3	1	0.57	2.03	3	1	0.41	3.17	6	1	1.26
1.6.2	Over provide storage area – Weatherproof store	1.60	3	1	0.56	2.03	3	1	0.49	3.23	6	1	1.36
1.6.3	Over provide storage area – Open store	1.70	3	1	0.53	2.10	3	1	0.40	3.50	6	1	1.28
1.7.1	Site layout plan is not tested for economic and efficient construction	1.83	3	1	0.79	3.80	5	3	0.48	6.93	12	3	3.15
1.7.2	Site layout plan is not sent to subcontractors and general foreman	2.27	4	1	1.08	1.87	3	1	0.51	4.23	9	1	2.30
1.7.3	Site layout plan is not placed on the notice board for information	2.63	4	1	1.03	1.90	2	1	0.31	5.03	8	2	2.24
1.7.4	Changes to the site layout plan are not notified immediately	2.47	4	1	1.07	1.93	4	1	0.64	4.67	9	2	2.38

Notes: *AR* average rating, *LR* largest rating, *SR* smallest rating, *SD* standard deviation

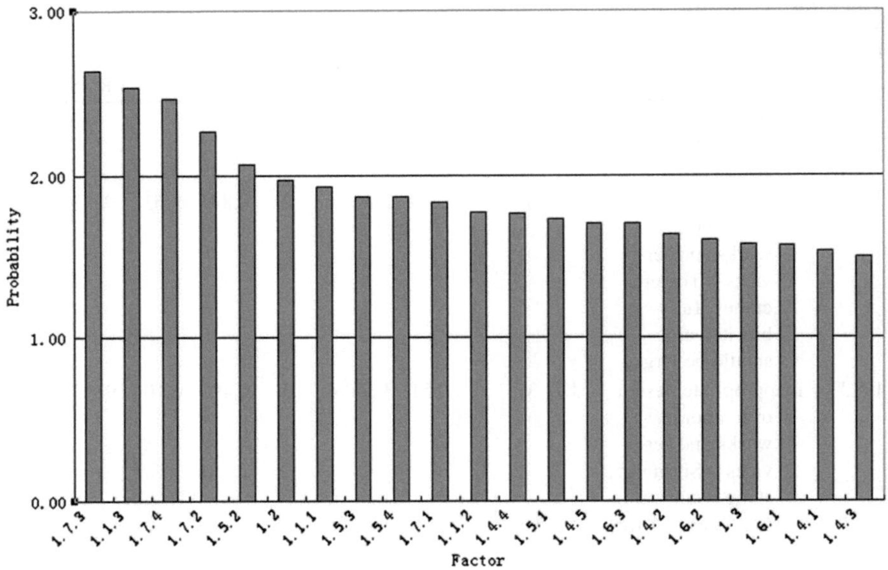

1.7.3 Site layout plan is not placed on the notice board for information
1.1.3 Does not think of green building materials
1.7.4 Changes to the site layout plan are not notified immediately
1.5.2 Inappropriate design of temporary works and services–Tower crane's fully blocked area should be large
1.2 No overall consideration of construction site facilities
1.1.1 Improper specifications of the precast concrete products
1.5.3 Inappropriate design of temporary works and services–Siting of static plants
1.5.4 Inappropriate design of temporary works and services–Parking of mobile plants
1.7.1 Site layout plan is not tested for economic and efficient construction
1.1.2 Inaccurate estimation of quantities required
1.4.4 Does not pay full attention to the construction requirements – The use of other equipment and plants
1.5.1 Inappropriate design of temporary works and services – Space for access
1.4.5 Does not pay full attention to the construction requirements – Services required
1.6.3 Over provide storage area – Open store
1.4.2 Does not pay full attention to the construction requirements – Office space
1.6.2 Over provide storage area – Weatherproof store
1.3 Does not comply with mandatory statutory requirements
1.6.1 Over provide storage area – Secure store
1.4.1 Does not pay full attention to the construction requirements – Duration
1.4.3 Does not pay full attention to the construction requirements – Maximum number of men on site

Fig. 9.1 The probability of non-value adding activities in site layout management

activities such as transferring and singling out activities happened regardless of the size of the storage area. Carbon emissions were emitted no matter which storage type was used.

2. The three most important factors that would increase the level of carbon emissions were:

   - 1.1.3 Does not think of the green building materials (4.37)
   - 1.7.1 Site layout plan is not tested for economic and efficient construction (3.80)
   - 1.1.2 Inaccurate estimation of quantities required (3.77)

Contractors agreed that the green building materials were very important if they aimed to achieve low carbon construction. In addition, the site layout

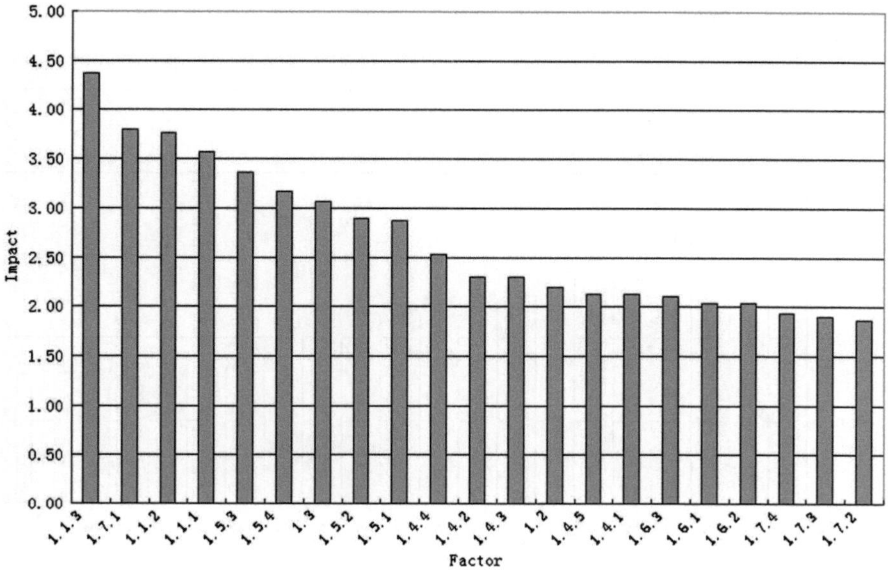

1.1.3 Does not think of green building materials
1.7.1 Site layout plan is not tested for economic and efficient production
1.1.2 Inaccurate estimation of quantities required
1.1.1 Improper specifications of the precast concrete products
1.5.3 Inappropriate design of temporary works and services–Siting of static plants
1.5.4 Inappropriate design of temporary works and services–Parking of mobile plants
1.3 Does not comply with mandatory statutory requirements
1.5.2 Inappropriate design of temporary works and services–Tower crane's fully blocked area should be large
1.5.1 Inappropriate design of temporary works and services–Space for access
1.4.4 Does not pay full attention to the construction requirements – The use of other equipment and plants
1.4.2 Does not pay full attention to the construction requirements – Office space
1.4.3 Does not pay full attention to the construction requirements – Maximum number of men on site
1.2 No overall consideration of construction site facilities
1.4.5 Does not pay full attention to the construction requirements – Services required
1.4.1 Does not pay full attention to the construction requirements – Duration
1.6.3 Over provide storage area – Open store
1.6.1 Over provide storage area – Secure store
1.6.2 Over provide storage area – Weatherproof store
1.7.4 Changes to the site layout plan are not notified immediately
1.7.3 Site layout plan is not placed on the notice board for information
1.7.2 Site layout plan is not sent to subcontractors and general foreman

Fig. 9.2 The impact of non-value adding activities in site layout management

should be designed to support economic and efficient construction, especially for transportation and erection activities which relied on the use of equipment and plants heavily. If a JIT delivery system was adopted in the construction sites, accurate estimation of the quantities of precast concrete products should be focused on. All contractors stated that the quantities could be accurately estimated based on a stable construction schedule (please refer to question 2.5.11 for more information). The impact of the non-value adding activities in site layout management is shown in Fig. 9.2.

3. When considering both the probability and impact, the severity of these non-value adding activities in site layout management is shown in Fig. 9.3. The most important factors included:

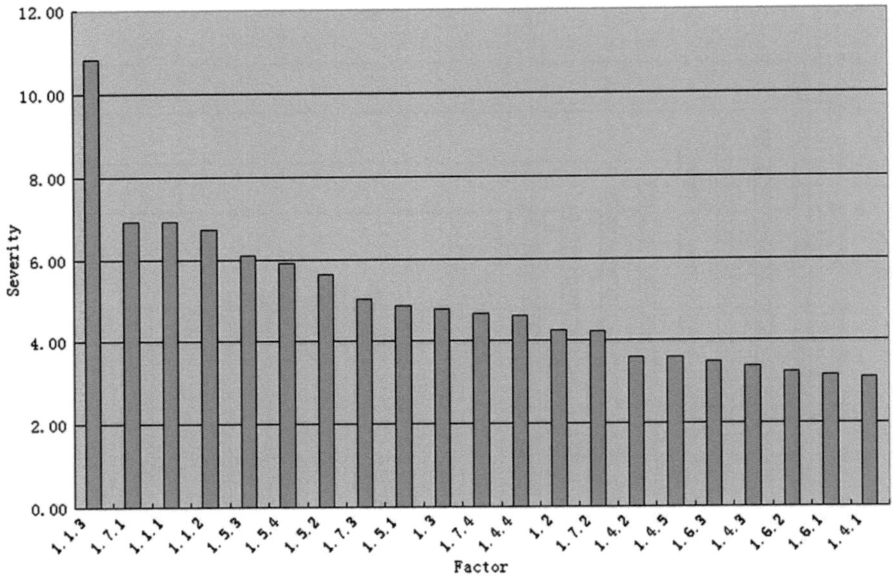

1.1.3 Does not think of green building materials
1.7.1 Site layout plan is not tested for economic and efficient production
1.1.1 Improper specifications of the precast concrete products
1.1.2 Inaccurate estimation of quantities required
1.5.3 Inappropriate design of temporary works and services–Siting of static plants
1.5.4 Inappropriate design of temporary works and services–Parking of mobile plants
1.5.2 Inappropriate design of temporary works and services–Tower crane's fully blocked area should be large
1.7.3 Site layout plan is not placed on the notice board for information
1.5.1 Inappropriate design of temporary works and services–Space for access
1.3 Does not comply with mandatory statutory requirements
1.7.4 Changes to the site layout plan are not notified immediately
1.4.4 Does not pay full attention to the construction requirements – The use of other equipment and plants
1.2 No overall consideration of construction site facilities
1.7.2 Site layout plan is not sent to subcontractors and general foreman
1.4.2 Does not pay full attention to the construction requirements – Office space
1.4.5 Does not pay full attention to the construction requirements – Services required
1.6.3 Over provide storage area – Open store
1.4.3 Does not pay full attention to the construction requirements – Maximum number of men on site
1.6.2 Over provide storage area – Weatherproof store
1.6.1 Over provide storage area – Secure store
1.4.1 Does not pay full attention to the construction requirements – Duration

Fig. 9.3 The severity of non-value adding activities in site layout management

- 1.1.3 Does not think of the green building materials (10.83)
- 1.7.1 Site layout plan is not tested for economic and efficient construction (6.93)
- 1.1.1 Improper specifications of the precast concrete products (6.93)
- 1.1.2 Inaccurate estimation of quantities required (6.73)

The contractors agreed on the importance of using green building materials to achieve low carbon construction, as well as the importance of an effective site layout plan to support green construction activities. The specifications of the precast concrete products should be completed at the very start and minimum alterations should be made to the specifications. According to one contractor, changes made from using off-site fabrications to on-site fabrications had happened in the past. Such changes would cause modifications to the site layout plan, as well as the erection method, causing more carbon emissions to be emitted.

The severity of the storage area in the construction sites was smaller compared to its severity in the precast concrete factories. By using the precast concrete products in the construction sites, there was no storage area needed for raw materials, such as cement and aggregates. Only open storage area for the precast concrete products was provided. Due to the small size of the storage area (which was typically designed to handle the quantity used for the following 1 or 2 days), contractors believed that this would not cause interruptions to the construction activities. However, as stated above, carbon emissions were emitted due to transferring and singling out activities despite the size of the storage area. If low carbon construction was targeted, the storage area should be eliminated by introducing the JIT delivery and the JIT installation process.

The routine considerations in designing the construction site layout were not very severe factors considering their low probability of occurrence, so much so that contractors overlooked these factors. These routine considerations included:

- 1.4 Construction requirements
- 1.5 Design of temporary works and services

However, a few factors in these routine considerations should be noted which included:

- 1.5.3 Inappropriate design of temporary works and services – Siting of static plants (6.10)
- 1.5.4 Inappropriate design of temporary works and services – Parking of mobile plants (5.90)
- 1.5.2 Inappropriate design of temporary works and services – Tower crane's fully blocked area should be large (5.63)

As these factors relied heavily on the use of equipment and plants, rigorous simulations should be conducted before choosing the type of static plants (e.g. tower crane, gantry) and mobile plants (e.g. crawler crane, forklift). The blocked area should be carefully designed so that direct use of the precast concrete products could be implemented. An case study was used to identify the importance of the site layout to achieve low carbon construction. The case study would be explained in the case study in Chap. 10.

For illustration purposes, the probabilities and impact of the non-value adding activities in site layout management were plotted in a P-I table (sometimes referred to as the P-I grid). As shown in Fig. 9.4, there are four categories of non-value adding activities, which were: unlikely and minor, rare catastrophe, frequent niggle and probable disaster. As can be seen from Fig. 9.4, most non-value activities fell into the category of "unlikely and minor" and "rare catastrophe".

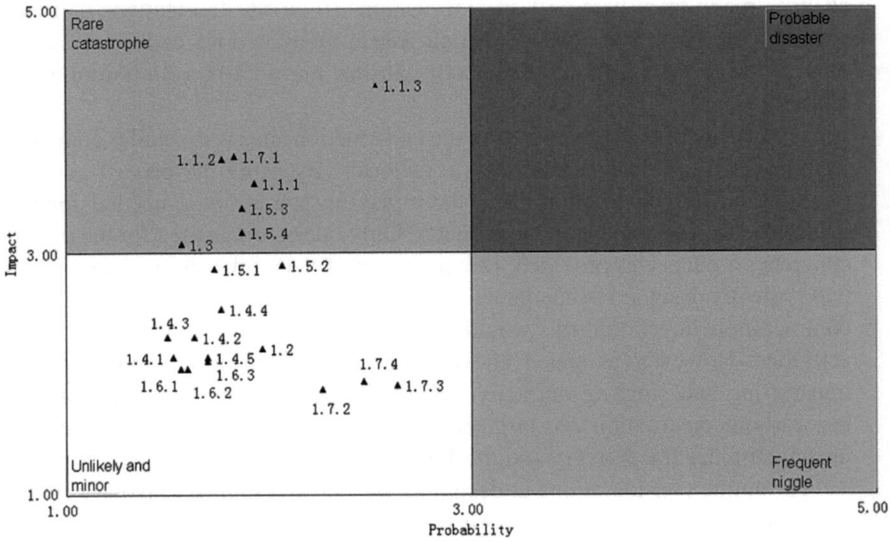

Fig. 9.4 P-I table for non-value adding activities in site layout management

"Transfer" actions should apply to low-probability, high-impact activities (rare catastrophe) and "accept" actions should apply to low-probability, low-impact activities (unlikely and minor). Details of the mitigation strategies would be explained in Sect. 9.7.

9.3.4 Parametric Tests

This section aimed to identify the non-value adding activities within the same group ranked by severity. Attention should be paid to those non-value adding activities with close degree of severities. The statistical significance between different groups should be identified. As the sample size was large (=30) with unknown distribution (normal or non-normal), parametric tests were adopted. A general procedure was followed:

1. Decide the sample type. In this study, parametric tests developed for two-paired samples were adopted to test the significance between two different non-value adding activities at 95 % confidence interval. As the sequence of the sample data could not be changed, paired sample t-test was adopted.
2. For example, factors 1.1.3 (Does not think of green building materials) and 1.1.1 (Improper specifications of the precast concrete products) were rated as the two most important non-value adding activities in site layout management in the construction sites. Paired sample t-test was conducted to test whether there was a significant difference between the means of these two factors. The null

Table 9.7 Test statistics for factors 1.1.3 and 1.1.1

		Paired differences					t	df	Sig. (2-tailed)
		Mean	Std. deviation	Std. error mean	95 % Confidence interval of the difference				
					Lower	Upper			
Pair 1	F1.1.3 and F1.1.1	3.900	5.047	0.9215	2.015	5.785	4.232	29	0.000

hypothesis was that the difference between the two mean values was zero. If p value (the minimum value that the null hypothesis could be rejected) is less than 0.05, the null hypothesis could be rejected and factor 1.1.1 would be significantly different from factor 1.1.3. The paired sample t-test of the two factors is shown in Table 9.7.

3. Following this procedure, more paired sample t-tests were conducted until the p value was less than 0.05. Since the p value for a factor with itself could not be identified, "Not Applicable" (N/A) was used in the paired sample t-test statistics shown in Table 9.8.

In this study, factor 1.1.1 was significantly different from factor 1.1.3 ($p = 0.0001 < 0.05$) and was therefore listed as the top factor of ranking group 2. Following the procedure, all non-value adding activities were grouped into four groups, as can be seen in Table 9.8.

As can be seen in Table 9.8, five ranking groups were identified by the paired sample t-tests. The most important factor which was listed in ranking group 1 was factor 1.1.3 (Does not think of green building materials). The least important factors that were listed in ranking group 4 and 5 included category 1.4 (Does not pay full attention to the construction requirements) and category 1.6 (storage area). However, it should be noted that as long as storage existed, the singling out and transferring activities would happen, causing more carbon emissions.

When allocating resources (time, human resources and costs) to eliminate the non-value adding activities, the ranking groups that were identified by the paired sample t-tests should be considered. For example, although factor 1.7.1 (Site layout is not tested on economic and efficient construction) had higher severity than factor 1.5.2 (Tower cranes' fully blocked area should be large), these two factors were statistically insignificantly different.

9.3.5 *Specific Analysis*

Specific qualitative analysis was conducted when a large difference exists between the largest rating (LR) and the smallest rating (SR) in the probability section. A large difference might imply that the contractors with high probability of

Table 9.8 Ranking and grouping of non-value adding activities in site layout management

			Severity		
Ranking	Factor no.	Description	AR	Sig. (2-tailed)	SD
1	1.1.3	Does not think of the green building materials	10.83	N/A	4.43
2	1.1.1	Improper specifications of the precast concrete products	6.93	0.0001 N/A	3.19
2	1.7.1	Site layout plan is not tested for economic and efficient construction	6.93	1.000	3.15
2	1.1.2	Inaccurate estimation of quantities required	6.73	0.709	2.68
2	1.5.3	Inappropriate design of temporary works and services – Sitting of static plants (e.g. hoists, tower cranes, concrete mixing plants)	6.10	0.232	1.56
2	1.5.4	Inappropriate design of temporary works and services – Parking of mobile plants	5.90	0.177	1.86
2	1.5.2	Inappropriate design of temporary works and services – Tower cranes' fully blocked area should be large	5.63	0.064	1.65
3	1.7.3	Site layout plan is not placed on the notice board for information	5.03	0.010 N/A	2.24
3	1.5.1	Inappropriate design of temporary works and services – Space for access	4.87	0.672	1.94
3	1.3	Does not comply with mandatory statutory requirements	4.77	0.676	2.49
3	1.7.4	Changes to the site layout plan are not notified immediately	4.67	0.317	2.38
3	1.4.4	Does not pay full attention to the construction requirements – The use of other equipment and plants	4.60	0.489	2.46
3	1.2	No overall consideration of construction site facilities	4.27	0.156	1.64
3	1.7.2	Site layout plan is not sent to subcontractors and general foreman	4.23	0.056	2.30
4	1.4.2	Does not pay full attention to the construction requirements – Office space	3.60	0.011 N/A	1.28
4	1.4.5	Does not pay full attention to the construction requirements – Services required	3.60	0.9999	1.40
4	1.6.3	Over provide storage area – Open store	3.5	0.748	1.28
4	1.4.3	Does not pay full attention to the construction requirements – Maximum number of men on site	3.37	0.243	1.47
4	1.6.2	Over provide storage area – Weatherproof store	3.23	0.239	1.36
4	1.6.1	Over provide storage area – Secure store	3.17	0.146	1.26
5	1.4.1	Does not pay full attention to the construction requirements – Duration	3.10	0.037	1.18

Notes: *AR* average rating, *SD* standard deviation

occurrence were not performing very well against those with low probability. Although there were no non-value adding activities which met the requirements for specific analysis (LR = 5 and SR = 1), a few non-value adding activities should be noted (LR = 4 and SR = 1). These factors included:

- 1.1.3 Does not think of green building materials

Most contractors stated that the use of green building materials was very common at the time of this study. The green building materials would be considered when disseminating the design specifications for approval by consultants. However, the decision would be subject to the client's approval. If the project aimed for a green building certification, the green building materials would be used. However, seven contractors (23.33 %) stated that the use of green building materials was not always considered (P = 4). They followed the design routines and when considering the use of the green building materials, their profit margins might be eroded.

- 1.7.2 Site layout plan is not sent to subcontractors and general foreman
- 1.7.3 Site layout plan is not placed on the notice board for information
- 1.7.4 Changes to the site layout plan are not notified immediately

The site layout management practices relating to the use of site layout plan varied among the contractors interviewed. 24 contractors (80.00 %) sent the site layout plan to the subcontractors. 23 contractors (76.67 %) placed the site layout plan on the notice board for information. Changes to the site layout plan rarely happened. However, due to changes in the construction methods (e.g. from off-site prefabrications to a combination of off-site prefabrications with on-site fabrications) and unsatisfactory performance of the precasters, the site layout plan might be changed. 22 contractors (73.33 %) stated that such changes would be notified to subcontractors immediately.

9.4 Lean Transportation Management in the Construction Sites

In this section, the transportation management practices of 30 contractors were investigated. Non-value adding activities in this section were derived from the literature review relating to JIT sourcing, TQM as well as JIT management practices. It should be noted that this section could also be named as supply chain management for contractors. However, in order to differentiate it from the supply chain management in the precast concrete factories, the term transportation management was used.

9.4.1 *Descriptive Analysis*

Similar to the section on site layout management, a few general questions relating to current transportation management were discussed with the contractors before

Table 9.9 General questions in the section of transportation management

Question no.	Description	Positive response	Positive response rate	Category
2.5.1	Is the transportation works subcontracted out?	29	96.67 %	Industry information
2.5.2	Does the contractor (or the subcontractors if the transportation job is subcontracted out) provide standing instructions to the drivers?	30	100 %	Human resources
2.5.3	Are the instructions made orally or through written documents?	27	90.00 %	Human resources
2.5.4	Is the operator aware of the typical damages to the precast concrete products during transportation?	30	100 %	Human resources
2.5.5	Is single sourcing adopted in the supply chain?	26	86.67 %	JIT sourcing
2.5.6	Is a long-term contract awarded to achieve loyalty and reduce the risk of disruption?	17	56.67 %	JIT sourcing
2.5.7	Is the contractor operating with a stable erection schedule?	30	100 %	Delivery considerations
2.5.8	What is the current delivery method adopted by the contractor?	30 (Direct delivery)		Delivery method
2.5.9	Has the company conducted any trade-off investigation relating to the small lot deliveries and reduced inventory?	0	0 %	Continuous improvement
2.5.10	Do you think that damages to the precast concrete products can be eliminated if sufficient care is provided?	30	100 %	Human Resource
2.5.11	Did re-delivery happen due to unclear delivery note?	22	73.33 %	Delivery consideration
2.5.12	Is the delivery note made orally or by written documents?	Oral 1	Written 29	Delivery considerations
2.5.13	Is the vehicle driver appropriately trained to provide sufficient care to the precast concrete products?	30	100 %	Human resources
2.5.14	Are routine inspections conducted before the release of the precast concrete products?	26	86.67 %	Delivery considerations

the non-value adding activities were provided for ranking. Details of the responses can be seen in Table 9.9.

A few implications can be identified from Table 9.9, including:

1. Most contractors (96.67 %) subcontracted the transportation works to the precasters. The transportation costs were included in the price structure of the precast concrete products. Similar with the price structure in the precast concrete

products, this was usually referred to as the "all-in-one" price. One contractor had his own precast concrete factory and the transportation works were conducted using his own trailers.

All contractors stated that trailer drivers were carefully trained by standing instructions. Most contractors (90.00 %) provided the standing instructions through written documents, while three contractors (10.00 %) stated that only on-the-job training was provided.

Although all contractors stated that the drivers were trained carefully so that sufficient care could be provided to the precast concrete products during transportation, they acknowledged the fact that 1–5 % damages to the precast concrete products would happen during transportation. The average percentage of damages to the precast concrete products was 3.17 %.

2. The evaluation and selection of the precasters was based on several criteria: a quantitative measure of quality (inspections and tests), certification programmes, delivery performance and most importantly, the price structure. These criteria were applied by all contractors when selecting suppliers (except the one who had his own precast concrete factory). Unlike the precasters who rarely conduct qualitative evaluation of the suppliers, all contractors stated that such evaluation of the precasters was conducted. This was to identify the reliability of the precasters. In addition, as the cost of transportation was integrated with the price of the products, the geographical location of the contractors seemed to be unimportant when selecting the precasters. For example, in one construction project, there were three precasters located within 3 km of the project. However, the contractor chose one precaster located 15 km away from the project site. An amount of 16.00 kg carbon emissions was emitted when a typical transportation arrangement of six precast concrete columns was used (without the empty return trip) due to the additional 12 km delivery distance. More importantly, using the "all-in-one" price instead of taking the transportation into consideration might lead to delay caused by traffic conditions, which could not be identified by comparing the prices. Of all the delivery methods provided (direct delivery, delivery with an interposed warehouse, milk round collection, suppliers supply a number of components), all contractors (100 %) were using the direct delivery method. It appears that if direct delivery was used and transportation cost was integrated with the product price, contractors in Singapore have less control to guarantee dependable deliveries.
3. If the precaster selected could provide all the precast concrete products used in the project, a single sourcing relationship was established between the contractor and the precaster. This single sourcing relationship was adopted by 26 contractors (86.67 %). However, only 17 contractors (56.67 %) established a long term relationship with the precasters. When selecting the precasters for the next construction project, these precasters would be preferred. Other contractors (43.33 %) evaluated and selected suppliers on a project-by-project basis. It should be noted that such long term relationship was largely dependent on the intervention of the clients. According to the contractors, the client's preference played an important role in selecting the precasters. Even if the long term

Table 9.10 The categories of non-value adding activities in transportation management

Category no.	Description
2.1	Damages during transportation
2.2	Selecting precasters
2.3	JIT management process
2.4	Delivery management

relationship between the precasters and the contractors was established, the selection of the suppliers could still be subject to the client's preference and approval. Small lot deliveries were adopted. However, it should be noted that such small lot deliveries covered the amount that would be used for the next 1 or 2 days and was not true small lot deliveries to support immediate usage in the context of the JIT delivery system.

4. It seems that contractors did not perform well on continuous improvement in terms of research at the time of this study. None of the precasters were conducting any research related to small lot deliveries and reduced inventory. They were satisfied with the current construction practices. According to the contractors, the small storage area did not pose problems to the construction activities.

9.4.2 Factors Description

A few categories of non-value adding activities in the transportation management were provided for the contractors to evaluate. These categories are shown in Table 9.10.

In each of these four categories, a few non-value adding activities were also identified. A general description for each non-value adding activity was provided as follows:

- *Category 2.1 Damages during transportation*

  2.1.1 No skilled attention to the details of supports and frames
  2.1.2 The driver is not aware of the few typical damages that may occur during transportation
  2.1.3 Inappropriate packings and supports
  2.1.4 No standing instructions

The precast concrete products should be delivered to the construction sites without damages. The most common damages that may occur during transportation included (Richardson 1973):

1. Damages due to the wrong placement of battens;
2. Damages caused by flexing the vehicle bed and torsion induced by road surfaces;
3. Cracks caused by vibration of the elements;
4. Damages to slender sections;
5. Broken nibs and bottom corners of panels, and
6. Damages caused by the use of slings and chains.

In order to prevent these damages from happening, standing instructions should be provided to the drivers, who should provide sufficient care to the precast concrete products during transportation. Appropriate packings, supports and frames should be provided in order to achieve defect-free transportation to reduce carbon emissions.

- *Category 2.2 Selecting precasters*

  2.2.1 Large quantity supply base
  2.2.2 No single sourcing supply with long-term contract
  2.2.3 Transportation is not taken into consideration
  2.2.4 No quality audits of the precasters prior to the award of the contract

Small lot-sizes were considered as the hallmark of JIT sourcing and were believed to be flexible enough to overcome the obstacles of higher delivery costs and loss of discount rates (Banerjee and Kim 1995). Achieving a long-term relationship was believed to have several benefits, such as easing the communications problems between the precasters and the contractors, obtaining discounts and building trust. As stated earlier, transportation cost was currently included in the overall product price. The geographical location of the precasters was hardly a factor to consider when selecting the precasters. However, many problems could be caused by sourcing based on the price structure alone, such as losing control of the delivery time. Although quality could be assured by many certification programmes, many benefits could still be achieved by conducting quality audits, especially when JIT delivery was scheduled. However, it seems that price was still the major consideration when selecting the precasters.

- *Category 2.3 JIT management process*

  2.3.1 Demand fluctuations
  2.3.2 Not fully prepared for the arrival of the precast concrete products
  2.3.3 No advance order and confirmation order
  2.3.4 Insufficient data exchange with the precasters

If a stable erection schedule could be anticipated, JIT deliveries could be applied more readily. However, this does not mean that JIT deliveries cannot be applied where there were demand fluctuations. More prompt communication was needed to meet demand fluctuations. Under extremely perfect conditions, the precast concrete products should be used immediately when delivered. However, due to various reasons, such as waiting for inspection and equipment, this was not always the case in the construction sites. Tommelein and Li (1999) provided the JIT delivery mapping alternatives for vertical supply chain integration. In this mapping strategy, both advance order and confirmation order were used to make sure that deliveries arrive on time. Cancellation of orders could be executed without undermining suppliers' benefits. As stated earlier, under demand fluctuations, prompt communication between the precasters and the contractors was of vital importance when applying the JIT delivery system. The data exchange methods included facsimile, email messages, mobile phones and some other electronic devices.

- *Category 2.4 Delivery*

  2.4.1 No accurate delivery notes
  2.4.2 Insufficient care
  2.4.3 Lack of routine inspection

It goes without saying that loading and unloading activities were vitally important for precast concrete products. The probability of damages to the precast concrete products increased with the frequency of such activities. Along with sufficient care when conducting loading and unloading activities, clear delivery notes should also be provided to avoid wrong delivery and routine inspections prior to the release of the products should be conducted to prevent double-handling, or double-delivering in the worst case scenarios.

9.4.3 Ranking Procedure

Similar to the section on site layout management, the non-value adding activities in transportation management were rated by both probability and impact using a five-point scale. The results are shown in Table 9.11. A few implications can be drawn from Table 9.11, including:

1. The most frequently occurring non-value adding activities in transportation management were:

   - 2.2.3 Transportation is not taken into consideration (3.17)
   - 2.3.3 No advance order and confirmation order (2.80)

   22 contractors (73.33 %) stated that transportation was not always a consideration when selecting the precasters (when P = 5, P = 4 and P = 3). The delivery performance of the precasters was very good in the aspects of managing to deliver on time with the right quantity and good quality, as rated by the 22 contractors. It was sometimes the contractors' fault that the delivery vehicles were left idling due to site congestion. The two-order system derived from the JIT delivery system and as proposed by Tommelein and Li (1999) was sometimes not adopted by contractors who believed that one order would be good enough because of the small quantity ordered every time.

   However, it should be noted that the two-order system would also help the precasters to deal with changes. It was not only the contractor who could benefit from this system. On the other hand, a few factors which did not frequently happen in transportation management included:

   - 2.3.4 Insufficient data exchange with the precasters (1.50)
   - 2.3.1 Demand fluctuations (1.60)
   - 2.4.1 No accurate delivery notes (1.67)

Table 9.11 Probability, impact and severity of the non-value adding activities in transportation management

Factor no.	Description	Probability (P)				Impact (I)				Severity (S)			
		AR	LR	SR	SD	AR	LR	SR	SD	AR	LR	SR	SD
2.1.1	No skilled attention to the details of supports and frames	2.43	3	1	0.68	3.43	4	2	0.68	8.20	12	3	2.75
2.1.2	The driver is not aware of the few typical damages that may occur during transportation	2.30	3	1	0.65	3.60	4	3	0.50	8.30	12	3	2.63
2.1.3	Inappropriate packings and supports	2.60	4	2	0.62	3.33	4	3	0.48	8.70	12	6	2.49
2.1.4	No standing instructions	1.90	3	1	0.66	3.40	4	2	0.72	6.43	9	2	2.36
2.2.1	Large quantity supply base	1.80	3	1	0.66	2.03	4	1	0.61	3.53	9	1	1.72
2.2.2	No single sourcing supply with long-term contract	2.60	4	1	1.13	2.87	4	2	0.73	7.40	16	3	3.79
2.2.3	Transportation is not taken into consideration	3.17	5	2	0.87	3.00	4	2	0.59	9.20	16	4	2.61
2.2.4	No quality audits of the precasters prior to the award of the contract	2.10	5	1	1.12	3.27	4	2	0.58	6.97	16	3	4.20
2.3.1	Demand fluctuations	1.60	3	1	0.77	3.40	4	2	0.67	5.53	12	2	3.19
2.3.2	Not fully prepared for the arrival of the precast concrete products	2.17	3	1	0.79	3.30	4	2	0.53	7.20	12	3	3.04
2.3.3	No advance order and confirmation order	2.80	5	1	1.35	2.80	4	1	0.55	7.57	15	3	3.76
2.3.4	Insufficient data exchange with the precasters	1.50	3	1	0.57	2.90	5	2	0.96	4.37	10	2	2.30
2.4.1	No accurate delivery notes	1.67	3	1	0.66	2.93	4	2	0.83	5.03	9	2	2.53
2.4.2	Insufficient care	2.60	4	2	0.62	3.87	5	3	0.63	10.07	16	6	2.88
2.4.3	Lack of routine inspection	2.40	4	1	0.93	3.57	4	2	0.57	8.50	16	4	3.51

Notes: *AR* average rating, *LR* largest rating, *SR* smallest rating, *SD* standard deviation

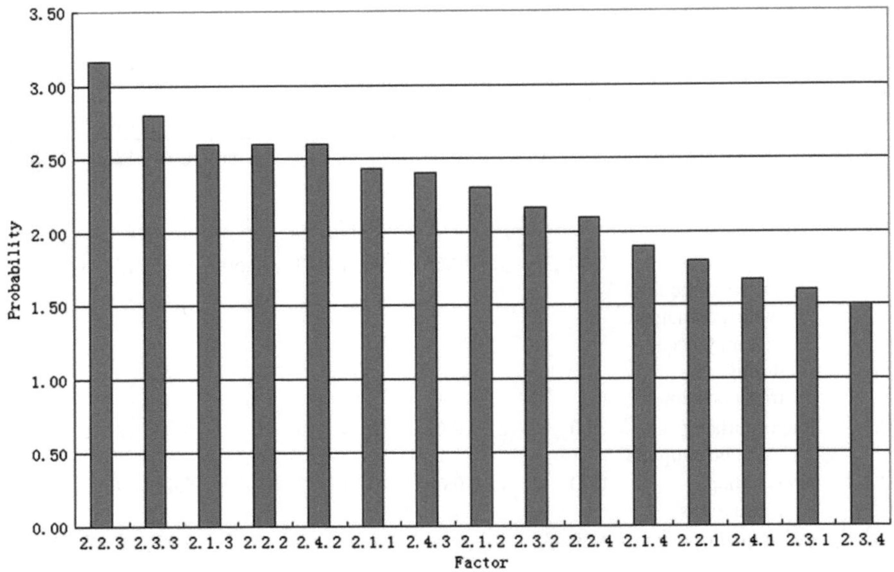

2.2.3 Transportation is not taken into consideration
2.3.3 No advance order and confirmation order
2.1.3 Inappropriate packings and supports
2.2.2 No single sourcing supply with long-term contract
2.4.2 Insufficient care
2.1.1 No skilled attention to the details of supports and frames
2.4.3 Lack of routine inspection
2.1.2 The driver is not aware of the few typical damages that may occur during transportation
2.3.2 Not fully prepared for the arrival of the precast concrete products
2.2.4 No quality audits of the precasters prior to the award of the contract
2.1.4 No standing instructions
2.2.1 Large quantity supply base
2.4.1 No accurate delivery notes
2.3.1 Demand fluctuations
2.3.4 Insufficient data exchange with the precasters

Fig. 9.5 The probability of non-value adding activities in transportation management

The data exchange between the precasters and the contractors seemed to be sufficient. The contractors were not facing demand fluctuations. Although the daily amount of usage might vary, such variation was very minimal. However, it was surprising to note that no immediate use of the delivered precast concrete products was conducted based on the stable erection schedule. This was better explained in question 4.6.9 in the erection management. The overall probability chart for non-value adding activities in transportation management is shown in Fig. 9.5.

2. The impact of the non-value adding activities in transportation management in the construction sites is shown in Fig. 9.6. The most important factors that could cause an increase in the level of carbon emissions during transportation included:

- 2.4.2 Insufficient care (3.87)
- 2.1.2 The driver is not aware of a few typical damages during transportation (3.60)
- 2.4.3 Lack of routine inspection (3.57)

It seems that the contractors agreed on the important role played by the drivers during transportation. All contractors stated that by providing sufficient care, the

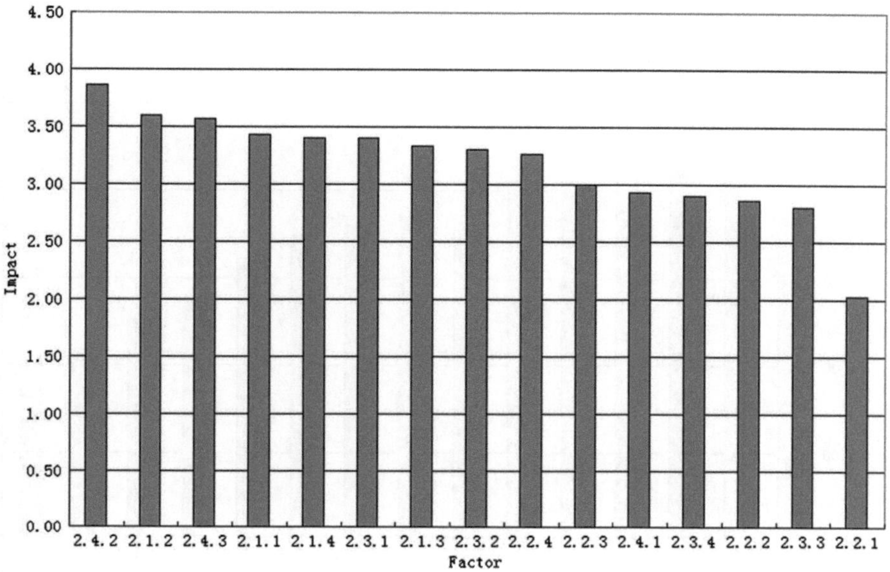

2.4.2 Insufficient care
2.1.2 The driver is not aware of the few typical damages that may occur during transportation
2.4.3 Lack of routine inspection
2.1.1 No skilled attention to the details of supports and frames
2.1.4 No standing instructions
2.3.1 Demand fluctuations
2.1.3 Inappropriate packings and supports
2.3.2 Not fully prepared for the arrival of the precast concrete products
2.2.4 No quality audits of the precasters prior to the award of the contract
2.2.3 Transportation is not taken into consideration
2.4.1 No accurate delivery notes
2.3.4 Insufficient data exchange with the precasters
2.2.2 No single sourcing supply with long-term contract
2.3.3 No advance order and confirmation order
2.2.1 Large quantity supply base

Fig. 9.6 The impact of non-value adding activities in transportation management

damages from transportation could be reduced. Routine inspection before the release of the precast concrete products was very important to reduce carbon emissions from transportation.

According to one contractor, during a 3-month period, two rejections of the precast concrete products had happened, causing re-delivery to be arranged.

On the other hand, the large quantity supply base (2.21) was not rated as an important factor by the contractors. Even if the quantity required for the next few days was delivered to the construction site, the precast concrete products could be stacked appropriately to avoid damages. In addition, as the erection schedule was very stable, the contractors would not face any demand fluctuation. It is a common practice to order the precast concrete products to be used in the next 1 or 2 days. Consequently, large quantity supply would not happen in current construction projects which used the precast concrete products.

3. The severity of the non-value adding activities is shown in Fig. 9.7. The most severe non-value adding activities included:

- 2.4.2 Insufficient care (10.07)
- 2.2.3 Transportation is not taken into consideration (9.20)
- 2.1.3 Inappropriate packings and supports (8.70)

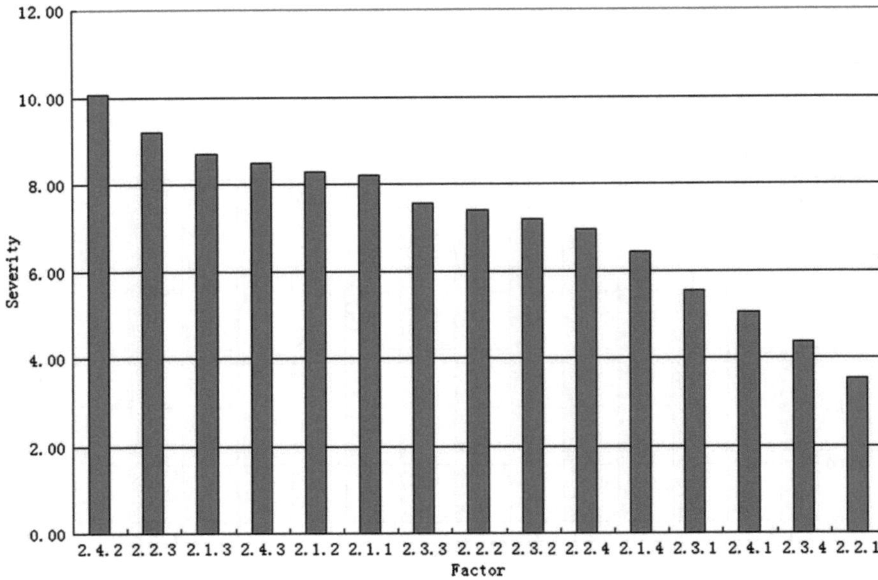

2.4.2 Insufficient care
2.2.3 Transportation is not taken into consideration
2.1.3 Inappropriate packings and supports
2.4.3 Lack of routine inspection
2.1.2 The driver is not aware of the few typical damages that may occur during transportation
2.1.1 No skilled attention to the details of supports and frames
2.3.3 No advance order and confirmation order
2.2.2 No single sourcing supply with long-term contract
2.3.2 Not fully prepared for the arrival of the precast concrete products
2.2.4 No quality audits of the precasters prior to the award of the contract
2.1.4 No standing instructions
2.3.1 Demand fluctuations
2.4.1 No accurate delivery notes
2.3.4 Insufficient data exchange with the precasters
2.2.1 Large quantity supply base

Fig. 9.7 The severity of non-value adding activities in transportation management

In order to achieve low carbon transportation for the precast concrete products, it seems that contractors should take transportation into consideration by ordering locally manufactured products. The price structure should not be the only consideration especially under the "all-in-one" price structure. For one contractor, there were three L6 (highest financial grade) precasters located 3 km away from the project site and yet the contractor chose one precaster that was located 15 km away. The selection would cause more carbon emissions to be emitted during transportation. In addition, by providing appropriate and sufficient packings and supports as well as sufficient care, damages during transportation could be eliminated.

For illustration purposes, the probabilities and impact of the non-value adding activities in transportation management were plotted in a P-I table, as shown in Fig. 9.8.

It seems that most non-value adding activities fell into the category of rare catastrophe with low probability of occurrence and high impact. Attention should be paid to the differences between the transportation management practices of the

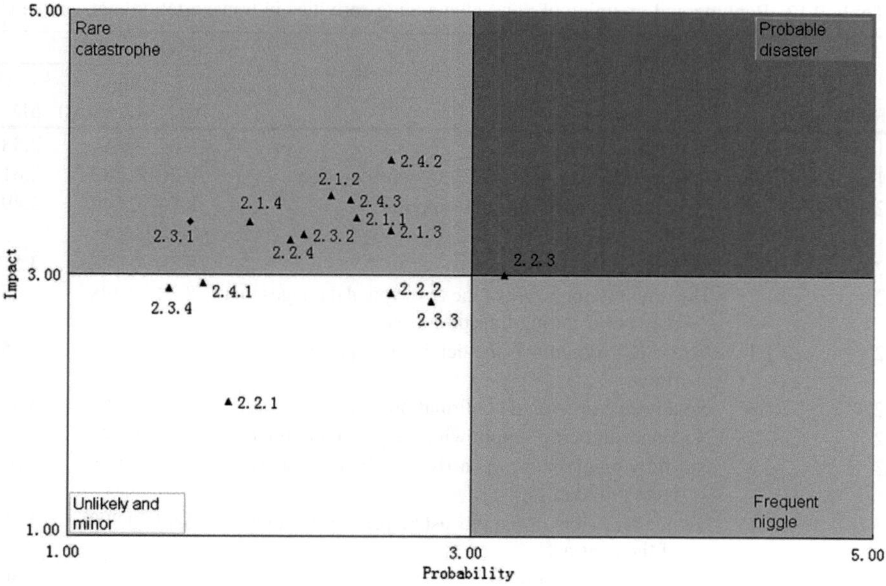

Fig. 9.8 P-I table for non-value adding activities in transportation management

precasters and the contractors. While a large quantity supply base was often used in precast concrete factories (4.12), this strategy was seldom adopted by the contractors (1.80). The precasters did not form long-term relationships with their suppliers (4.29) while many contractors did build a long-term relationship with the precasters (2.60). Both the precasters and the contractors were performing well in being prepared for the arrival of either the raw materials or the precast concrete products. Data exchange between the suppliers and the precasters, as well as between the precasters and the contractors seemed to be sufficient and adequate.

9.4.4 Parametric Tests

Similar to Sect. 9.3.4, parametric tests (paired sample t-tests) were conducted to rank the non-value adding activities into groups. The parametric tests followed the general procedure described in Sect. 9.3.4.

The results of the paired sample t-tests for the non-value adding activities in transportation management are presented in Table 9.12. Five ranking groups were identified by the paired sample t-tests. Factors in the ranking group one had the highest level of severity and should be eliminated immediately. On the other hand, factors in the ranking group five had the lowest level of severity and should not be focused on unduly because the contractors appeared to be performing very well in these areas.

Table 9.12 Ranking and grouping of non-value adding activities in transportation management

			Severity		
Ranking	Factor no.	Description	AR	Sig. (2-tailed)	SD
1	2.4.2	Insufficient care	10.07	N/A	2.88
1	2.2.3	Transportation is not taken into consideration	9.20	0.183	2.61
2	2.1.3	Inappropriate packings and supports	8.70	0.0495 N/A	2.49
2	2.4.3	Lack of routine inspection	8.50	0.798	3.51
2	2.1.2	The driver is not aware of the few typical damages that may occur during transportation	8.30	0.498	2.63
2	2.1.1	No skilled attention to the details of supports and frames	8.20	0.471	2.75
2	2.3.3	No advance order and confirmation order	7.57	0.199	3.76
2	2.2.2	No single sourcing supply with long-term contract	7.40	0.142	3.79
3	2.3.2	Not fully prepared for the arrival of the precast concrete products	7.20	0.008 N/A	3.04
3	2.2.4	No quality audits of the precasters prior to the award of the contract	6.97	0.800	4.20
3	2.1.4	No standing instructions	6.43	0.220	2.36
4	2.3.1	Demand fluctuations	5.53	0.041 N/A	3.19
4	2.4.1	No accurate delivery notes	5.03	0.336	2.53
5	2.3.4	Insufficient data exchange with the precasters	4.37	0.021 N/A	2.30
5	2.2.1	Large quantity supply base	3.53	0.077	1.72

Notes: *AR* average rating, *SD* standard deviation

9.4.5 Specific Analysis

Examined by the criterion that LR equals 5 and SR equals 1, two non-value adding activities were chosen for specific analysis. These two factors were:

- 2.2.4 No quality audits of the precasters prior to the award of the contract

Twenty-five contractors (83.33 %) stated that quality audits of the precasters were conducted. This was conducted to relate more to the reliability of the precasters rather than the quality of the precast concrete products, which could be observed from many certification programmes. Five contractors (16.67 %) stated that the quality audits were not conducted because by examining their previous working experiences with the precasters, the quality and reliability of the precasters were found to be satisfactory. One contractor (3.33 %) operated his own precast concrete factory and no quality audit was deemed necessary.

- 2.3.3 No advance order and confirmation order

Seven contractors (23.33 %) admitted that the two-order system was used when purchasing the precast concrete products (when P = 1, 2 and 3). Most contractors placed one order several days before the erection date. By using both the advance order and the confirmation order, contractors would help the precasters to manage changes, especially when the JIT delivery was scheduled. Both the contractors and the precasters were less vulnerable to changes under the two-order system, although it seemed that the precasters might benefit more by doing so.

9.5 Lean Stock Management in the Construction Sites

In this section, the stock management practices at the construction sites were examined by the lean production philosophy. The non-value adding activities in this section were derived from a literature review (Ohno 1988; Low and Mok 1999; Alwi et al. 2002).

9.5.1 Descriptive Analysis

A few general questions related to the current stock management practices were discussed with the contractors. These general questions were shown in Table 9.13.

The implications from Table 9.13 are explained as follows:

1. All contractors provided working instructions to the employees who were conducting loading and unloading activities. Twenty-two contractors (73.33 %) adopted on-the-job training by providing working instructions orally. On the other hand, eight contractors (26.67 %) offered written documents as working instructions. Twenty-five contractors (83.33 %) encountered damaged products caused by inappropriate stacking at the stockyard. The percentage of

Table 9.13 General questions in the section on stock management

Question no.	Description	Positive response	Positive response rate	Category
3.5.1	Have the precast concrete products been damaged due to inappropriate stacking?	25	83.33 %	Inventory
3.5.2	Are working instructions provided to employees who are in charge of loading and unloading activities?	30	100 %	Human resources
3.5.3	Are the working instructions provided by on-the-job trainings or through written documents?	On-the job trainings 22 (73.33 %)	Written documents 8 (26.67 %)	Human resources
3.5.4	Are periodic stock checks conducted?	26	86.67 %	Inventory
3.5.5	Is a computer stock control system adopted for stock management?	0	0 %	Inventory

Table 9.14 Four major categories of non-value adding activities in stock management

Category no.	Description
3.1	Layout and equipment
3.2	Staffing arrangements
3.3	Stacking arrangements
3.4	Stock records

the damages was very small, ranging from 1 to 2 %. Minimal repair was needed when such damages happened.

2. Periodic stock checks were conducted by 26 contractors (86.67), mostly on a daily basis. However, four contractors (13.33 %) believed that such stock checks were unnecessary given that the precast concrete products had been checked at the precast concrete factories before deliveries and would be used in the following 1 or 2 days. However, it was recommended that the stock checks be conducted by all contractors in case damages that cannot be repaired happened to the precast concrete products. This was a proactive measure to achieve a smooth work flow. No special computer softwares were used in stock control. Only Windows Excel was used.

9.5.2 *Factors Description*

Four categories of non-value adding activities were identified through a literature review (Ohno 1988; Low and Mok 1999; Alwi et al. 2002). The four categories of non-value adding activities in this context are shown in Table 9.14.

General description for each non-value adding activity is provided as follows:

- *Category 3.1 Layout and equipment*

3.1.1 Lack of well ordered stockyard
3.1.2 Lack of secured stockyard
3.1.3 Inappropriate selection of equipment

To prevent damages to the precast concrete products, a well ordered stockyard was required. Bumps along the delivery path could cause damages, especially in the contact area between the delivery vehicle and the precast concrete products. Loss of products might lead to replacements of the precast concrete products. In addition, inappropriate selection of equipment might also damage the products.

- *Category 3.2 Staffing arrangements*

3.2.1 Inappropriate staffing arrangements

In the construction sites, an integrated site team which included a crane driver, a banksman and one charge hand should be provided to make sure that the loading and unloading activities were correctly and efficiently conducted.

- *Category 3.3 Stacking arrangements*

 3.3.1 Unclear identification marks
 3.3.2 Inappropriate battens
 3.3.3 Unclear working instructions

Damages due to inappropriate stacking happened in the construction sites of most contractors. Battens should be inserted correctly to prevent such damages. In addition, unclear identification marks and work instructions might lead to double-handling, which was also a category of waste examined by the lean concept.

- *Category 4.4 Stock records*

 4.4.1 Lack of periodic stock checks
 4.4.2 Lack of computerized stock control

The main purpose of stock records was to identify the qualities and quantity of the precast concrete products at the stockyard in the construction sites. Based on the quality and quantities, subsequent deliveries could be arranged. However, if the JIT delivery system was used and there was no inventory in the construction sites, no periodic stock checks would be necessary.

9.5.3 Ranking Procedure

The probability, impact and severity of the non-value adding activities in stock management in the construction sites are listed in Table 9.15. A few implications can be drawn from Table 9.15, including:

1. The most frequent non-value adding activities that occurred in stock management were:

   - 3.4.2 Lack of computerized stock control (3.40)
   - 3.1.1 Lack of well ordered stockyard (2.70)
   - 3.4.1 Lack of periodic stock checks (2.63)

   It seems that the contractors were satisfied with using Windows Excel to conduct stock control. Some contractors only estimated the quantity of the precast concrete products and no records were kept. Although a well ordered stockyard could support a smooth workflow, the stockyard was not always well maintained, especially during heavy rain.

   In addition, the stock checks were not always conducted by the contractors as noted earlier. The precast concrete products had been checked at the precast concrete factories before deliveries and would be used in the following 1 or 2 days, which rendered the stock checks unnecessary. The inappropriate use of battens happened to some of the contractors. Examples in this context were given in the case study in Chap. 8. The probability chart is shown in Fig. 9.9.

Table 9.15 Probability, impact and severity of the non-value adding activities in stock management

Factor no.	Description	Probability (P)				Impact (I)				Severity (S)			
		AR	LR	SR	SD	AR	LR	SR	SD	AR	LR	SR	SD
3.1.1	Lack of well ordered stockyard	2.70	3	1	0.60	3.10	4	2	0.40	8.40	12	2	2.04
3.1.2	Lack of secured stockyard	2.00	3	1	0.95	3.23	4	2	0.86	7.13	12	2	4.58
3.1.3	Inappropriate selection of equipment	1.60	3	1	0.56	3.73	4	3	0.45	6.10	12	3	2.51
3.2.1	Inappropriate staffing arrangements	1.87	3	1	0.57	3.83	5	3	0.46	7.30	15	3	2.77
3.3.1	Unclear identification marks	2.20	3	1	0.48	3.10	4	2	0.40	6.90	12	3	2.17
3.3.2	Inappropriate battens	2.40	3	1	0.56	3.60	4	3	0.50	8.53	12	3	1.83
3.3.3	Unclear working instructions	1.97	3	1	0.41	3.83	4	3	0.38	7.53	12	3	1.66
3.4.1	Lack of periodic stock checks	2.63	4	1	0.96	2.23	4	2	0.57	5.70	12	2	2.17
3.4.2	Lack of computerized stock control	3.40	4	1	1.13	1.43	3	1	0.57	4.50	8	1	2.21

Notes: *AR* average rating, *LR* largest rating, *SR* smallest rating, *SD* standard deviation

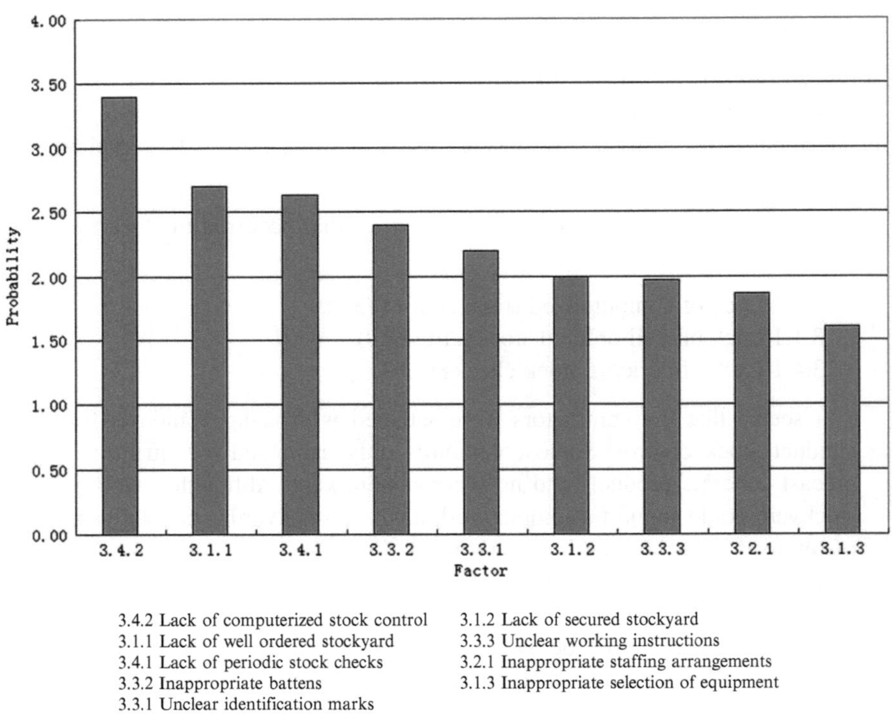

Fig. 9.9 The probability of non-value adding activities in stock management

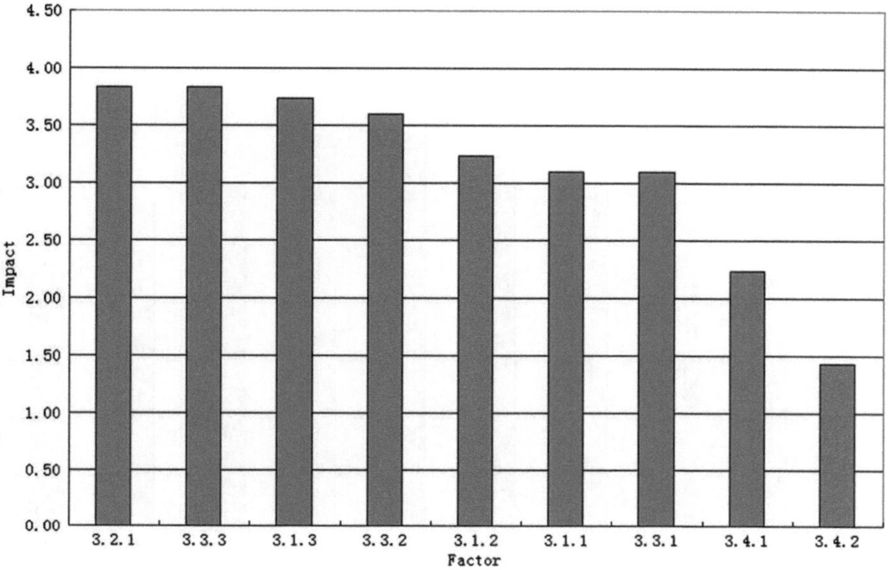

Fig. 9.10 The impact of non-value adding activities in stock management

Factors 3.1.3 (inappropriate selection of equipment) and 3.2.1 (inappropriate staffing arrangements) had very low probability of occurrence. It seemed that the contractors were performing well in selecting equipment and providing appropriate staffing arrangements.

2. The three most important factors with high negative impact on carbon emissions included:

   - 3.2.1 Inappropriate staffing arrangements (3.83)
   - 3.3.3 Unclear working instructions (3.83)
   - 3.1.3 Inappropriate selection of equipment (3.73)

   It seems that the contractors agreed that facilities managers and operators should have the greatest potential to reduce carbon emissions. This was because in the construction sites, it was the equipment which consumed energy and generated carbon emissions. On the other hand, computerized stock control (factor 3.4.2) and periodic stock checks (factor 3.4.1) were not very important as rated by the contractors. The impact of the non-value adding activities in stock management is shown in Fig. 9.10.

3. As shown in Fig. 9.11, the three most severe factors in stock management were:

   - 3.3.2 Inappropriate battens (8.53)
   - 3.1.1 Lack of well ordered stockyard (8.40)
   - 3.3.3 Unclear working instructions (7.53)

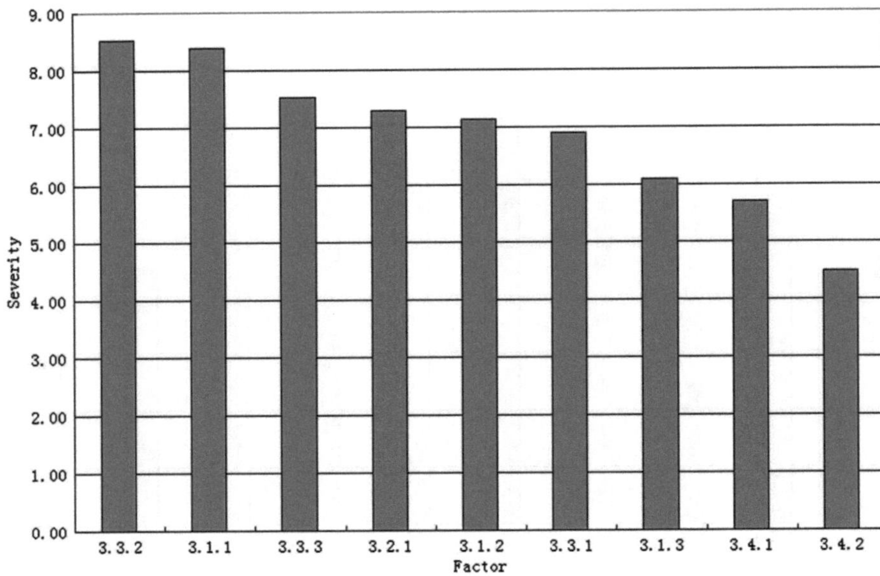

3.3.2 Inappropriate battens
3.1.1 Lack of well ordered stockyard
3.3.3 Unclear working instructions
3.2.1 Inappropriate staffing arrangements
3.1.2 Lack of secured stockyard
3.3.1 Unclear identification marks
3.1.3 Inappropriate selection of equipment
3.4.1 Lack of periodic stock checks
3.4.2 Lack of computerized stock control

Fig. 9.11 The severity of non-value adding activities in stock management

Inappropriate battens inserted and lack of well ordered stockyard could lead to damages to the precast concrete products. Figures and charts relating to the stacking requirements should be provided to the employees who conducted the loading and unloading activities. The working instructions could be provided in terms of both on-the-job trainings and written instructions. Although on-the-job trainings were adopted for most contractors, it should be noted that written instructions should be provided as well so that a defined procedure could be referred to. Computerized stock control (factor 3.4.2) and periodic stock checks (factor 3.4.1) had the lowest severity due to their relatively low impact on the level of carbon emissions.

Similarly, these non-value adding activities were plotted in a P-I table to identify the category which each factor belonged to. The P-I table is shown in Fig. 9.12. As can be seen from Fig. 9.12, most non-value adding activities fell into the category of rare catastrophe due to their low probability of occurrence and high impact. Special attention should be paid to activities that are near the boundary of each category. Mixed strategies should be taken when eliminating such activities. Detailed mitigations strategies would be explained in Sect. 9.7.

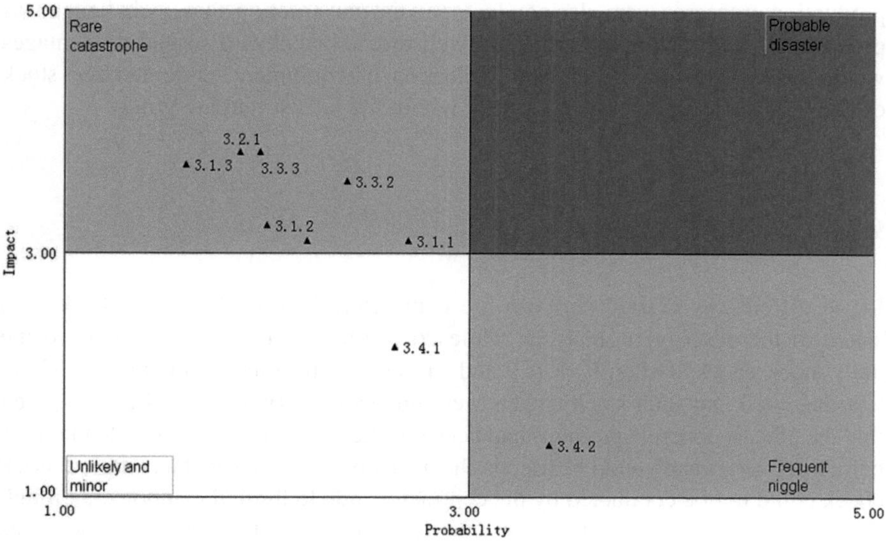

Fig. 9.12 P-I table for non-value adding activities in stock management

Table 9.16 Ranking and grouping of non-value adding activities in stock management

			Severity		
Ranking	Factor no.	Description	AR	Sig. (2-tailed)	SD
1	3.3.2	Inappropriate battens	8.53	N/A	1.83
1	3.1.1	Lack of well ordered stockyard	8.40	0.797	2.04
2	3.3.3	Unclear working instructions	7.53	0.030 N/A	1.66
2	3.2.1	Inappropriate staffing arrangements	7.30	0.719	2.77
2	3.1.2	Lack of secured stockyard	7.13	0.629	4.58
2	3.3.1	Unclear identification marks	6.90	0.242	2.17
3	3.1.3	Inappropriate selection of equipment	6.10	0.010 N/A	2.51
3	3.4.1	Lack of periodic stock checks	5.70	0.360	2.17
4	3.4.2	Lack of computerized stock control	4.50	0.038	2.21

Notes: *AR* average rating, *SD* standard deviation

9.5.4 Parametric Tests

Four ranking groups were identified by the paired sample t-tests, as shown in Table 9.16. Factors which were in ranking group 1 included:

- 3.3.2 Inappropriate battens
- 3.1.1 Lack of well ordered stockyard

The maintenance of the stockyard was especially important when using a combination of both on-site fabrications with off-site fabrications. As the production processes were also conducted in the construction sites, a challenge was posed to the contractors to organize a well ordered stockyard so that no damages would happen. On the other hand, selection of equipment, computerized stock control and periodic stock checks were within the lowest ranking group.

9.5.5 Specific Analysis

Large differences existed between the contractors in periodic stock checks. As stated in the descriptive analysis, while 26 contractors (86.67 %) conducted the daily stock checks (when P = 1, 2 and 3), there were four contractors (13.33 %) who believed that such stock checks were unnecessary (when P = 4 and 5) given that the precast concrete products had been checked at the precast concrete factories before deliveries and would be used in the next 1 or 2 days. Even if a complete stock check could not be conducted by the contractors due to limited manpower, a rough check relating to the quantities and quality of the precast concrete products was recommended.

9.6 Lean Erection Management in the Construction Sites

Erection management was the last value stage in the value chain of the precast concrete products after production. The precast concrete products were singled out from the storage area, lifted to the respective floor and installed at the right place. Tower cranes and crawler cranes might be used for erection activities in the construction sites. Due to the involvement of such equipment, erection efficiency was important to achieve low-carbon erection.

9.6.1 Descriptive Analysis

A few general questions related to the current erection management practices were discussed with the contractors. These general questions covered many areas, including erection methods, regulations, equipment and human resources. These questions are elaborated in Table 9.17. A few implications could be inferred from Table 9.17, including:

1. All contractors provided erection method statements (or erection manual) to guide erection activities. The erection method statement was published as written documents and sent to the consultants for approval. It seemed that by sending the erection method statement to the consultants for approval, the erection efficiency could be assured. However, it should be noted that if such

Table 9.17 General questions in the section of erection management

Question no.	Description	Positive response	Positive response rate	Category
4.6.1	Is a detailed erection method statement provided by the contractor?	30	100 %	Erection method
4.6.2	Is the erection method statement published as written documents?	30	100 %	Erection method
4.6.3	Is the erection method designed by professionals who are familiar with erection activities?	30	100 %	Erection method
4.6.4	Does the contractor emphasize on erection accuracy?	27	100 %	Erection method
4.6.5	During the pre-erection stage, are there laws and regulations that the contractor must abide with?	30	100 %	Regulations
4.6.6	Has it happened before that due to a breach of law and regulations, the erection activities were ceased?	5	16.67 %	Regulations
4.6.7	Are there standard crane operating manuals which the operator can refer to?	30	100 %	Human resources
4.6.8	Does the maintenance schedule of the equipment match the rate of use?	30	100 %	Equipment
4.6.9	How did the contractor use the precast concrete products (deliver-store-use or deliver-use)?	Deliver-store-use 30	Deliver-use 0	Erection method

erection method statement was not examined by the lean professionals, a few non-value adding activities could not be identified, e.g. the singling out activities, transferring activities and storage. No contractors were using the precast concrete products based on a "deliver-use" method. However, one of the contractors stated that a combination of both "deliver-store-use" (around 70 % of the time) and "deliver-use" (around 30 % of the time) was adopted. This validated the applicability of the JIT delivery system in construction projects which used precast concrete products.

2. During erection, there were regulations that the contractors must abide with, mainly those relating to health and safety regulations, e.g. Workplace Safety and Health Act 2006. Five contractors (16.67 %) stated that their erection activities had been ceased because of a failure to meet such regulations. Stopping erection activities caused interruptions to the delivery schedule and more importantly the production schedule of the precasters.
3. The equipment (e.g. tower cranes, crawler cranes, forklifts) were well maintained according to the contractors. The maintenance schedule matched the rate of use. For some contractors, the maintenance schedule was placed on the notice board for information.

Table 9.18 Five major categories of non-value adding activities in erection management

Category no.	Description
4.1	Erection method statement
4.2	Pre-erection preparation
4.3	Erection operations
4.4	Maintenance of equipment and plants
4.5	Legal aspects

The maintenance schedule included: item number, year manufactured, date of the inspection, expiry date of the inspection, date of maintenance, date to schedule for servicing and operating status. Most crane drivers were hired from the crane leasing company (the costs of the crane driver were included in the rental price of the crane), while some crane drivers were provided by the contractor. In both cases, the crane drivers needed to obtain an operating licence from the Building and Construction Authority by attending training programmes.

9.6.2 Factors Description

Five categories of non-value adding activities were identified in erection management, as shown in Table 9.18. General description for each non-value adding activity is provided as follows:

- *Category 4.1 Erection method statement*

  4.1.1 No erection method statement
  4.1.2 The erection method statement is not recorded and modified with the erection process
  4.1.3 The erection method statement is not disseminated to the construction site
  4.1.4 No illustrations with sketches and diagrams
  4.1.5 Lack of feedback system about the erection method statement

The role of the erection method statement was to ensure that appropriate control was exercised in the construction sites. Descriptions of the erection method and the equipment involved in the erection process should be provided in the statement. Other information such as the sequence of the erection and available crane time should also be provided. In addition, it was recommended that the erection method statement be recorded and modified alongside with the erection process as the erection environment changed. This should also be disseminated to every employee involved in the erection process and based on the feedback from the employees, the erection method statement could be revised.

- *Category 4.2 Pre-erection preparation*

  4.2.1 No pre-erection surveys
  4.2.2 Lack of preparatory work in the construction sites
  4.2.3 Cease of erection by laws and regulations before installation
  4.2.4 Inappropriate crane selection

It is important that the appropriate crane was selected to carry out the erection activities, which could be achieved by conducting an in-depth discussion with the manager of the crane hire company. Pre-erection surveys (e.g. site conditions, straightness and verticality of the precast concrete products, etc.) and preparatory works (e.g. provision of holdings) should be completed before the erection started. This was to ensure that accurate erections could be conducted without further alterations.

- *Category 4.3 Erection operations*

4.3.1 Inappropriate siting of tower crane
4.3.2 Inappropriate crane operations
4.3.3 No special attention paid to joints and connections
4.3.4 Inexperienced employees at the construction sites
4.3.5 No process to learn from experienced employees
4.3.6 Inadequate supervision during erection

The siting of tower crane should be determined in the erection method statement. An appropriately trained operator should be hired to carry out the erection activities and a banksman should be involved to supervise such activities. Besides the erection accuracy, timing was one important factor which the supervisors should consider if the JIT installation method was adopted.

- *Category 4.4 Maintenance of the plants and equipment*

4.4.1 Lack of regular planned maintenance of the plants and equipment
4.4.2 The maintenance schedule does not match the rate of use
4.4.3 Lack of inspections and tests conducted at the frequency required

As observed in the construction sites, most tower cranes (sometimes crawler cranes) were switched on at 7 am and switched off at 9 pm. The maintenance schedule should match the rate of use, which was not always normal. Inspections and tests should be conducted regularly in case the equipment broke down and caused delays.

- *Category 4.5 Legal aspects*

4.5.1 Cease of erection due to a breach of law and regulations during installation

The relevant laws and regulations should be complied with during erection. These laws and regulations included: Building Control Act 1989, Clean Air Act 1990, Environmental Public Health Act 1987, Water Pollution Control and Drainage Act 1975 and so on. One contractor had to stop the construction activities because of a breach of the Environmental Public Health Act 1987 (Control of Noise from Construction Sites). The stoppage would cause interruptions to the erection and more importantly bring uncertainties to the precasters.

9.6.3 Ranking Procedure

The probability, impact and severity of non-value adding activities in erection management can be seen in Table 9.19. Based on Table 9.19, a few implications can be inferred. These implications were:

1. The non-value adding activities with high probability of occurrence in erection management included:

   - 4.3.5 No process to learn from experienced employees (3.10)
   - 4.3.4 Inexperienced employees at the construction sites (2.57)
   - 4.3.3 No special attention paid to joints and connections (2.53)

   Although internal meeting were often organized by the contractors to discuss issues relating to further improvements, these internal meetings lacked the presence of the experienced employees. Most internal meetings were organized every week. In addition, all contractors admitted that the competence of the employees could be improved.

   If such improvements were obtained, the damages to the precast concrete products could be reduced. Not enough attention was paid to the joints and connections, leading to damages to the precast concrete products. On the other hand, the non-value adding activities with low probability of occurrence included:

   - 4.1.1 No erection method statement (1.13)
   - 4.5.1 Cease of erection due to a breach of law and regulations during installation (1.23)
   - 4.4.1 Lack of regular planned maintenance of the plants and equipment (1.30)

   Erection method statement was always provided by the contractors. Although stoppage of erection happened to several contractors due to a breach of the law and regulations, the probability was extremely low. The equipment used at the construction sites were carefully maintained. The maintenance schedule matched the rate of use. The probability chart is shown in Fig. 9.13.
2. Factors with high impact on carbon emission levels included:

   - 4.3.2 Inappropriate crane operations (4.00)
   - 4.3.1 Inappropriate siting of tower crane (3.83)
   - 4.1.5 Lack of feedback system about the erection method statement (3.77)

   The importance of factors 4.3.2 and 4.3.1 validated the role of the tower crane in reducing carbon emissions in the construction site. The tower crane should be sited with large coverage area. Crane operations should be conducted carefully by the operator because this had the largest impact on the level of carbon emissions. According to some contractors, the erection method statement was the procedure that the erection activities should follow. It should be continually improved as the project progressed and when the project was completed.

Table 9.19 Probability, impact and severity of the non-value adding activities in erection management

Factor		Probability (P)				Impact (I)				Severity (S)			
no.	Description	AR	LR	SR	SD	AR	LR	SR	SD	AR	LR	SR	SD
4.1.1	No erection method statement	1.13	2	1	0.35	3.70	5	3	0.79	4.37	10	3	2.33
4.1.2	The erection method statement is not recorded and modified with the erection process	1.90	3	1	0.61	3.40	4	3	0.50	5.97	8	3	1.61
4.1.3	The erection method statement is not disseminated to the construction site	1.70	2	1	0.47	3.00	4	2	0.53	5.10	8	2	1.56
4.1.4	No illustrations with sketches and diagrams	1.50	3	1	0.73	3.03	4	2	0.56	4.50	9	2	2.29
4.1.5	Lack of feedback system about the erection method statement	1.90	3	1	0.61	3.77	4	2	0.50	7.20	12	3	2.64
4.2.1	No pre-erection surveys	1.70	2	1	0.47	2.97	4	2	0.49	5.03	8	2	1.61
4.2.2	Lack of preparatory work in the construction sites	1.80	2	1	0.41	2.97	4	2	0.56	5.33	8	2	1.56
4.2.3	Cease of erection by laws and regulations before installation	1.73	3	1	0.83	3.37	5	2	0.89	6.10	12	2	3.77
4.2.4	Inappropriate crane selection	1.60	2	1	0.50	3.73	4	3	0.45	6.03	8	3	2.14
4.3.1	Inappropriate siting of tower crane	1.73	2	1	0.45	3.83	5	3	0.46	6.67	10	3	1.97
4.3.2	Inappropriate crane operations	2.40	4	2	0.56	4.00	5	3	0.53	9.53	16	6	2.26
4.3.3	No special attention paid to joints and connections	2.53	4	2	0.57	2.90	4	2	0.61	7.47	12	4	2.54
4.3.4	Inexperienced employees at the construction sites	2.57	3	2	0.50	3.03	4	2	0.49	7.83	12	4	2.09
4.3.5	No process to learn from experienced employees	3.10	4	2	0.76	3.03	4	2	0.49	9.40	16	4	2.66
4.3.6	Inadequate supervision during erection	2.30	3	1	0.65	3.17	4	2	0.59	7.17	9	3	2.00

(continued)

Table 9.19 (continued)

Factor no.	Description	Probability (P)				Impact (I)				Severity (S)			
		AR	LR	SR	SD	AR	LR	SR	SD	AR	LR	SR	SD
4.4.1	Lack of regular planned maintenance of the plants and equipment	1.30	3	1	0.53	3.67	5	3	0.66	4.73	12	3	2.07
4.4.2	The maintenance schedule does not match the rate of use	1.63	3	1	0.61	3.13	4	2	0.43	5.10	9	3	2.01
4.4.3	Lack of inspections and tests conducted at the frequency required	2.23	4	1	1.04	2.83	4	2	0.59	6.07	12	2	2.45
4.5.1	Cease of erection due to a breach of law and regulations during installation	1.23	2	1	0.43	3.23	4	2	0.73	4.13	8	2	2.13

Notes: *AR* average rating, *LR* largest rating, *SR* smallest rating, *SD* standard deviation

On the other hand, non-value adding activities with relatively low impact on the level of carbon emissions included:

- 4.4.3 Lack of inspections and tests conducted at the frequency required (2.83)
- 4.3.3 No special attention paid to joints and connections (2.90)

However, it should be noted that these factors had moderate impact on the level of carbon emissions ad indicated using the five-point scale, even though they had relatively low impact than the other factors described above. The impact chart is shown in Fig. 9.14.

3. As shown in Fig. 9.15, the non-value adding activities with high severity included:

- 4.3.2 Inappropriate crane operations (9.53)
- 4.3.5 No process to learn from experienced employees (9.40)
- 4.3.4 Inexperienced employees in the construction sites (7.83)

These factors validated the importance of effective crane operations, competent employees and continuous improvement to reduce carbon emissions in the construction sites. On the other hand, failures to comply with regulations during erection (factor 4.5.1) and lack of erection method statement (factor 4.1.1) should not be unduly focused on, because the contractors appeared to be performing well in these areas.

As can be seen from Fig. 9.16, most non-value adding activities fall into the category of rare catastrophe, or the intersection area of rare catastrophe and unlikely and minor, which seems reasonable based on the high impact of these non-value

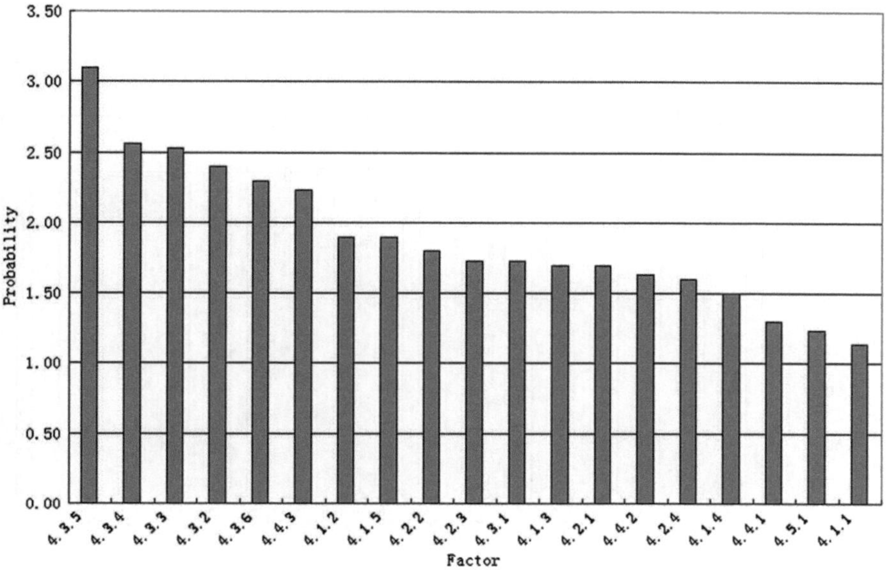

4.3.5 No process to learn from experienced employees
4.3.4 Inexperienced employees at the construction sites
4.3.3 No special attention paid to joints and connections
4.3.2 Inappropriate crane operations
4.3.6 Inadequate supervision during erection
4.4.3 Lack of inspections and tests conducted at the frequency required
4.1.2 The erection method statement is not recorded and modified with the erection process
4.1.5 Lack of feedback system about the erection method statement
4.2.2 Lack of preparatory work in the
4.2.3 Cease of erection by laws and regulations before installation
4.3.1 Inappropriate siting of tower crane
4.1.3 The erection method statement is not disseminated to the construction site
4.2.1 No pre-erection surveys
4.4.2 The maintenance schedule does not match the rate of use
4.2.4 Inappropriate crane selection
4.1.4 No illustrations with sketches and diagrams
4.4.1 Lack of regular planned maintenance of the plants and equipment
4.5.1 Cease of erection due to a breach of law and regulations during installation
4.1.1 No erection method statement

Fig. 9.13 The probability of the non-value adding activities in erection management

adding activities. These activities had relatively low probability of occurrence and high impact. Attention should be paid to factors near the boundaries. Mixed strategies should be taken when eliminating such activities. The mitigation strategies will be discussed in Sect. 9.7.

9.6.4 Parametric Tests

Four ranking groups were identified by conducting the paired sample t-tests for non-value adding activities in erection management. As can be seen from Table 9.20, the most severe non-value adding activities that should be focused on included:

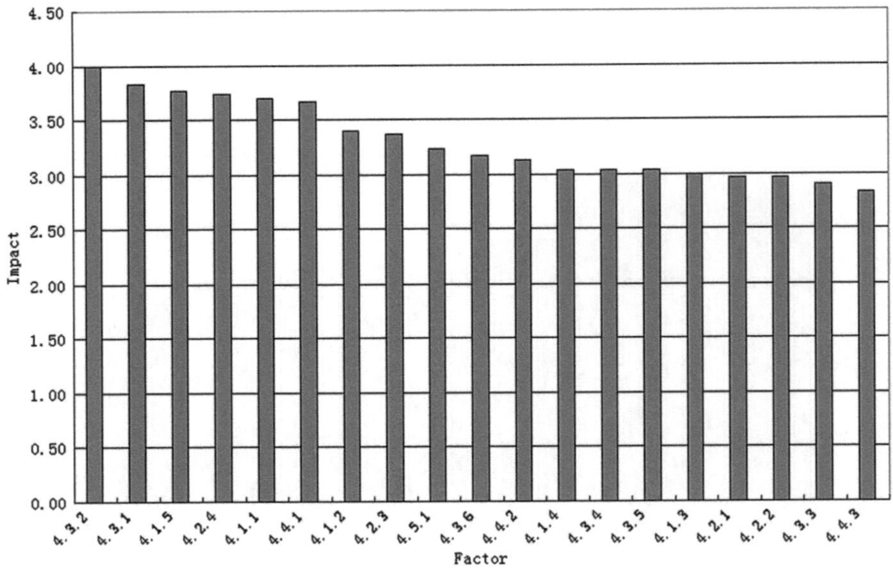

4.3.2 Inappropriate crane operations
4.3.1 Inappropriate siting of tower crane
4.1.5 Lack of feedback system about the erection method statement
4.2.4 Inappropriate crane selection
4.1.1 No erection method statement
4.4.1 Lack of regular planned maintenance of the plants and equipment
4.1.2 The erection method statement is not recorded and modified with the erection process
4.2.3 Cease of erection by laws and regulations before installation
4.5.1 Cease of erection due to a breach of law and regulations during installation
4.3.6 Inadequate supervision during erection
4.4.2 The maintenance schedule does not match the rate of use
4.1.4 No illustrations with sketches and diagrams
4.3.4 Inexperienced employees at the construction sites
4.3.5 No process to learn from experienced employees
4.1.3 The erection method statement is not disseminated to the construction site
4.2.1 No pre-erection surveys
4.2.2 Lack of preparatory work in the construction sites
4.3.3 No special attention paid to joints and connections
4.4.3 Lack of inspections and tests conducted at

Fig. 9.14 The impact of non-value adding activities in erection management

- 4.3.2 Inappropriate crane operations
- 4.3.5 No process to learn from experienced employees

It seemed that the contractors should focus on the crane operators to ensure a smooth work flow in order to reduce carbon emissions. A continuous improvement process should be developed to learn from the experienced employees.

As stated previously, facility managers and operators should have the greatest potential to reduce carbon emissions. This was because the equipment generated carbon emissions in the construction site. The facility managers and operators should be involved in the continuous improvement plans.

The areas where the contractors were performing well included:

- Comply with laws and regulations during installation (factor 4.5.1)
- Provide the erection method statement (factor 4.1.1)

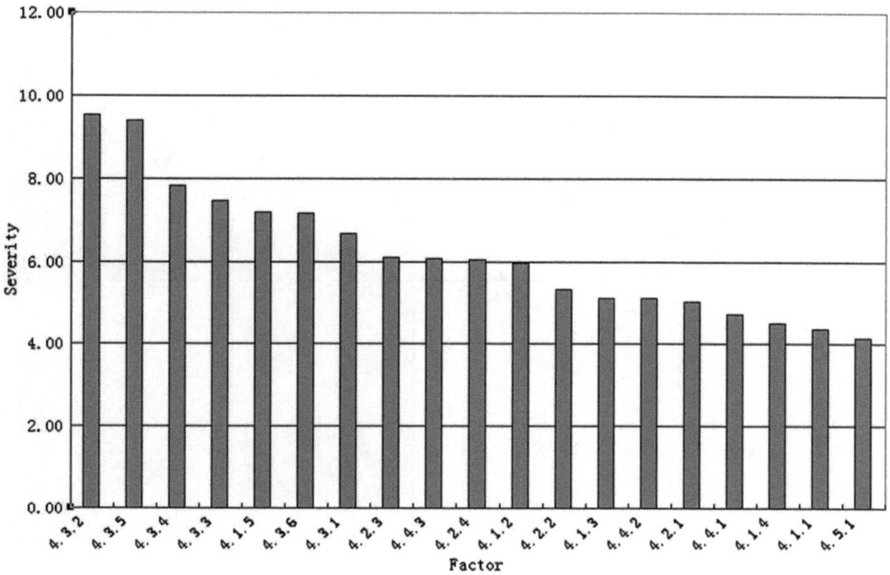

4.3.2 Inappropriate crane operations
4.3.5 No process to learn from experienced employees
4.3.4 Inexperienced employees at the construction sites
4.3.3 No special attention paid to joints and connections
4.1.5 Lack of feedback system about the erection method statement
4.3.6 Inadequate supervision during erection
4.3.1 Inappropriate siting of tower crane
4.2.3 Cease of erection by laws and regulations before installation
4.4.3 Lack of inspections and tests conducted at the frequency required
4.1.2 The erection method statement is not recorded and modified with the erection process
4.2.2 Lack of preparatory work in the construction sites
4.1.3 The erection method statement is not disseminated to the construction site
4.4.2 The maintenance schedule does not match the rate of use
4.2.1 No pre-erection surveys
4.4.1 Lack of regular planned maintenance of the plants and equipment
4.1.4 No illustrations with sketches and diagrams
4.1.1 No erection method statement
4.5.1 Cease of erection due to a breach of law and regulations during installation

Fig. 9.15 The severity of non-value adding activities in erection management

- Provide illustrations with sketches and diagrams to the erection method statement (factor 4.1.4)
- Regularly maintain the equipment (4.4.1)

These factors belonged to the ranking group 4 with lower level of severity. Due to the relatively low severity, the contractors should not overly address these problems but should instead allocate reasonable resources (e.g. time, human resources, costs) to address the problems.

9.6.5 *Specific Analysis*

No specific analysis was required here, following the screening principle (when LR = 5 and SR = 1).

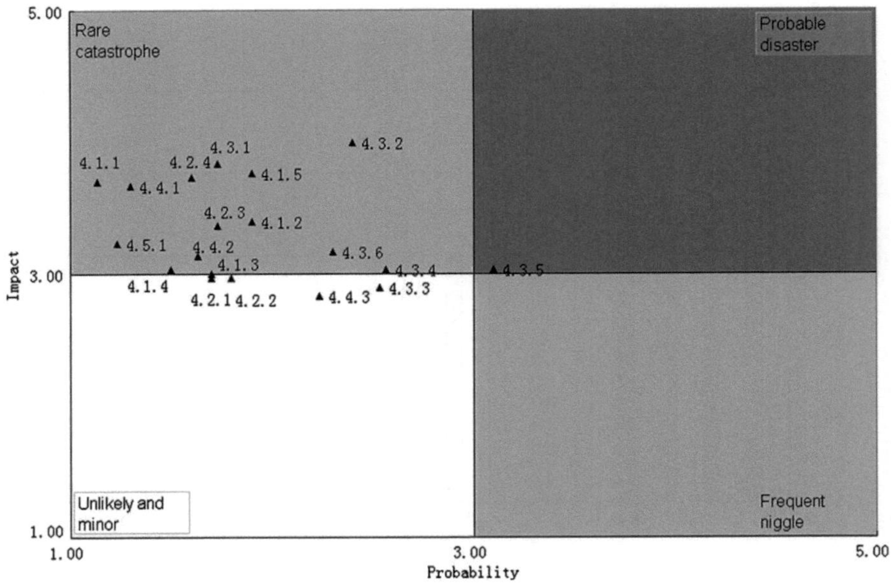

Fig. 9.16 P-I table for non-value adding activities in erection management

9.7 Mitigation Strategies and Actions for Contractors

In this section, mitigation strategies and mitigation actions were developed for the contractors to manage the non-value adding activities examined previously. It should be noted that mitigation strategies define the type of mitigation efforts while mitigation actions form the detailed mitigation plans. The type of the mitigation efforts (i.e. accept, transfer, control, avoid or mixed strategies) should be identified before mitigation plans were developed.

9.7.1 General Procedure to Develop Mitigation Actions

The development of mitigation strategies for the contractors to eliminate non-value adding activities followed the standard procedure, which was described in Sect. 9.7.1. These steps were:

- **Step 1**: Identifying categories. In this step, the activity was plotted into the P-I table to obtain the category it belonged to. Four categories were identified in the P-I table, which are unlikely and minor, rare catastrophe, frequent niggle and probable disaster.
- **Step 2**: Identifying mitigation strategies. Detailed mitigation strategies (i.e. accept, transfer, control, avoid or mixed strategies) were developed in this stage.

Table 9.20 Ranking and grouping of non-value adding activities in erection management

			Severity		
Ranking	Factor no.	Description	AR	Sig. (2-tailed)	SD
1	4.3.2	Inappropriate crane operations	9.53	N/A	2.26
1	4.3.5	No process to learn from experienced employees	9.40	.0825	2.66
2	4.3.4	Inexperienced employees at the construction sites	7.83	0.013	2.09
				N/A	
2	4.3.3	No special attention paid to joints and connections	7.47	0.245	2.54
2	4.1.5	Lack of feedback system about the erection method statement	7.20	0.334	2.64
2	4.3.6	Inadequate supervision during erection	7.17	0.141	2.00
3	4.3.1	Inappropriate siting of tower crane	6.67	0.039	1.97
				N/A	
3	4.2.3	Cease of erection by laws and regulations before installation	6.10	0.393	3.77
3	4.4.3	Lack of inspections and tests conducted at the frequency required	6.07	0.335	2.45
3	4.2.4	Inappropriate crane selection	6.03	0.132	2.14
3	4.1.2	The erection method statement is not recorded and modified with the erection process	5.97	0.070	1.61
4	4.2.2	Lack of preparatory work in the construction sites	5.33	0.012	1.56
				N/A	
4	4.1.3	The erection method statement is not disseminated to the construction site	5.10	0.472	1.56
4	4.4.2	The maintenance schedule does not match the rate of use	5.10	0.647	2.01
4	4.2.1	No pre-erection surveys	5.03	0.231	1.61
4	4.4.1	Lack of regular planned maintenance of the plants and equipment	4.73	0.220	2.07
4	4.1.4	No illustrations with sketches and diagrams	4.50	0.321	2.29
4	4.1.1	No erection method statement	4.37	0.250	2.33
4	4.5.1	Cease of erection due to a breach of law and regulations during installation	4.13	0.067	2.13

Notes: *AR* average rating, *SD* standard deviation

- **Step 3**: Allocating resources. This step was conducted by comparing the relative priority using the paired sample t-tests. Factors within the same category of severity but with higher average rating (AR) should be allocated with more resources, such as more training programmes, more competent supervisors, etc.
- **Step 4**: Defining mitigation actions. In this step, the mitigation strategies identified above were elaborated into detailed mitigation actions. If training programmes were proposed as mitigation strategies, the number of persons who should be trained, as well as the frequency of such training would form the detailed mitigation actions.
- **Step 5**: Evaluating mitigation results. The non-value adding activities eliminated after mitigation actions should be plotted on the P-I table to identify the results. A looped application of the process tree should be monitored on a continuous basis.

9.7.2 Developing the Mitigation Actions for the Contractors

Four main mitigation strategies have been identified in Sect. 9.7.1, which are accept, control, transfer and avoid. Following the process tree, the non-value adding activities in the construction sites were categorized into 7 groups by their relative severity. Mitigation strategies are provided in Table 9.21 as well. A few implications can be obtained from Table 9.21, including:

1. The most severe activities in the construction sites included:

    - 1.1.3 Does not think of green building materials (10.83)
    - 2.4.2 Insufficient care during transportation (10.07)
    - 4.3.2 Inappropriate crane operations (9.53)
    - 4.3.5 No process to learn from experienced employees (9.40)

    Green building materials played a very important role to achieve low carbon construction, even though the current usage of green building materials was limited. The contractors agreed that by providing sufficient care during transportation, the damages (3.17 %) to the precast concrete products could be reduced. Inappropriate crane operations contributed to increasing the level of carbon emissions. For example, due to lack of communications, the precast concrete products were held in mid-air for several minutes. This was very common in the construction sites. Internal periodic meeting should include facilities managers and operators. These personnel had the greatest potential to reduce carbon emissions, which were usually generated by the use of equipment.
2. Non-value adding activities in ranking group 2 to group 5 had intermediate impact on the carbon emission levels in the construction sites. The mitigation actions varied among these activities, depending on the nature of the activity. Most actions fell into the category of "transfer" and "control" and "mixed strategies". As stated above, "control" mainly applied to high-probability and low-impact activities by improving the overall competency of employees. On the other hand, "transfer" applies to high-impact but low-probability activities by improving the competence of the supervisors or the specific persons in charge. "Mixed strategies" applied to the non-value adding activities that required both the inputs of employees and supervisors. For example, employees should be trained on the general procedures to provide appropriate packings and supports to the precast concrete products. After such packings and supports were provided, the supervisors should conduct a quality check (factor 2.1.3).
3. Non-value adding activities in ranking groups 6 and 7 have little impact on the carbon emission levels in the construction sites. "Accept" actions could be taken because it would not be efficient to focus on these activities too much. Similarly, attention should be paid to factors in the overlapping area. However, there were three exceptions that should be noted, which were:

    - 4.5.1 Cease of erection due to a breach of law and regulations during installation

Table 9.21 Ranking, grouping and mitigation actions for non-value adding activities in the construction sites

		Severity						Severity			
R	Factor	AR	Sig. (2-tailed)	SD	MA	R	Factor	AR	Sig. (2-tailed)	SD	MA
1	1.1.3	10.83	N/A	4.43	T	2	2.1.3	8.70	0.019 N/A	2.49	MS
1	2.4.2	10.07	0.350	2.88	T	2	3.3.2	8.53	0.783	1.83	MS
1	4.3.2	9.53	0.201	2.26	T	2	2.4.3	8.50	0.798	3.51	T
1	4.3.5	9.40	0.195	2.66	T	2	3.1.1	8.40	0.613	2.04	MS
1	2.2.3	9.20	0.097	2.61	T	2	2.1.2	8.30	0.498	2.63	T
						2	2.1.1	10.38	0.471	4.44	MS
3	3.3.3	7.53	0.044 N/A	1.66	T	2	4.3.4	7.83	0.194	2.09	C
3	4.3.3	7.47	0.913	2.54	T	2	2.3.3	7.57	0.199	3.76	T
3	2.2.2	7.40	0.853	3.79	T						
3	3.2.1	7.30	0.719	2.77	T	4	4.3.1	6.67	0.008 N/A	1.97	T
3	2.3.2	7.20	0.644	3.04	MS	4	2.1.4	6.43	0.678	2.36	T
3	4.1.5	7.20	0.517	2.64	T	4	1.5.3	6.10	0.268	1.56	T
3	4.3.6	7.17	0.425	2.00	T	4	3.1.3	6.10	0.293	2.51	T
3	3.1.2	7.13	0.629	4.58	T	4	4.2.3	6.10	0.393	3.77	T
3	2.2.4	6.97	0.504	4.20	T	4	4.4.3	6.07	0.335	2.45	T
3	1.1.1	6.93	0.350	3.19	T	4	4.2.4	6.03	0.132	2.14	T
3	1.7.1	6.93	0.303	3.15	T	4	4.1.2	5.97	0.070	1.61	T
3	3.3.1	6.90	0.242	2.17	T	4	1.5.4	5.90	0.166	1.86	T
3	1.1.2	6.73	0.165	2.68	T	4	3.4.1	5.70	0.104	2.17	T
						4	1.5.2	5.35	0.061	3.57	T
5	4.2.2	5.33	0.012 N/A	1.56	MS	4	2.3.1	5.53	0.122	3.19	T
5	4.1.3	5.10	0.472	1.56	T						
5	4.4.2	5.10	0.647	2.01	T	6	2.3.4	4.37	0.170	2.30	AC
5	1.7.3	5.03	0.393	2.24	T	6	4.1.1	4.37	0.999	2.33	AC
5	2.4.1	5.03	0.617	2.53	T	6	1.2	4.27	0.865	1.64	AC
5	4.2.1	5.03	0.231	1.61	T	6	1.7.2	4.23	0.745	2.30	AC
5	1.5.1	4.87	0.143	1.94	T	6	4.5.1	4.13	0.607	2.13	MS
5	1.3	4.77	0.277	2.49	T	6	1.4.2	3.60	0.124	1.28	AC
5	4.4.1	4.73	0.220	2.07	MS	6	1.4.5	3.60	0.115	1.40	AC
5	1.7.4	4.67	0.106	2.38	T	6	2.2.1	3.53	0.077	1.72	MS
5	1.4.4	4.60	0.206	2.46	T	6	1.6.3	3.50	0.059	1.28	MS
5	3.4.2	4.50	0.122	2.21	MS						
5	4.1.4	4.50	0.125	2.29	MS						
7	1.4.3	3.37	0.033 N/A	1.47	AC	7	1.6.1	3.17	0.501	1.26	AC
7	1.6.2	3.23	0.684	1.36	AC	7	1.4.1	3.10	0.147	1.18	AC

Notes: *R* ranking group, *AR* average rating, *SD* standard deviation, *MA* mitigation actions, *AC* accept, *T* transfer, *AV* avoid, *C* control, *MS* mixed strategies

Although the severity of this factor to the contractors (on the level of carbon emissions) was small due to its low probability of occurrence, it would cause huge disruptions to the production processes in the precast concrete factories. If the JIT delivery system was adopted between the precasters and the contractors, the contractors should have the responsibility to manage the stability of the demand.

- 2.2.1 Large quantity supply base and 1.6.3 Over provide storage area – open store

According to the contractors, large quantity supply base was not very common for contractors. Quantities required in the following 1 or 2 days would be ordered. The interruptions to other construction activities were minimal due to the small inventory. However, it should be noted that the transferring and singling out activities for each precast concrete product remained even though a small inventory was provided. Although it might be difficult to apply JIT delivery and installation for all precast concrete products, partial applications would help to reduce inventory and carbon emissions.

4. The delivery performance of the precasters in Singapore was satisfactory, as stated by most contractors. According to one contractor, it was not the delivery performance of the precasters which impeded the application of JIT delivery system. It was the contractors' unsatisfactory planning performance which made the JIT delivery impossible, or at best, difficult. If overlapping of construction activities happened, the delivery trailers would be kept waiting outside the entrance until other construction activities were completed.

   According to Waters-Fuller (1996), one of the most important factors that might impede the adoption of JIT sourcing was related to the huge demand fluctuations. However, all contractors were experiencing demand stability. A few JIT sourcing activities could consequently be tested on feasibility, including immediate installation, long-term relationship, the two-order system, etc. Ansari and Modarress (1987) stated that the lack of supplier support was the most significant barrier to success. Government intervention with tax benefits might help to push both the precasters and the suppliers to move towards lean production, which could be supported by the current production practices.

9.8 Summary

The non-value adding activities in the construction sites were described, evaluated and ranked in this chapter. After an analysis, non-value adding activities in each value stage, from site layout management, transportation management, stock management to erection management, were classified into several ranking groups categorized by severity. This ranking procedure was also conducted for all non-value adding activities. Seven ranking groups were identified by the paired sample t-tests. The most important ones were related to the use of green building materials, insufficient care during transportation, inappropriate crane operations and no process to learn from experienced employees. The appropriate actions to reduce these non-value adding activities were also recommended in this section. In addition, many other non-value adding activities with moderate impact on the carbon emission levels were identified in this chapter. It was believed that by reducing the probability of occurrence of such activities, the contractors could achieve low carbon construction. A case study is provided in the following chapter for the contractors to identify the potential improvements that could be obtained by applying the lean production concept.

References

Alwi S, Hampson K, Mohamed S (2002) Waste in the Indonesian construction project. In: Proceedings of the 1st international conference of CIB W107 – creating a sustainable construction industry in developing countries, Spier, South Africa

Ansari A, Modarress B (1987) Potential benefits of JIT purchasing for US manufacturers. Prod Invent Manag J 28(2):30–33

Banerjee A, Kim SL (1995) An integrated JIT inventory model. Int J Oper Prod Manage 15 (9):237–244

Calvert RE (1995) Introduction to building management. Butterworth-Heinemann Ltd., Oxford

Low SP, Mok SH (1999) The application of JIT philosophy to construction: a case study in site layout. Constr Manage Econ 17(5):657–668

Neil JM (1980) Teaching site layout for construction. In: ASCE meeting, Portland, OR, 14–18 April, pp 1–11

Ohno T (1988) Toyota production system: beyond large-scale production. Diamond, Inc., Tokyo

Richardson JG (1973) Precast concrete production. Cement and Concrete Association, London

Stalk GJ, Hout TM (1990) Competing against time. Free Press, New York

Tommelein ID, Li EY (1999) Just-in-time concrete delivery mapping alternatives for vertical supply chain integration. In: Tommelein ID (ed) Proceedings of IGLC-7, Berkeley, CA

Waters-Fuller N (1996) The benefits and costs of JIT sourcing: a study of Scottish suppliers. Int J Phys Distrib Logist Manag 26(4):35–50

Xin N (2010) A construction site layout planning decision-making system for dynamic, unequal-area and multi-objective optimization problem. VDM Verlag Publishing, Saarbrucken

Chapter 10
Case Study of a Construction Site

10.1 Introduction

The applications of the lean production philosophy in precast concrete factories have been discussed in Chaps. 7 and 8. Significant results can be obtained when the lean production philosophy is applied to reduce carbon emissions in precast concrete factories. The non-value adding activities from the lean perspective in the construction sites were identified in Chap. 9. The purpose of this case study is therefore to investigate how the lean production principles can be used in construction sites to identify and eliminate wastes. These wastes are also quantified using carbon emissions as the indicator in this case study.

The life cycle of precast concrete products involves five stages, namely production, delivery, erection, operation and demolition. Since the stage relating to operation is usually very long, the operational stage will be excluded from this study. Similarly, there are a lot of uncertainties when demolition happens in many years later. Demolition stage will not therefore be considered in this study. A normal value stream for precast concrete products in construction sites is illustrated by Fig. 10.1. The four stages listed in Fig. 10.1 are the value stages that will be considered in this study. The carbon emissions that can be reduced by applying the lean production philosophy in each stage will be evaluated.

Three contractors in the Singapore construction industry (which will be referred to as Contractors A1, A2 and A3 in the following context) were chosen for case studies to validate the lean improvements used in reducing carbon emissions. An ideal case study would be a construction project which used the precast concrete column described in Chap. 7. However, due to the time lag between the case study of the precaster and the contractor, such an ideal case study would be difficult to identify. This chapter therefore combined the observations from the three construction projects to generate a typical construction project that used the precast concrete column. The three contractors were chosen after they agreed to participate in the study.

P. Wu and S.P. Low, *Lean and Cleaner Production: Applications in Prefabrication to Reduce Carbon Emissions*, DOI 10.1007/978-3-642-42062-7_10,

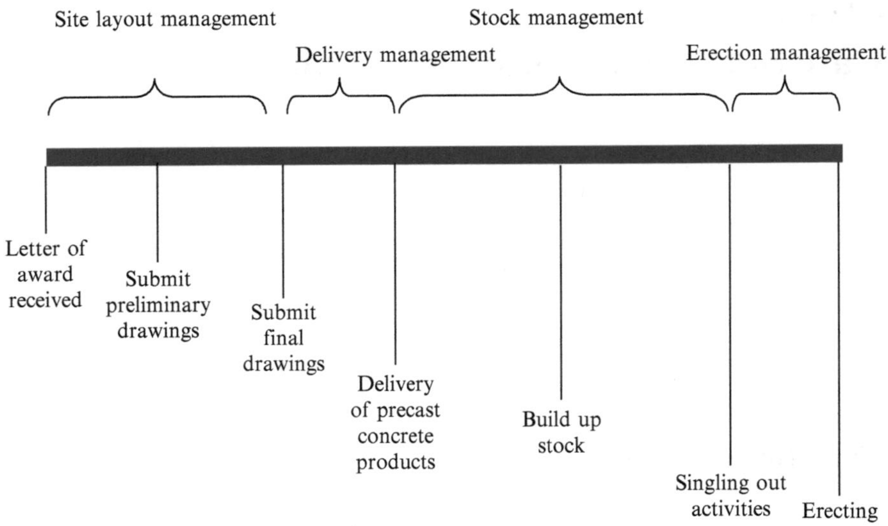

Fig. 10.1 The value stream of precast concrete products in construction sites

10.2 General Procedure to Generate the Case Study

A general procedure to quantify the lean improvements included two steps, which were the procedure to develop the case study and the procedure to calculate the carbon emissions values arising from the lean improvements. It should be noted that due to the time constraint, it was difficult to identify an ideal case study where the precast concrete columns (16HPC1) were used. In addition, for large public housing projects in Singapore, a combination of on-site fabrication and off-site fabrication was often adopted to achieve economic benefits. This would impose challenges to the contractors to organize the delivery well, especially at the main entrance. A total of three contractors were investigated. A general description of the three contractors is provided as follows:

- Contractor A1: Contractor A1 was constructing a small public housing project (only one low-rise block) where only off-site fabrication was used.
- Contractor A2: Contractor A2 was constructing a medium public housing project (four low-rise blocks) where a combination of on-site fabrication with off-site fabrication was used.
- Contractor A3: Contractor A3 was constructing a large private condominium project (three high-rise blocks) where on-site fabrication was used.

10.2.1 Observations from Contractor A1

Contractor A1 was constructing a small public housing project, with only one low-rise block. Due to the small size of the project, only off-site fabrication was

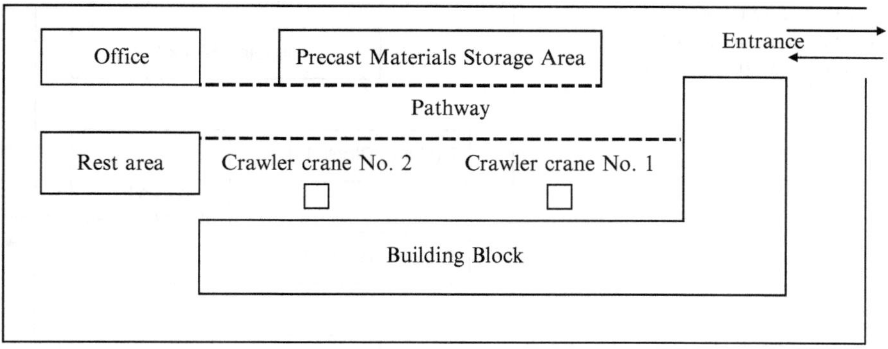

Fig. 10.2 Site layout plan of the project carried out by Contractor A1 (not to scale)

adopted. The precast concrete products were transported from the precast concrete factory to the construction site, unloaded to the storage area and hoisted to the respective floor for installation.

The installation process carried out by the crawler crane was very straightforward. The crawler crane unloaded the delivered precast concrete products to the storage area and loaded the precast concrete products to the floor for installation. The unloading time per precast concrete column was 3–5 min while the loading time per precast concrete column (10th storey) was 10–15 min. The site layout plan of the project is shown in Fig. 10.2.

As can be seen from the site layout plan, the storage area covered a small proportion of the total site area. Only the precast concrete columns that would be used in the following 1 or 2 days were stored. It should be noted that despite the small proportion of the storage area, unloading activities occurred for each precast concrete column. Thirty-two percentage of the crane energy was consumed by the transferring and singling out activities. The contractor needed to provide sufficient care to the precast concrete columns at the storage area to ensure there were no damages caused to the precast concrete columns.

The precaster for this project was located 15 km away from the site. However, it was discovered that there were three precasters (all with high financial grade) located 3 km away from the site. By choosing these precasters, an amount of 16.00 kg CO_2 could be reduced on a per delivery basis (for six precast concrete columns). Two rejections of the precast concrete columns had happened in a 3-month period. Although the probability of the rejections was very minimal, it should be noted that this would cause interruptions to the erection activities and might cause the crane to idle.

The erection activities were carefully conducted. Contractor A1 focused on erection accuracy and ensured that all the accuracy checks were conducted before releasing the hooks. Such performance would prevent double-handling.

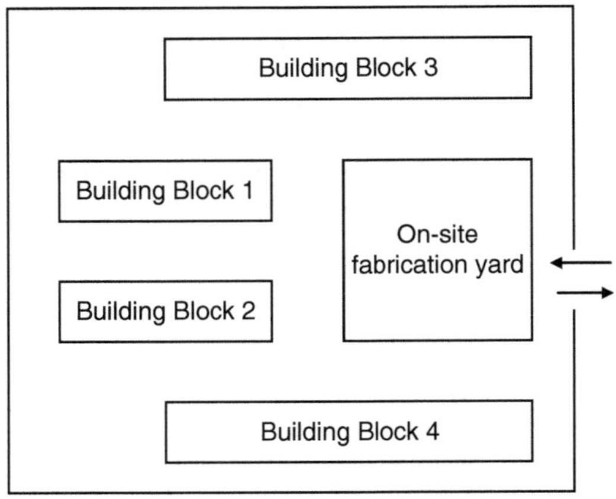

Fig. 10.3 Site layout plan of the project carried out by Contractor A2 (not to scale)

10.2.2 Observations from Contractor A2

The site layout of the project conducted by Contractor A2 is shown in Fig. 10.3. Due to the involvement of both on-site fabrication and off-site fabrication, Contractor A2 faced problems in transportation management.

One problem was the overlapping of the delivery times. Many materials were required by Contractor A2, including concrete, steel, precast concrete products, finishing sands, plastic decorations, etc. Due to the overlapping the delivery times, Contractor A2 did not perform very well in managing a smooth transportation flow, especially for the precast concrete products. Queuing outside of the main entrance was very common. As observed at the main entrance, the transportation vehicle of the precast concrete products had to wait for 5–30 min before the precast concrete products could be unloaded.

More importantly, these equipment were left idling while queuing. The drivers were not even in the vehicle when the idling happened. According to the American Trucking Association (1998), if a truck was left idling for one hour, an amount of 10.397 kg CO_2 would be emitted. This was very large compared to the effective carbon emissions of 8.53 kg per erection (which will be explained in Sect. 10.6). No traffic controller was provided at the main entrance which caused the traffic congestion to deteriorate further.

10.2.3 Observations from Contractor A3

Contractor A3 was using only on-site fabrication. The project was a large condominium project where green building materials (green concrete) were used. Contractor A3 faced a few problems in site layout management, stock management and erection management. The site layout could be improved if the lean concept was adopted. Contractor A3 was chosen as the main target of the case study. The problems that Contractor A3 was facing were explained in details in Sect. 10.4.

10.2.4 The Case Study – Contractor A3

A case study was generated using the observations from the three contractors described above. The case study included:

- The site layout management, stock management and erection management practices that were derived from Contractor A3.
- The geographical locations of the project and the precasters were derived from Contractor A1.
- The delivery management arrangement was derived from Contractor A2.

Such arrangements were made based on several reasons including:

1. The aim of the hypothetical case study was to represent a typical housing project in the Singapore construction industry. For this type of projects, in order to achieve economic savings, the use of both on-site fabrication and off-site fabrication was very common.
2. The main information source was based on what Contractor A3 had provided. Contractor A3 was adopting only on-site fabrication and assumptions should therefore be made relating to off-site fabrication.
3. By managing both on-site fabrication with off-site fabrication, Contractor A3 might face similar problems as Contractor A2.
4. The project was located at a place near the construction project of Contractor A1, which made the ordering of manufactured products in close geographical proximity possible.

10.3 General Procedure to Calculate the Lean Improvements

Similarly to Sect. 8.2, a general procedure to calculate the lean improvements was developed for Contractor A3. The general procedure included two major subprocesses, which were the screening process and the estimation process. As seen in Fig. 10.4, the screening process included three steps:

1. Screening step 1: In this step, the relative importance of the factor identified for Contractor A3 was rated by the probability of occurrence and impact. Factors with no probability of occurrence (P = 1) were dropped from the assessment. It should be noted that absolute care should be paid here because operational activities may be different from what Contractor A3 has rated. Screening step 1 should be based mainly on observations made in Contractor A3, along with ratings on the probability of occurrence.
2. Screening step 2: Following Step 1, factors that required assessment were re-categorized into different originating groups. Carbon emissions in precasters were generated by several sources, which included equipment, electricity, waste of raw materials (only when on-site fabrication is adopted), waste of finished

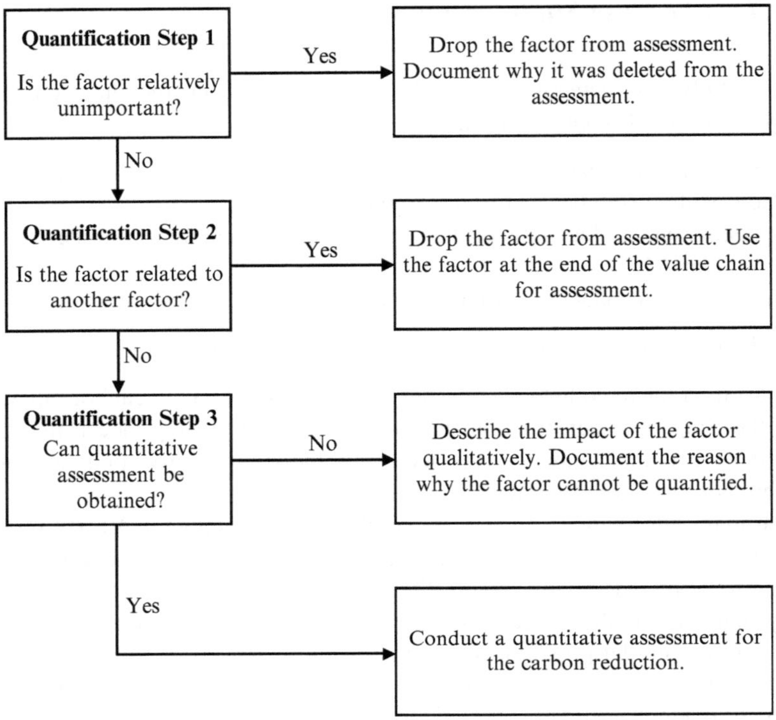

Fig. 10.4 Screening procedure to identify factors that could be estimated

products and capital facilities (such as site offices and employee dormitories). Factors under different originating groups were assessed by different equations.

3. Screening step 3: Factors which could not be categorized in any of the groups mentioned in step 2 might not be eligible for a quantitative assessment. Qualitative descriptions of the impact of such activities to the level of carbon emissions were then provided for Contractor A3.

When factors could be categorized into different originating groups and quantitative assessments could be obtained, the estimation process was then conducted. The estimation of the improvements was conducted for each originating category. The equations used in the estimation process included:

- For equipment:

$$\text{Emissions}_{\text{GHG, fuel}} = \text{Fuel Consumption}_{\text{fuel}} * \text{Emission Factor}_{\text{GHG, fuel}} \quad (10.1)$$

Where:

$\text{Emissions}_{\text{GHG, fuel}}$ = emissions of a given GHG by type of fuel (kg GHG)
$\text{Fuel Consumption}_{\text{fuel}}$ = amount of the fuel combusted (TJ)
$\text{Emission Factor}_{\text{GHG, fuel}}$ = default emission factor of a given GHG by type of fuel (kg gas/TJ)

- For electricity:

$$\text{Emissions}_{\text{GHG,electricity}} = \text{Electricity usage} * \text{Emission Factor}_{\text{GHG,electricity}} \quad (10.2)$$

Where:

$\text{Emissions}_{\text{GHG, electricity}}$ = emissions of a given GHG by electricity (kg GHG)
Electricity usage = amount of the electricity used (KWh)
$\text{Emission Factor}_{\text{GHG, electricity}}$ = default emission factor of a given GHG by electricity (kg GHG/kWh)

- For waste of raw materials or waste of finished products

$$\text{Emissions}_{\text{materials,products}} = \text{Embodied carbon} * \text{Quantity} \quad (10.3)$$

Where:

$\text{Emissions}_{\text{materials, products}}$ = carbon emissions by type of materials/products (kg CO_2)
Embodied carbon = embodied emission factors by type of materials/products (kg CO_2/kg)
Quantity = amount of the materials/products consumed (kg)

Capital facilities were normally supported by electricity. The amount of carbon emissions in this category was therefore calculated by Eq. 10.2.

10.4 Methodology

Similar to Sect. 8.3, LCA was adopted to evaluate the environmental impact on the precast concrete construction sites. A four-step evaluation procedure was used, including:

1. Life cycle goal and scope definition.
2. Life cycle inventory analysis which includes data collection and description of data.
3. Life cycle impact assessment.
4. Life cycle interpretation.

The LCI data was investigated in each of the four value stages after production, which were: site layout management, delivery management, stock management and erection management, as explained earlier in Fig. 10.1. A few estimation criteria and assumption were made so that the calculation process could follow a standard procedure. The estimation criteria included:

1. Compliance with approved methodologies and standards. Similar with the case study of Precaster A, a few general standards of life cycle assessment should be followed. For example, the system boundaries of the case study should be defined at the very start so that every step that consumed energy and generated carbon emissions would be included in the life cycle assessment.

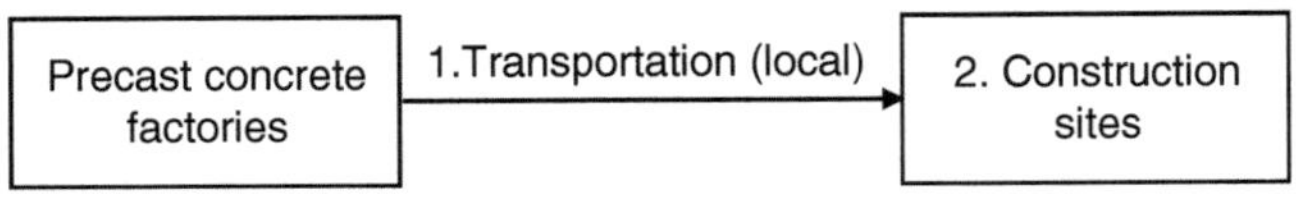

Fig. 10.5 The systems boundaries of this case study

Table 10.1 Energy consumption and emissions factors used in this case study

Energy consumption and emissions factors	Sources
Carbon emissions from road transportation	Peyroteo et al. (2007)
Carbon emissions from the idling of delivery vehicles	U.S. DOE (2010)
Energy consumption of tower crane	Provided by contractors
Carbon emissions of tower crane	Derived from energy consumption of tower crane

2. System boundaries. The system boundaries of this case study are shown in Fig. 10.5.

   Two major processes were included in this study, which included the transportation of the precast concrete columns from the precast concrete factories to the construction sites and the erection activities in the construction sites. The transportation could be integrated to the "cradle-to-gate" boundary of the precast concrete factories so that a "cradle-to-site" estimation of the precast concrete products could be obtained. The transportation and erection can also be integrated to the "cradle-to-gate" boundary of the precast concrete factories to obtain a complete life cycle before the operational stage.
3. Data selection. Data that originated from the Singapore construction industry would be preferred, such as the emissions factors for electricity generation. However, foreign data (such as the European and world-wide average data) could be used where local data was not available. Many LCI studies were referred to when conducting this case study. These LCI studies are shown in Table 10.1.

In accordance with the estimation criteria, a few estimation assumptions were made to calculate the carbon emissions generated in a complete erection cycle. These estimation assumptions included:

1. Road transportation was adopted for the delivery of the precast concrete columns from the precast concrete factories to the construction sites. Round trip local transportation distances were assumed to be 15 miles (24.15 km) based on the geographical location of the precast concrete factories.
2. Energy consumption per erection cycle (which included transportation and erection) was calculated by the following procedure:

   - Six precast concrete columns were transported from the precast concrete factory to the construction site. The energy consumption per delivery was divided by six to obtain the energy consumed to deliver one precast concrete column to the construction site.

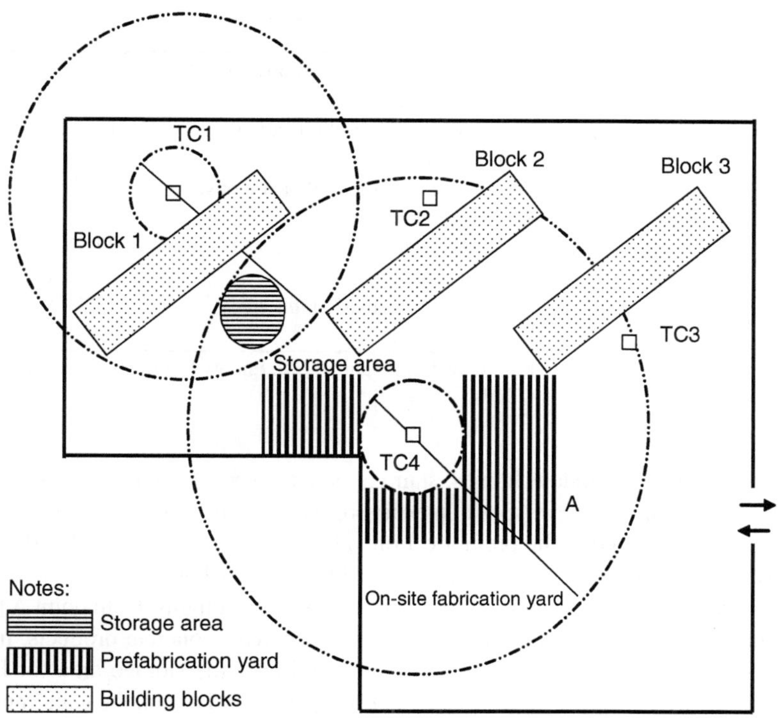

Fig. 10.6 Site layout of the project (not to scale)

- Similarly, if the delivery vehicles were left idling during transportation and erection, this idling time was divided by six to obtain the energy consumed for one precast concrete column.
- Tower crane was used to erect the precast concrete columns. Mobile cranes might be adopted in construction sites for the transportation of other materials, which included plastics, finishing materials, etc. These materials were not within the scope of this study and their carbon emissions would not be calculated.
- One-hour energy consumption of the tower crane was observed, as well as the number of the precast concrete columns that were erected during this time. The energy consumption was divided by the number and the energy consumption per precast concrete column could be obtained. This procedure was adopted based on the reason that the energy consumption per column was very difficult to record in construction sites.

3. The site layout of the project conducted by Contractor A3 was shown in Fig. 10.6. As shown in Fig. 10.6, four tower cranes were used to build three building blocks. A combination of both on-site fabrication and off-site fabrication was adopted in this project. The precast concrete columns as well as some other precast concrete

Table 10.2 Fuel consumption and emissions factors of the tower crane

Tower crane number	TC1	TC4
Fuel consumption (normal operations) Litre gasoline per hour	7	6
Fuel consumption (full operations) Litre gasoline per hour	14	12.5
Emissions factor (normal operations) Kg CO_2 per hour	18.76	16.08
Emissions factor (full operations) Kg CO_2 per hour	37.52	33.5
Emissions factor used in this study Kg CO_2 per hour	18.76	16.08

Emissions factor source: U.S. Energy Information Administration (2010)

products such as window frames, stair cases, etc., were ordered from the precast concrete factories. A few types of precast concrete products such as planks, beams, etc. were fabricated in the prefabrication yard. Tower crane (TC) 1, 2 and 3 were used to conduct erection activities for each building block while TC 4 was used to conduct transferring activities for the precast concrete columns from point A to the storage area and production activities for other precast concrete products. In this study, only TC1 and TC4 were considered and only the storage area for Block 1 was shown in the site layout plan. Contractor A3 could obtain the lean improvements for the whole project by adding up the following activities:

- Erection activities conducted by TC2 and TC3;
- Transferring activities conducted by TC4 to the storage area for Block 2 and Block 3; and
- Production activities of other precast concrete products carried out by TC4.

4. The details of the cranes used in the erection processes are shown in Table 10.2. This study chose Building Block 1 as the research objective. Only TC1 and TC4 were used in the erection processes of Building Block 1. TC2 and TC3 were not considered in this case study. TC1 was used to conduct erection activities for block 1 while TC 4 was used to conduct transferring activities for the precast concrete columns from point A to the storage area and production activities for other precast concrete products.

10.5 The Screening and Estimation Process

Non-value adding activities identified in each value stage of precast concrete production were tested and rated by Contractor A3. Factors with no probability of occurrence (P = 1) were dropped from the assessment due to their relatively low importance, since Contractor A3 was performing well in these areas.

Table 10.3 Carbon emissions in one complete erection cycle without non-value adding activities

Category	Carbon emissions kg CO_2/column
Transportation of the precast concrete columns from the precast concrete factories to the construction site	5.40
Erection of the precast concrete columns from the delivery vehicle to the building block	3.13
Total	8.53

10.5.1 Carbon Emissions in One Complete Erection Cycle

Before estimating how much carbon emissions could be saved by applying the lean production philosophy in construction sites, the carbon emissions that occurred in one complete erection cycle, including transportation and erection in this study, should be calculated. It should be noted that this amount of carbon emissions should be based on the assumptions that there were no non-value adding activities in both transportation and erection. Just-in-time installations were carried out when the precast concrete columns arrived at the construction sites.

It should be noted that the energy consumption in the complete erection cycle was related to the geographical location of the precaster who arranged the delivery of the precast concrete columns and the height of the floors where the erection works were conducted. In this case, the precaster was located 24.15 km away from the construction site and the erection works were conducted on the 31st floor of the building with a height of 103.6 m.

As explained earlier in Table 10.1, an emissions factor of 0.12 kg CO_2/km/t was adopted to calculate the carbon emissions during transportation. A total of 5.40 kg ($0.12 \times 24.15 \times 1.864$) CO_2 was emitted during transportation.

In addition, a total of six precast concrete columns were erected in 1 h. This assumption was based on the 2-h observations made on the construction site every day for four consecutive days. A total of 3.13 kg (18.76/6) CO_2 was emitted when installing one precast concrete column for a height of 103.6 m. The breakdown of the carbon emissions in one complete erection cycle is shown in Table 10.3. It should be noted that this amount of carbon emissions (8.53 kg CO_2/column) was referred to as effective carbon in the following context because it was caused only by value-adding activities. An immediate use of the precast concrete products was assumed when these arrived at the construction site. The installation works of the precast concrete columns could not be completed without generating this amount of carbon emissions.

10.5.2 Site Layout Management

The site layout of the construction project was carefully planned by Contractor A3. According to the project manager, this site layout was sent to the consultants for approval. The siting of the tower crane and parking of the mobile plants were also carefully planned to achieve economic and efficient erection.

Unlike the production process in the precast concrete factories, Contractor A3 did not over provide for material storage area in terms of secured store, weather-proof store and open store. The storage area was designed for erection activities in the next 1 or 2 days.

However, several non-value adding activities should be highlighted. These non-value adding activities included:

1. Does not think of green building materials. In accordance with previous findings in Sect. 9.3.3, green building materials were not usually considered in precast concrete projects, although they had a relatively high impact on the level of carbon emissions. The carbon emissions that could be reduced by using green building materials (e.g. green concrete) were highly dependent on the materials used. In this project, green concrete was adopted from the 1st to 20th storey. Normal concrete was used from 20th storey to the roof. This decision was made based on two reasons:

    - Columns from the 1st to 20th storey required a high strength concrete of Grade 50 (G50) and above and columns from the 20th storey to the roof required a concrete strength of G35. Current green concrete design might not be very stable under high strength, as expressed as a concern by the project manager in Contractor A3. More tests should be conducted using green concrete.
    - Using the green concrete from the 20th storey to the roof would achieve the objective that 60 % of the total volume of concrete used the green concrete. This would enable the project to obtain the Green Mark point which required that 50 % of the total concrete volume to use green concrete.

    A detailed discussion with the green concrete supplier was conducted to validate the above two points. As for the first point, G50 green concrete was not widely used at the time of this study; this was used only in a few projects. The performance of the green concrete would not be a problem as stated by the supplier. It was the price of the green concrete that might impede its full use throughout the project. It should be noted that the G35 green concrete per cubic metre was more expensive than the G35 normal concrete by almost S$5, which would cause revenue to decrease by about S$170,000. If green concrete of higher grade was adopted, the profitability of the project could be reduced sharply.

    When calculating the carbon reduction by using the green concrete throughout the project, it was assumed that G50 green concrete was used in the precast concrete columns from the 21st to the 39th storey (G35 green concrete was used in the precast concrete columns from the 1st to 20th storey by Contractor A3). Fifty percentage slag cement was used to replace Portland cement. An amount of 7.22 kg CO_2 could be reduced for the precast concrete columns used from the 21st to the 39th storey, based on the emissions factors proposed by Prusinski et al. (2006).

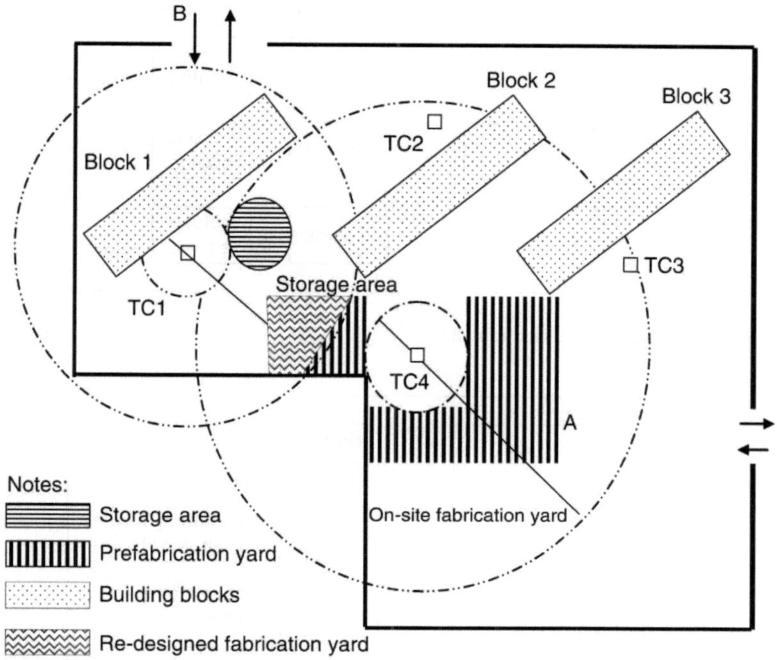

Fig. 10.7 Re-allocating TC1 to achieve smooth work flow

2. The site layout is not carefully planned to achieve economic and efficient production. As stated in Sect. 9.3.3, factor 1.7.1 had a relatively high severity on the level of carbon emissions. The site layout should be carefully designed to achieve economic and efficient production. Contractor A used both on-site fabrication and off-site fabrication to achieve economic savings. However, the design of the area for on-site fabrication should be conducted carefully. In the construction site of Contractor A3, the following procedure could be taken to improve the workflow to achieve low-carbon installation:

   - Step 1 – Re-allocate TC1 to achieve immediate usage of the precast concrete products produced in the on-site fabrication yard. As can be seen in Fig. 10.4, the overlapping area of TC1 and TC4 was the storage area for the precast concrete products which would be used in Block 1. The precast concrete products produced in the on-site fabrication yard would be transferred to the storage area for installation. Such transferring activities were not adding value to the erection process. It was proposed that TC1 be re-allocated so that direct usage of the precast concrete products could be achieved. This re-allocation can be seen in Fig. 10.7. It should be noted that direct usage of the precast concrete products produced in the on-site fabrication yard could also be achieved by choosing a tower crane with a longer jib. However, this might increase the costs by using such tower cranes.

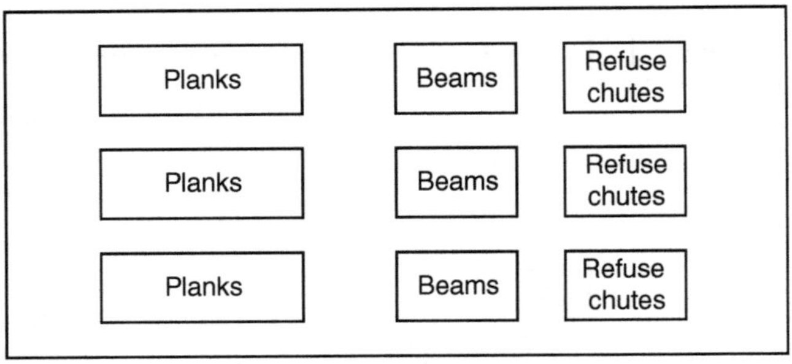

Fig. 10.8 Previous on-site fabrication yard

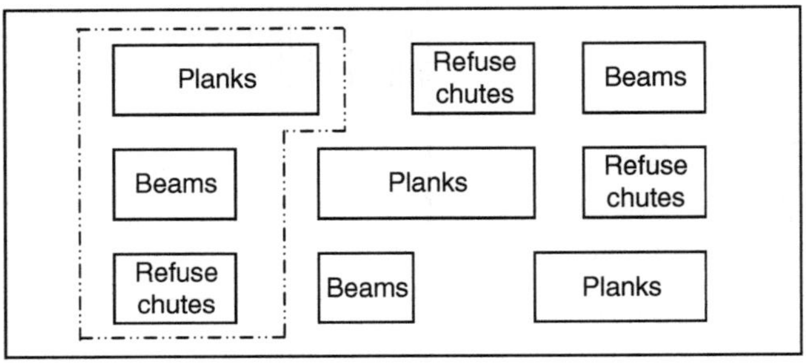

Fig. 10.9 Re-designed fabrication yard

- Step 2 – Re-design the on-site fabrication yard. Following the re-allocation of TC1, the on-site fabrication yard should also be re-designed. This area is shown in Fig. 10.7. The precast concrete products used for Block 1 could be grouped into this area. As illustrated in Fig. 10.8, under normal circumstances, the production of the precast concrete products would be well organized so much so that each type of product would be produced in a separate section of the site layout. This was the strategy that Contractor A3 adopted for on-site fabrication of the precast concrete products. However, it was proposed that another grouping strategy be adopted to achieve low inventory and smooth work flow, as shown in Fig. 10.9.

  The strategy adopted in Fig. 10.9 was to group the precast concrete products which were used in Building Block 1 into an area so that TC1 could conduct the installation activities once the components were produced. Unlike the traditional production arrangement shown in Fig. 10.8, this strategy would enable the contractor to conduct immediate installation without transferring activities and inventory, thus reducing carbon emissions.
- Step 3 – Open an additional entrance at location B. An additional entrance should be opened in location B to facilitate the delivery of the precast concrete products to Building Block 1. The precast concrete products would not be transferred by TC4 to the storage area for installation, thus reducing carbon emissions. Immediate installation could be arranged for Building Block 1 if such an entrance was created.

Table 10.4 Quantification of the lean improvements in site layout management

Factor no.	Description	Primary calculations kg CO_2/column	Secondary calculations kg CO_2/column	Qualitative assessments
1.1.3	Does not think of green building materials	7.22[a]	N/A	N/A
1.8.1	The site layout is not carefully planned to achieve economic and efficient production	2.01	N/A	Yes[b]

Notes:
[a]The amount of carbon emissions that can be reduced by using green building materials was highly dependent on the materials and technologies used. This study showed one example proposed by Prusinski et al. (2006) to replace 50 % of the Portland cement by slag cement
[b]Please refer to problem 2 in this section for more information. It should be noted that an efficient site layout did not only help to reduce the carbon emissions calculated in this section. It was also related to the following sections, such as transportation and erection

- Step 4 – Grouping the precast concrete products used for Building Block 2 and 3. Similar to the site re-design for Building Block 1, as shown Fig. 10.7, the production of the precast concrete products for Building Block 2 and 3 (using TC2 and TC3 respectively) could be re-arranged. Such arrangement would make the precast concrete products available for installation once produced, thus reducing transferring activities.

An amount of 2.01 kg carbon emissions was emitted by transferring the precast concrete column from location A to the storage area. This amount of carbon emissions could be reduced if the above site arrangements were made.

The overall assessment of the impact of site layout management to the amount of carbon emissions produced is shown in Table 10.4. As can be seen from Table 10.4, by using green building materials, the carbon emissions could be significantly reduced. The amount of carbon emissions caused by not using green building materials represented 84.6 % of the total carbon emissions in a complete erection cycle. It seems that contractors were interested only in using green building materials to obtain Green Mark certifications. Once the minimum requirements were met, it appears that things would revert back to where they were.

10.5.3 Transportation Management

The transportation of the precast concrete columns from the precast concrete factories to the construction site was operated by a subcontractor. An "all-in-one" price was adopted so that Contractor A3 could leave the transportation job to the subcontractor and focus on erection and other construction activities. The quantities required in the following 1 or 2 days would be estimated. The precast concrete columns were delivered to the construction site, unloaded to the storage area and

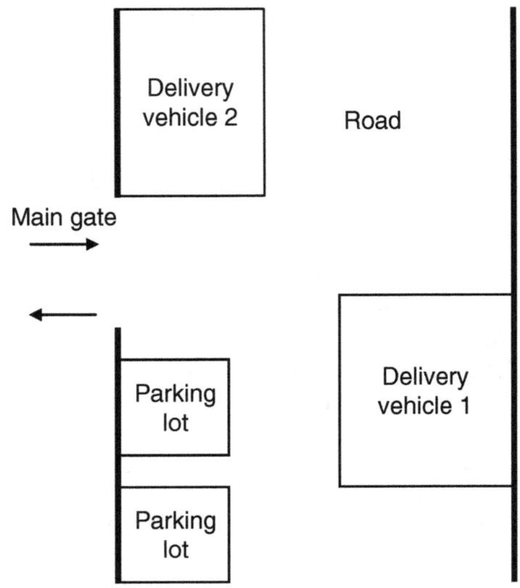

Fig. 10.10 The delivery vehicles that were left idling caused by overlapping of the delivery times

installed when needed. Contractor A3 was not facing demand fluctuations. According to the project manager interviewed, although there would be small changes in the quantities of the precast concrete columns from day to day, such changes were very small and a stable erection schedule could be anticipated. In addition, wrong deliveries have never happened before which showed that the precaster was performing well in making accurate delivery notes. However, following the screening procedure, Contractor A3 appeared to face several problems in transportation management, which included:

1. Transportation was not taken into consideration. In according with previous findings in Sect. 9.4.3, Contractor A did not take transportation into consideration, because the "all-in-one" price was adopted for the precast concrete columns used in this project. As the transportation job was subcontracted to the precasters, Contractor A3 might lose control of the delivery time, which might cause interruptions to the preparation for the arrival of the precast concrete columns. As observed in the building site of Contractor A3, the delivery time had caused many interruptions to the erection process in construction sites. The delivery time of ready mix concrete (for joining the sections of different precast concrete products) and the delivery time of the precast concrete columns overlapped and caused the delivery vehicles to idle on the road from 5 to 30 min. In the 4 days of observation in Contractor A3, these idling times often happened and generated a significant amount of carbon emissions. As can be seen in Fig. 10.10, since the delivery times of different delivery vehicles, including in-transit mixers, were not well organized, delivery vehicle 1 was left idling for 10 min before it was driven through the main gate for precast installation. The idling time for delivery vehicle 2 was even worse. It was kept idling for 30 min before the arrangement for installation was completed. An average idling time of 15 min was assumed for each delivery of precast concrete columns, which would generate 2.60 kg CO_2 emissions for the six precast concrete columns delivered.

Table 10.5 Quantification of the lean improvements in transportation management

Factor no.	Description	Primary calculations kg CO_2/column	Secondary calculations kg CO_2/column	Qualitative assessments
2.1.2	Typical damages during transportation	12.19	N/A	N/A
2.2.3	Transportation is not taken into consideration	0.43[a]	N/A	N/A
2.3.2	Not fully prepared for the arrival of finished products	0.43[a]	N/A	N/A

Notes:
[a]0.43 kg CO_2 was emitted because the delivery vehicle was left idling for 15mins. The idling could be caused by both factors 2.2.3 and 2.3.2

2. Not fully prepared for the arrival of finished products. One important reason that contributed to the carbon emissions identified in problem 1 was that Contractor A3 was not fully prepared for the arrival of the precast concrete columns. The human resources allocated to facilitate such activities were not enough. No traffic controllers were even provided at the main gate to facilitate the movement of the delivery vehicles. According to Richardson (1991), it was advisable that one person, or in the case of large contracts one group of people, at the precast works be made responsible for the arrangements for delivery of elements and equipment so that errors caused by wrongly passed messages could be avoided. In addition, the site layout was not designed well which would cause traffic congestion at the main gate. The drawbacks of the site layout design have been explained earlier.
3. Typical damages during transportation. Damages to the precast concrete columns were common during transportation. However, such damages were very small and can be rectified by repairs. These typical damages included:

   (a) Broken nibs and corners
   (b) Damages due to the wrong placement of battens
   (c) Damages to slender sections
   (d) Damages caused by the use of slings and chains

   According to the project manager interviewed, a 2 % waste of finished products could be assumed. If the same precast concrete column as used in the computation in Sect. 8.3 was adopted in this project, another 12.19 kg of CO_2 would be emitted by repairing activities.

The overall assessment of the non-value adding activities in transportation management is shown in Table 10.5. As shown in Table 10.5, the most important factor during transportation related to the damages to the precast concrete columns, which would represent 142 % of the effective carbon as calculated in Sect. 10.4.1. Actions should be taken by the precasters (if the delivery was subcontracted out to the precasters) to reduce such damages identified above. If the delivery was instead arranged by the contractor, the contractor should then investigate the delivery process.

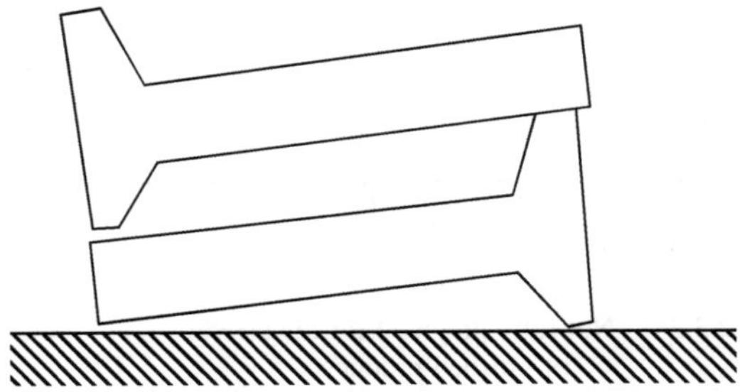

Fig. 10.11 Inappropriate stacking of the precast concrete products in the storage area

More importantly, sufficient care should be provided in the delivery process by the vehicle drivers. According to the project manager interviewed, such damages could be reduced if sufficient care was provided when carrying out loading and unloading activities, as well as during transportation. It was recommended that a certification programme be instituted to examine whether or not the employees would be able to handle the precast concrete columns without waste when carrying out loading and unloading activities (e.g. the Freight Operator Recognition Scheme in the UK).

An amount of 0.43 kg CO_2 was emitted because of idling, which was 5.04 % of the effective carbon. It was recommended that the supply chain management practices be reviewed for Contractor A3. Overlapping of delivery times should be avoided. In addition, at least one traffic controller should be provided at the main gate to facilitate the deliveries. It should be noted that the carbon emissions caused by vehicle idling is highly related to the idling time. If the delivery vehicle was left idling for 30 min, 10 % more carbon emissions would be generated.

10.5.4 Stock Management

According to Contractor A3, appropriate equipment (tower crane in this case) was selected to manage the storage and appropriately trained staff were arranged to manage the storage. There was no wrong use of the precast concrete columns due to unclear identification marks. The main non-value adding activities in Contractor A3 included:

1. Lack of well ordered stockyard. The stockyard was not well maintained, especially the on-site fabrication yard. The stockyard was worse when there was a heavy rain during the visit. It seemed that the drainage of the site could be improved to avoid potential damages to the precast concrete products.
2. Inappropriate battens. Some precast concrete products were stacked without using the appropriate battens. This omission would cause damages to the precast concrete products, as shown in Fig. 10.11.

   The working instructions provided to the employees who were in charge of stock management were made in terms of on-the-job training. A detailed programme including the stacking requirements was not provided. While it

Table 10.6 Quantification of the lean improvements in stock management

Factor no.	Description	Primary calculations kg CO_2/column	Secondary calculations kg CO_2/column	Qualitative assessments
3.1.1	Lack of well ordered stockyard	6.10[a]	N/A	N/A
3.3.2	Inappropriate battens		N/A	N/A
3.4.1	Lack of periodic stock checks		N/A	N/A

Notes:
[a]This amount of carbon emissions was caused by damages to the precast concrete columns during storage, which was 1 % according to Contractor A3

was proposed that the use of more than two battens along the length should be avoided except for extremely flexible prestressed planks (Richardson 1991), this was not always followed by Contractor A3. Some other stacking requirements were not followed as well, which were (Richardson 1991):

- The top surface of the upper element in a stack to be swept prior to adding a further element.
- Both faces of a panel undergo similar conditions of temperature and weather exposure during the storage.

3. Lack of periodic stock checks. According to Contractor A3, the quantity of the precast concrete columns used in this project was carefully estimated. The quality of the precast concrete columns was checked in the precast concrete factories before these were arranged for delivery. Periodic stock checks were unnecessary on the construction site because the precast concrete products would be used in the following 1 or 2 days. However, such stock checks were still recommended for several reasons:

- Potential damages that could be caused by inappropriate stacking.
- Unanticipated rejection of the precast concrete columns which might cause interruptions to the installation process.
- Installation delays caused by adverse weather conditions.

According to Contractor A3, the damages that happened during stacking were very minimal. Small repairs would be conducted if such damages were found. One percentage of the damages to the precast concrete columns could be anticipated during storage. The overall assessment of the lean improvements in stock management is shown in Table 10.6.

10.5.5 Erection Management

Contractor A3 provided a written erection method statement and a video tape relating to the erection activities. This written erection method statement was prepared by Contractor A3 and submitted to the consultants for approval. According to Contractor

A3, such an erection method statement was prepared so that issues relating to health and safety could be properly addressed. In fact, the erection method statement could help Contractor A3 reduce energy consumption and carbon emissions by following the procedures listed in the erection method statement strictly.

Erection activities were carried out properly in the construction site of Contractor A3. Four persons were allocated to carry out the erection activities, including two charge-hands, a banksman and a crane driver. According to the project manager, erection accuracy was satisfactory. Once the precast concrete column was placed at the right position that was properly marked, there would be no further movement of the column. No additional energy was consumed because of erection inaccuracy.

The continuous improvement plan of Contractor A3 seemed to be good to support further improvements. Internal periodic meetings were organized every week to discuss further improvements. Consultants were involved in the periodic meetings every 2 weeks. However, it should be noted that lean experts should also be involved in such periodic meetings. There were many activities which were considered as normal activities, but could be categorized into non-value adding activities when examined by the lean production philosophy, such as storage, multi-handling, equipment idling, etc.

Similar to Sect. 10.5.4, activities in erection management in Contractor A3 were investigated to identify the improvements in carbon reduction. These activities included:

1. Inappropriate crane operations. The precast concrete columns were used based on a "deliver-store-use" method. The precast concrete columns were stored in the storage area located at the side of the building block before these were erected in the following 1 or 2 days. Unloading the precast concrete columns and stacking at the storage area would cause double handling and would possibly damage the products due to the double handling or site storage. According to the project manager of Contractor A3, a 1 % damage to the precast concrete columns could be anticipated during the storage, which would cause a 6.10 kg increase in the level of carbon emissions.

   The other source of carbon emissions caused by a storage area was the slinging out and transferring activities. As has been observed in Contractor A3, TC4 was used to conduct the singling out and transferring activities, as well as the casting activities for some on-site prefabricated products, such as window frames, staircases, etc. A total of eight precast concrete products were transferred in one hour and an amount of 16.08 kg of CO_2 was emitted. The carbon emissions emitted by one transfer was 2.01 kg.
2. Unqualified employees at construction site. As observed in Contractor A3, a smooth work flow was not always followed. There were a few interruptions to the work flow which were caused by unqualified employees. These interruptions included:

   - When transferring the precast concrete column from the pick-up point (location A as shown in Fig. 10.4) to the storage area, the precast concrete column has been suspended in the air for almost 5 min when the employees were trying to make a place for the column in the storage area. This non-value adding activity happened every day in the four conservative days of observations. It was assumed that this frequency would remain for the rest of the work on 31st floor. An amount of 0.08 kg of CO_2 was emitted by this non-value adding activities.

- Due to the on-site fabrication of other precast concrete products, TC4 was heavily used. The use of TC4 did not follow a smooth work flow. As observed in the construction site, when TC4 was about to pick up the precast concrete column at the pick-up point, the crane operator was told to conduct the demoulding work first. Such unnecessary swing of the jib would cause an increase in the level of carbon emissions. Every 10-s swing would cause an increase of 0.045 kg of CO_2 emissions. It should be noted that due to the low probability of occurrence about this non-value adding activity (which happened two times during the investigation), this amount of CO_2 was recorded as 0.045 kg CO_2 per 10-s swing.
- The position where the precast concrete column should be installed was marked with a clear line in the construction site. Erection accuracy should be checked before releasing the precast concrete column. However, as observed in Contractor A3, some parts of the erection accuracy checks were conducted after the release of the precast concrete column. For example, a ruler was used to check whether the surface of the precast concrete column and the marked point followed a straight line. This accuracy check was conducted when the precast concrete column was released from the tower crane. This might cause double-handling when the result of the accuracy check was not satisfactory.

3. Inadequate supervision during erection processes. The supervision on the construction site was not enough to support a smooth work flow. Some activities caused interruptions to the work flow. These non-value adding activities included:

    - The erection activities on the construction site were not always conducted by the integrated site team, which should usually include one banksman, one crane operator and two charge-hands. As observed on the construction site, the banksman who should direct the operations of a crane was conducting the works that should be handled by the charge-hands. Many sudden accelerating and braking were caused if only one person was involved in such loading and unloading activities.
    - The precast concrete column has been suspended in mid-air for 3 min before erection activities were conducted. The banksman and charge-hands were asked to conduct some other works in the construction site. This interrupted the work flow and caused an increase in the level of carbon emissions. During the 2-h site visit for four conservative days, this happened everyday when 16 precast concrete columns were erected. An amount of 0.80 kg of CO_2 was generated because of the 3-min suspension in mid-air. This amount was evenly distributed to the 16 precast concrete columns erected during the 2-h observation.

Table 10.7 Quantification of the lean improvements in erection management

		Primary calculations	Secondary calculations	
Factor no.	Description	kg CO_2/column	kg CO_2/column	Qualitative assessments
4.3.1	Inappropriate crane operation	2.01	6.10	N/A
4.3.4	Unqualified employees at construction site	0.08	0.045[a]	Yes[b]
4.3.6	Inadequate supervision during erection processes	0.05	N/A	Yes[c]

Notes:
[a]The amount of 0.045 kg of CO_2 was based on a per 10-s swing basis
[b]Please refer to problem 2 in this section. There were a few activities that might increase the level of carbon emissions. However, due to the low probability of occurrence of such activities, their impacts on the level of carbon emissions were not calculated but were qualitatively described
[c]Please refer to problem 3 in this section. Supervisors who were aware of the lean philosophy seemed to be one of the factors for successful lean erection in the construction site

It should be noted that adequate supervision should be provided by supervisors who were aware of the lean principles. Otherwise, the activities that were non-value adding activities as examined by the lean philosophy would be considered as necessary construction activities based on their previous experiences.

The overall assessment for erection management is shown in Table 10.7. It should be noted that the amount of 2.01 kg of CO_2 emissions was caused by transferring activities which were the same source of carbon emissions for factor 1.8.1 (The site layout is not carefully planned to achieve economic and efficient production) as identified in Sect. 10.5.2.

10.6 Results

By reducing the non-value adding activities in each value stage in the precast concrete construction site of Contractor A3, a few benefits could be achieved, including:

- Reduced inventory (which might lead to less working capital investments because fewer employees were involved in managing inventory);
- Reduced floor area (which might lead to less fixed capital investments and period expenses);
- Short installation time;
- Decreased probability of defects and re-work; and
- Increased efficiency.

Table 10.8 Carbon reduction achieved by applying the lean production philosophy

Category	The amount of carbon emissions ($kg\ CO_2$/column)
Waste of finished products	18.29
2.1.2 Typical damages during transportation	12.19
3.1.1 Lack of well ordered stockyard	6.10
3.3.2 Inappropriate battens	
3.4.1 Lack of periodic stock checks	
4.3.1 Inappropriate crane operations	
Inappropriate installation arrangement	2.57
1.8.1 The site layout is not carefully planned to achieve economic and efficient productions	2.01
4.3.1 Inappropriate crane operation	
2.2.3 Transportation is not taken into consideration	0.43
2.3.2 Not fully prepared for the arrival of the finished products	
4.3.4 Unqualified employees at the construction site	0.08
4.3.6 Inadequate supervision during erection processes	0.05
Other savings	7.22
1.1.3 Does not think of green building materials	7.22
Other qualitatively described lean improvements	Please refer to the qualitative assessment for each factor
1.8.1 The site layout is not carefully planned to achieve economic and efficient production	
4.3.4 Unqualified employees at the construction site	
4.3.6 Inadequate supervision during erection processes	

The carbon reduction that could be achieved by applying the lean production philosophy is shown in Table 10.8. A few implications could be obtained from Table 10.8, including:

1. The carbon emissions during the erection processes could be represented by the following formula:

$$\text{The amount of } CO_2 = \underbrace{8.53}_{\text{Effective carbon}} + \underbrace{18.29 + 2.57}_{\text{Wastes}} \quad (\text{kg})$$

   Effective carbon was obtained when there were no non-value adding activities in the erection cycle, as has been calculated in Sect. 10.5.1. A total amount of 20.86 kg of CO_2 was emitted in terms of waste of finished products and inappropriate installation arrangements. This amount of carbon emissions was 2.45 times that of the effective carbon. An amount of 18.29 kg of CO_2 was emitted in terms of wastes of finished products, either during transportation or during storage. This amount of carbon emissions represented 2.14 times that of effective carbon. Another amount of 2.57 kg of CO_2 was emitted in terms of inappropriate installation arrangements, such as inefficient site layout, inappropriate crane operations, unqualified employees and inadequate supervision. This amount of carbon emissions represented 30.13 % of the effective carbon. The details of the breakdown are shown in Table 10.9.

Table 10.9 The breakdown of carbon reduction when applying the lean production philosophy

Category	The amount of carbon emissions kg CO_2/column	Percentage (%)
Effective carbon	8.53	100
Wastes of finished products	18.29	214
Inappropriate installation arrangements	2.57	30.13
Carbon savings when using green building materials	7.22	84.64

2. The decision to use green building materials to reduce carbon emissions in construction projects was highly dependent on the green building certification programmes. It seems that such decision was not based on the intention to reduce carbon emissions but on the intention to obtain certain scores in the green building certification programmes. If the minimum requirements were met (i.e. 50 % of the total concrete was replaced by green concrete in this case), no further usage would be adopted.

   Green building materials (green concrete in this case) was slightly higher in price than normal building materials. Incentives should therefore be provided to encourage contractors, developers and clients to use green building materials to achieve green building certifications. The GM GFA (Green Mark Gross Floor Area) Incentive Scheme has been provided for contractors, developers and clients who aim to develop buildings that attain higher tier Green Mark ratings. This scheme will grant additional floor area over and above the Master Plan Gross Plot Ratio (GPR) control (BCA 2010). More incentive schemes should be provided to influence the behavior of contractors, developers and clients towards using green building materials.
3. As stated previously, facilities managers and employees who were working closely with equipment should have the greatest potential to reduce carbon emissions in construction sites. However, the competence of the facilities managers and employees could be improved, as observed in the construction sites. Many non-value adding activities happened in the construction sites because the employees lacked a vision of smooth work flow. All preparatory work should be completed before the installation work could start. For example, when observing the delivery of finishing sands (contained in fabric bags) to the construction site, it was found that the wooden supports for the bags (so that these bags of sand could be transferred using forklifts) were only provided while the delivery vehicle was left idling. In addition, during 1 day of the site observations when there was a heavy rain, the air compressor was left running for almost 1 h until the rain stopped.

   These non-value adding activities were not within the scope of this study. Their impact on the level of carbon emissions were therefore not assessed in this study. However, this does not mean that these activities were not important. When a typical 2HP air compressor was left idling for 1 h, almost 0.8 kg of CO_2

would be emitted (calculated based on the emission factor for electricity generation of Singapore). This amount of carbon emissions was 29.9 % of the effective carbon when installing the precast concrete columns to the 31st storey in this study. As more project management and environmental management practices would be considered in green building rating systems (such as the BCA Green Mark in Singapore), such non-value adding activities might affect the certification of the construction project.

4. Equipment idling was a serious problem in this project. This might be caused by using on-site fabrication with off-site fabrication. While on-site fabrication could bring economic savings, the site arrangements and transportation arrangements should be more carefully prepared than when using only off-site fabrication. In this case, both site arrangements and transportation arrangements could be improved to prevent equipment idling to reduce carbon emissions. In fact, if equipment idling was listed as one consideration in green building certification, this project might fail in this criterion.
5. Unlike the production processes in the precast concrete factories, the erection processes in construction sites were very straightforward. Immediate installation could be arranged if a JIT delivery system was organized. However, this does not mean that a complete re-examination of the erection processes is not necessary. Waste of the finished precast concrete products could cause two times more carbon emissions than the amount caused by the erection cycle. Inappropriate erection arrangements could cause another 30 % increase in carbon emissions in this project. The carbon emissions caused by these non-value adding activities could be reduced if the lean production philosophy was applied in the construction site.
6. Attention should be paid to the results of the case studies. Although the "ideal" case would be a construction project using the same type of precast concrete column, this was difficult at the time of this study because of the time lag when the research studies with the contractors were conducted sometime after the research studies with precasters were completed. The amount of carbon emissions that could be reduced by applying the lean production philosophy might vary significantly depending on the frequency of the on-site non-value adding activities. However, as stated in Sect. 6.3, the case study was based on "analytic generalization" rather than "statistical generalization" to test the empirical results of the survey. The contractors could use the same procedures presented in this case study to identify their own on-site non-value adding activities and quantify the carbon reduction that could be achieved by applying the lean concept.
7. Different contractors might face different non-value adding activities. The improvements in this case study were identified for Contractors A1, A2 and A3. Although there were many non-value adding activities identified in Chap. 8, it should be noted that the non-value adding activities in each precast concrete project should be identified case by case.

10.7 Summary

This chapter validated the applicability of the lean production philosophy to reduce carbon emissions in construction sites. Three contractors in Singapore was chosen for the case study and the erection cycle of precast concrete columns carried out by this contractor was examined. It was found that the erection cycle, including site layout management, transportation management, stock management and erection management, could be improved to reduce carbon emissions in the construction site. Two major categories that might cause an increase in the level of carbon emissions were waste of the finished products and inappropriate installation arrangements. It was found that the delivery performance of the precasters was satisfactory by delivering the precast concrete columns to the construction site on time. However, the site arrangements were not efficient enough.

The lean production philosophy can also help the contractors to improve their project management and environmental management practices. If such improvements were achieved, contractors might obtain better scores in the green building certification programmes, such as the BCA Green Mark Scheme. Other benefits associated with low-carbon erection could be achieved as well, including low inventory, reduced floor area, short installation time and increased efficiency.

References

American Trucking Association (1998) Green truck website. [Online] Available at: http://www.truckline.com/air_emissions/1420.html. Accessed 9 Aug 2010

Building Construction Authority (2010) Green Mark Groos Floor Area (GM GFA) incentive scheme. Available from: http://www.bca.gov.sg/greenmark/gmgfa.html. Cited 29 Nov 2013

Peyroteo A, Silva M, Jalali S (2007) Life cycle assessment of steel and reinforced concrete structures: a new analysis tool. In: Portugal SB 2007 sustainable construction, material and practices. IOS Press, Amsterdam, pp 1–6

Prusinski JR, Marceau ML, Van Geem MG (2006) Life cycle inventory of slag cement concrete. In: Naik T (ed) Proceedings of the 8th international conference on fly ash, silica fume, slag and natural pozzolans in concrete. CANMET/ACI, pp 1–26

Richardson JG (1991) Quality in precast concrete: design, production and supervision. Longman Scientific & Technical, New York

U.S. Department of Energy (2010) National idling reduction network news. [Online] Available at: http://www1.eere.energy.gov. Accessed 9 Aug 2010

U.S. Energy Information Administration (2010) Emission factors and global warming potential. Available from: http://www.eia.gov/oiaf/1605/emission_factors.html. Cited 29 Nov 2013

Chapter 11
Carbon Labelling and Reflections

11.1 Introduction

The preceding chapters demonstrated that the lean production philosophy could be applied in both the precast concrete factories and the construction sites (using precast concrete components) to reduce carbon emissions. The level of carbon emissions that could be reduced was related to the frequency of many non-value adding activities that have been identified in previous chapters.

However, there remained a few questions that needed to be discussed in this study. These questions included:

- How would the lean concept contribute to the LCA approach when evaluating the environmental impacts?
- Could the lean concept be applied to other construction materials to achieve low-carbon production?
- How could the value concept in lean production be broadened to meet the challenge of sustainable development?

The implications of this study to the precasters, the contractors and regulatory authorities should be addressed as well so that appropriate strategies could be taken to achieve either low-carbon production or low-carbon construction. In addition, the validation of results is also provided in this chapter. Three sources of validation are provided which are: review of the calculation process, comparison with previous research and external review.

11.2 Lean in Carbon Labelling Programmes

11.2.1 Introduction

As the largest source of carbon emissions (AIA 2007), more and more efforts were directed to help the construction industry to achieve sustainability, including

P. Wu and S.P. Low, *Lean and Cleaner Production: Applications in Prefabrication to Reduce Carbon Emissions*, DOI 10.1007/978-3-642-42062-7_11,

designing many environmental labelling programmes. According to Hemmelskamp and Brockmann (1997), environmental labels that served as quality marks for products according to selected criteria could be used to verify their environmental compatibility. While an overall assessment for all the environmental impacts might be difficult, many environmental labelling programmes focused on a single criterion, such as carbon emissions (e.g. carbon labelling programmes) and energy consumption (e.g. the Energy Star Programme in the U.S.). According to Ball (2002), most labelling programmes followed the Life Cycle Assessment (LCA) rules and could offer an absolute measurement of the single criterion that was selected for the assessment. For example, the carbon labelling programme focused on the inputs of materials and energy to estimate the carbon level of the finished products. The carbon levels of different products could therefore be compared to identify their contributions to the global climate change.

However, when these environmental labelling programmes were examined, it was found that a relative comparison between the current operation and the best operation (which is referred to as lean benchmark in this study) was overlooked. As observed in the precast concrete factories, many non-value adding activities which consumed energy and resources happened frequently and might adversely affect the labelling score of the products. The lean concept indicated a new area which the traditional LCA approach did not cover.

11.2.2 LCA in Environmental Labelling Programmes

Driven by the pressing pressure of environmental challenges, there have been a number of attempts to initiate environmental labelling or eco-labelling schemes (Ball 2002). Environmental labelling programmes might provide one or several pieces of environment-related information, such as modelling of energy consumption, water consumption, carbon emissions and wastes. These pieces of information were aggregated into a single score for making decisions when selecting materials with the best environmental performance. In the building and construction industry, the labelling programmes could be used to assess the whole building performance as well as the performance of construction materials. Trusty (2001) divided the labelling programmes into three levels, which were:

- Level 1: Product comparison tools (e.g. UK Ecopoints, Blue Angel, NF Environment Mark)
- Level 2: Whole building design or decision support tools (e.g. Whole Life Cycle Costing, Multi-Criteria Decision Making)
- Level 3: Whole building assessment frameworks (BREEAM, LEED®, Green Globes)

For example, LEED® (the Leadership in Energy and Environmental Design) was a voluntary consensus standard developed by the U.S. Green Building Council (USGBC) for developing sustainable buildings that had superior performance in the areas of sustainable site development, water savings, energy efficiency, materials selection, and indoor air environmental quality (Vijayan and Kumar 2005). The Green Globes offered a simpler methodology and employed a user-friendly interactive guide for assessing and integrating green design principles for buildings (Smith et al. 2006). Both labelling programmes were known as the whole building performance assessment tools. On the other hand, the BRE methodology for environmental profiles for construction materials, components and buildings offered a standardized method to identify and assess the environmental effects associated with building materials over their life cycle – that included their extraction, processing, use, maintenance and eventual disposal (BRE 2010). Based on the methodology, the UK Ecopoints was initiated by BRE to measure the total environmental impacts of a particular product or process (Huovila and Curwell 2007). Environmental labelling programmes of construction materials should be completed in close cooperation with manufacturers, as information related to inputs of raw materials, energy as well as the detailed design was mostly provided by the manufacturers. The Whole Life Cycle Costing approach was a technique which enables comparative cost assessments to be made over s specified period of time, taking into account all relevant economic factors both in terms of initial costs and future operational costs (Gluch and Baumann 2004). According to Balcomb and Curtner (2000), the Multi-Criteria Decision-Making (MCDM) technique was designed to guide design teams in a way that would make sustainable building design easy and inexpensive. Both methods belonged to the Level 2 assessment (i.e. whole building design or decision support tool) and could offer the design team a good evaluation of the proposed building to achieve ultimate building sustainability. In accordance with the research aim, this study focuses on the review of environmental labelling methods for construction materials rather than for the whole building assessment and the decision support tools.

According to Ball (2002), the first eco-labelling scheme for buildings was initiated by the Building Research Establishment (BRE) in the UK, which was known as the BRE Environmental Assessment Method (BREEAM). In addition, the BRE methodology for environmental profiles for construction materials, components and buildings was established to assess the environmental impacts and provide reliable environmental information. In the U.S., the Energy Star Programme provided energy assessment tools for buildings. However, although the Energy Star Programme could help to assess the energy consumption of buildings, it was usually considered as an incomplete eco-labelling scheme (Ball 2002). According to Ball (2002), energy was the commodity in the building and construction industry that could be easily measured and quantified, but it was by no means the only sustainable factor that should be assessed. Based on this, the eco-labelling programmes for building materials could therefore be divided into two groups: one group that offered a complete assessment tool for all environmental aspects and the other one that provided information on a single assessment area. For example, the UK Ecopoints measured the total environmental impact of a particular product or process while the Energy Star Programme only accounted for the energy section.

Table 11.1 Some carbon labelling practices in current environmental labelling programmes

Environmental labelling programmes	Carbon labelling practices under the programme
EU – Ecolabel	Assess the GHG emissions of the products based on key life cycle assessment techniques and principles
UK – The Ecopoints	Use LCA assessment and techniques to transfer to GHG emissions into ecopoints – 1 kg of CO_2 equivalent gets 0.0029 ecopoints
Singapore – Singapore green labelling scheme	Use LCA assessment and techniques to address the carbon emissions values at the product's point of production (e.g. kg CO_2/t or kg CO_2/m^3)
US – BEES® 4.0	Use LCA assessment and techniques to transfer the GHG emissions into an environmental performance score

There were no isolated carbon labelling schemes for construction materials. Most carbon labelling schemes were integrated in the environmental labelling programmes. As shown in Table 11.1, the most commonly adopted strategy in developing the carbon emissions values for products was to use LCA assessment and techniques, similar to current environmental labelling programmes, which were designed and tested under life cycle analysis (LCA) – a method to evaluate the environmental impacts in the life cycle of the products (ISO 14040–14043).

According to Hertwich et al. (1997), LCA was the most prominent and most comprehensive approach used in environmental labelling which classified emissions into categories reflecting the environmental impacts they caused, such as acidification or ozone depletion, and aggregated the emissions in each category to an equivalency potential based on how much each emission contributed to the respective impact. However, environmental labelling programmes which used LCA as the assessment approach might lead to a few problems, including:

1. The crucial point of environmental labelling was the credibility of the ecolabel information (Karl and Orwat 1999). However, a LCA was only a snapshot of a product/system at a point in time under specified assumptions (Grant and Macdonald 2009). For example, both wastes of raw materials and damages to finished products were very common when producing precast concrete products. Whether or not the wastes and damages would be included in the calculation of the embodied carbon were subject to the analyst's own LCA assumptions.
2. The comprehensiveness of the ecolabel information was currently represented by a single sign. Although a single sign could offer the customers an intuitive explanation of the products' environmental compatibility, it might suppress other information when evaluating the products' environmental quality. According to Grant and Macdonald (2009), LCA had little to say about the adaptability of the system, its limits, risks or potential, which were all necessary information to evaluate the products' environmental compatibility.
3. New innovative technologies often looked inefficient in the early design stage and might fare poorly in LCA terms even if they were potentially of great benefit to the environment. It seems that LCA lacked a long-term view and analysis of the products' environmental performance.

Table 11.2 The carbon emissions value of the precast concrete product (modified)

Raw materials		Emission factors	Carbon emissions	
1. Cement	320.1300 kg	0.4970 kg CO_2/kg	159.1046 kg CO_2	24.59 %
2. Aggregates	540.6600 kg	0.0050 kg CO_2/kg	2.7033 kg CO_2	0.42 %
3. Reinforcement	178.0000 kg	1.7000 kg CO_2/kg	302.6000 kg CO_2	46.76 %
Energy inputs				
4a. Transportation (cement) (international)	320.1300 kg	104.2000 kg CO_2/t	33.3575 kg CO_2	5.15 %
4b. Transportation (aggregate) (international)	540.6600 kg	121.6000 kg CO_2/t	65.7443 kg CO_2	10.16 %
4c. Transportation (reinforcement) (international)	178.0000 kg	35.7000 kg CO_2/t	6.3546 kg CO_2	0.98 %
5. Concrete plant operation	68.6000 KWh/m^3	0.5233 kg CO_2/KWh	25.5200 kg CO_2	3.94 %
6a. Transportation (concrete)(local)	24.1500 km	0.1200 kg CO_2/km/t	9.7721 kg CO_2	1.51 %
6b. Transportation (reinforcement) (local)	24.1500 km	0.1200 kg CO_2/km/t	1.0317 kg CO_2	0.16 %
7. Precast concrete production	6.5000 KWh	0.5233 kg CO_2/KWh	3.4015 kg CO_2	0.53 %
		Total	609.5895 kg CO_2	94.20 %
8. Waste of raw materials – 2 %			9.2900 kg CO_2	1.44 %
9. Waste of finished products – 3 %			18.2900 kg CO_2	2.83 %
10. Inappropriate production arrangements			9.9330	1.53 %
		Total	647.1025 kg CO_2	

11.2.3 Lean in Environmental Labelling Programmes

In order to calculate the carbon score, a process tree of the product should be defined at the very start. Based on the estimation in Chap. 8, the carbon emissions value of this product was shown in Table 11.2. A total of 647.1025 kg CO_2 was emitted in the life cycle (cradle to gate) of the precast concrete product, including the inputs of waste and inappropriate production arrangements. This life cycle included the extraction of raw materials, the transportation of raw materials (both international and local) and the production processes in both the concrete plant and precast concrete factory. It should be noted that the estimation should not be based solely on design specifications because wastes of either raw materials or finished products do happen frequently in precast concrete factories.

Table 11.3 Non-value adding activities in the precast concrete production process

Category	The amount of carbon emissions ($kg\ CO_2$/column)
Waste of finished products	18.29
Too much inventory in factory	
Damaged products during inventory	
Damaged products when handling	
Double-handling or delivery due to unsatisfied quality or specifications	
Illumination savings	13.11
Over provide material storage	
Large quantity supply base	
Too much inventory in factory	
Waste of raw materials	9.29
Over provide material storage	
The site layout is not carefully planned to achieve economic and efficient production	
Waste of raw materials in the production process	
Materials damaged during handling	
Unnecessary materials handling	
Inappropriate production arrangements	9.933
Improper specification of building materials	7.25
Over provide material storage	0.58
The site layout is not carefully planned to achieve economic and efficient production	0.96
Transportation is not taken into consideration	0.58
Raw materials do not meet specifications	0.47
Unnecessary materials handling	0.003
Double-handling or delivery due to unsatisfied quality or specifications	0.09
Delivery performance	0.14
Unclear identification marks	0.07
Unclear delivery notes	0.07

When examined by the lean concept, not all processes listed in the process tree were value adding to the overall production. As observed in the precast concrete factory, there were a few non-value adding activities that consumed energy and generated emissions. These activities and their carbon emissions value were calculated and shown in Table 11.3.

Carbon labelling programmes could offer accurate estimation of the embodied carbon of construction materials. However, the industry lacked a benchmark to identify how efficient the production was. Current carbon labelling programmes could offer a short term benchmark between different precasters and the industry leaders. The comparison could be intuitively interpreted by the scores achieved from the carbon labelling programmes. For example, according to the UK Ecopoints, 1 kg of CO_2 equivalent could be translated into 0.0029 ecopoints. In this case, the carbon emissions value of this product (which was 647.1025 kg CO_2 per column) could be translated to 1.88 ecopoints. Consumers could therefore obtain the products' carbon performance through the ecopoints achieved. Products with the lower ecopoints would be chosen. However, the long term benchmark was currently missing for construction materials, or at least in the precast concrete sector. In other words, a lean benchmark which might represent tomorrow's world class company was not provided for comparison at all.

In this case, when all the non-value adding activities were eliminated, a total amount of 609.5895 kg CO_2 was emitted per precast concrete column, which could be translated into 1.77 ecopoints. It should be noted that illumination savings and delivery performance of the finished products were not within the boundary of the LCA study of this product, which was cradle to gate. The carbon emissions from these two sources were not used. However, such sources do represent a potential improvement that could be achieved by applying the lean concept. In this case, 0.11 ecopoints were saved by applying the lean concept in the precast concrete factory.

Unlike normal carbon labelling programmes which highlighted the inputs and outputs of resources and energy by design specifications, the lean score/benchmark advocated issuing the carbon label based on a refined production process. In practice, it was recommended that a lean score was provided, associated with the carbon score obtained from the carbon labelling programme. In this case, the 0.11 ecopoints was the lean score when the production process was refined. The lean score was the improvements that can be achieved by the precasters and should be determined case by case. In addition, as the lean score represented the wastes in the production process, the lower the score, the more efficient the production process was.

When issuing the lean score for the precast concrete products, there were several areas that should be considered. Firstly, the efficiency of current production compared to the leanest case where all the non-value adding activities were eliminated should be calculated. This was designed to address the problem that LCA had little to say about the potential of the systems. Secondly, a score should be assigned to new innovative technology, which might fare poorly in the LCA rating, but could bring about potential benefits to the environment in the future. In addition, production processes with "continuous improvement" plans should have better scores than those without such plans. This offered a long-term view in the benchmarking process. Otherwise, both production processes would be issued with the same score under normal carbon labelling programmes, where the limits and risks of the production systems were not reflected at all.

The applicability of the lean concept in carbon labelling programme for precast concrete products could therefore be summarized into the four phases as illustrated in Fig. 11.1. These four phases were:

1. Identification. In this phase, the production process tree is refined by the lean concept to identify the true valuable inputs. Non-value adding activities are identified and eliminated in order to provide the lean benchmark.
2. Analysis. Carbon reduction by eliminating the non-value adding activities identified in phase one is calculated. Both quantitative and qualitative method should be adopted. For non-value adding activities which involve the use of energy consumption and raw materials, quantitative calculation should be provided.

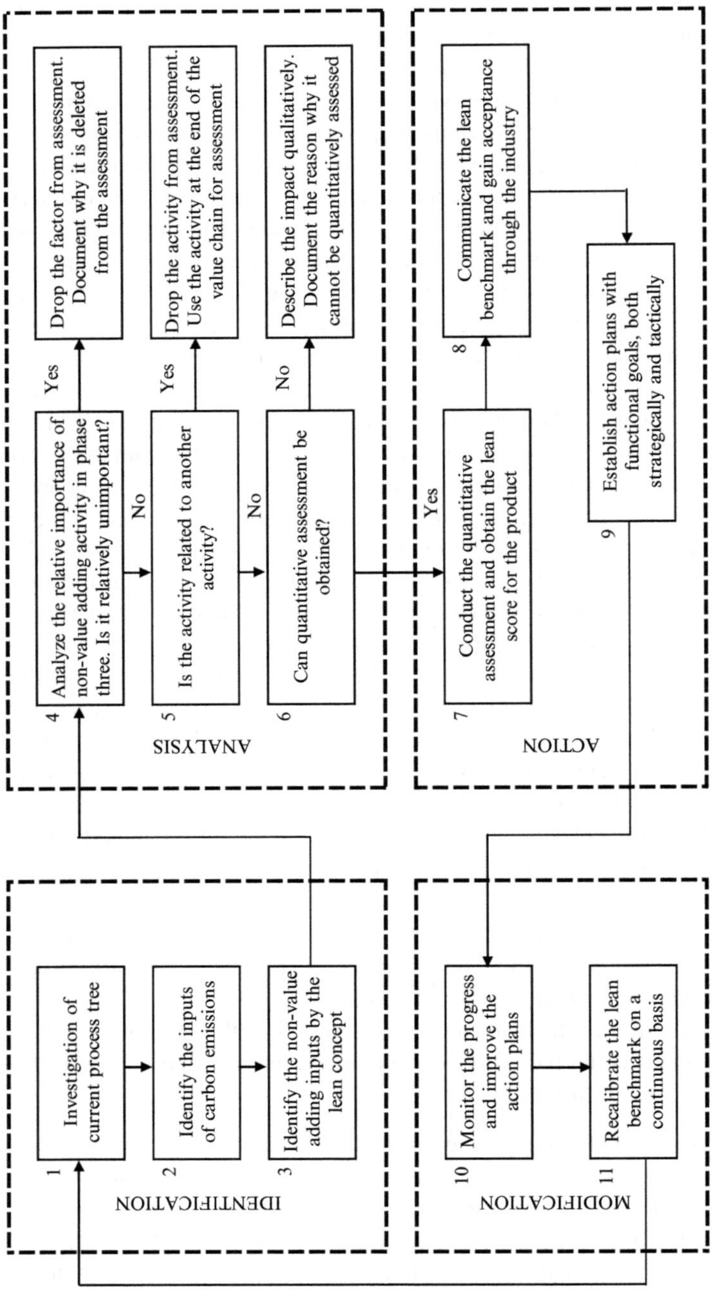

Fig. 11.1 Lean benchmarking process in carbon labelling programmes

On the other hand, for evaluating the impacts of issues that are difficult to quantify (e.g. continuous improvement and top management commitment), qualitative evaluation of the impacts should be conducted.

3. Action. A corresponding lean benchmark associated with the lean score should be obtained in this stage to provide a relative measurement about the performance of the materials towards environmental sustainability. A "continuous improvement" plan to help the manufacture and use of materials improve towards the benchmark should be prepared as well.
4. Modification. As technology improves and the production process refines, the lean score calculated in phase 3 may change. Recalibration of the lean benchmark on a continuous basis is therefore proposed in this stage to offer up-to-date information for customers.

11.3 Applying the Lean Concept to Other Construction Materials

In an industry as complex as construction, a wide range of raw materials were used in a number of different applications (CIRIA 1995). The lean benchmarking process was first designed for precast concrete products because of the origin of the philosophy i.e. the manufacturing industry. The lean production philosophy originated from the automobile industry and has been applied in the manufacturing industry for decades. The production process of precast concrete products had many similarities with the manufacturing industry so that the application of lean in the precast concrete industry would require less modifications of the lean concept.

However, the origin of the lean concept does not preclude applying the lean concept to the carbon labelling programmes of other construction materials. The production process of construction materials can be viewed as a manufacturing process. According to Groover (2010), manufacturing was the transformation of materials into items of greater value by means of one or more processing and/or assembly operations. In fact, according to Groover (2010), the terms production and manufacturing could be used interchangeably. The construction materials industry belonged to the secondary manufacturing industry which used natural resources and transformed them into consumer and capital goods.

Viewed as a manufacturing process, the production of other construction materials was very similar to the production of precast concrete products. CIRIA (1995) identified the life cycle of construction materials into three stages: production, in-service use and after-use. Production included the extraction of raw materials, storage, transportation, process and packaging, all of which were very common in the production of both precast concrete products and other construction materials.

In addition, the carbon labelling programmes designed for precast concrete products and other construction materials were the same. These programmes used LCA as the evaluation approach. Therefore, the carbon labelling programmes for other construction materials might face the same problems caused by LCA, such as the lack of evaluation about the limits, risks and potential of the production systems.

The lean concept was appropriate to address the problems caused by LCA for construction materials other than precast concrete products. A production tree could still be obtained and refined, based on which a lean score could be calculated. With both the carbon score and the lean score, the true environmental performance of the products could be more transparently and accurately indicated.

11.4 The Value Concept in the TFV Framework

Due to the rising recognition of the environmental impacts, more and more research has been conducted to develop and explain the value of the environment, which was usually referred to as environmental values. Many people argued that the environment did not only have instrumental values (i.e. to provide support to human beings), it had intrinsic moral values (Callicott 1984, 1986, 1995; Nash 1989; Norton 1991; Rolston 1988). According to Satterfield (2001), even if due to consciousness, only humans were moral agents (and thus could evaluate things); that was not to say that ecosystems, organisms and species were not morally good or possessed certain kinds of value in and of themselves. The argument of the value of the environment had developed into a research area, which was environmental ethics.

The value of the buildings (how buildings add value) was improved by considering the environment as one of the values. Ofori (1992) stated that the environment should be added as the fourth value to the construction industry along with quality, time and cost. Huovila and Koskela (1998) stated that the building industry had to adapt to new and emerging construction markets which had environmental and social dimensions. However, the consideration of the environment in the construction industry was in its infancy and there remained many problems to be solved, such as it being hard to quantify, difficult to monitor, etc. This discussion here is therefore not to assess the success and failure of such consideration but rather to take tentative steps towards a new approach to describe environmental values in the construction industry.

O'Neill and Spash (2000, p. 522) provided a few assumptions about the valuing agent (customer), including:

- Agents' values are expressions of their preferences.
- Their preferences are ordered and have a certain structure – they are transitive, reflexive, complete and continuous.
- The strength of agents' preferences for marginal changes in a bundle of goods is expressed in their willingness to pay for their satisfaction.
- Agents have subjective probabilities about the likelihood of different possible outcomes.
- Agents are instrumentally rational. They act so as to realize the greatest expected satisfaction of preferences, given budget constraints and assignments of probabilities of different possible states of the world.

It seems that the environment had certain characteristics of being a potential customer. The environment had preferences, which might include diversity, preservation of species, etc. These preferences were represented by the environment's impact on human beings. This was why many professionals argued that the environment had intrinsic values and should be respected. Dickson (2000) stated that the environmental problems might happen because people had ethically misguided attitudes towards the natural environment and these problems could be solved by getting people to appreciate the ethical significance of the natural world. Callicott (2002) broadened the concept of intrinsic value to animals and living beings other than humans.

The normally recognized values in the construction industry were quality, time and cost. Kaplan and Norton (1996) proposed a balanced scorecard to understand how buildings added value to clients. There were four aspects in the scorecard, which were financial value, indoor environmental quality, spatial quality and symbolism. Winch (2002) developed a new process based on Porter's value system concept (Porter 1985) to capture the value generated through the project life-cycle – in terms of both profits and learning. More and more researchers started to include the environment as one more pillar of values of the buildings (Ofori 1992; Huovila and Koskela 1998; Lapinski et al. 2006).

The lean concept had proven to be effective in increasing environmental benefits by eliminating waste, preventing pollution and maximizing the owners' value (Riley et al. 2005; Ferng and Price 2005; Luo et al. 2005; Lapinski et al. 2006; Huovila and Koskela 1998). Nahmans (2009) stated that it was a natural extension to apply the lean concept to achieve green construction. By applying the lean concept to a production line, 9 to 6.5 people (labor waste), 12 % space (equipment waste) and 10 % wallboard (material waste) could be reduced. EPA (2003) found that lean produced an operational and cultural environment that was highly conducive to waste minimization and pollution prevention, and that lean provided an excellent platform for environmental management tools such as life cycle assessment and design for environment.

However, when the value concept in these lean applications was examined, it was found that there were several problems. These problems included:

1. Unitary concept of value. Norton and Steinemann (2001) stated that it was useful to have a variety of ways of expressing values and a variety of ways of measuring values. Huovila and Koskela (1998) identified the contribution of the lean construction principles to sustainable development. The values included minimization of resource depletion, minimization of pollution and matching business and environmental excellence (Huovila and Koskela 1998). Luo et al. (2005) applied the lean concept to prefabrication and stated that lean could contribute to improve quality and supply chain and reduce waste. Bae and Kim (2007) found that different lean applications might have different results on the three pillar of sustainable development. For example, lean supply (the JIT system) might have influence on economic and environmental impacts rather than social impacts. In these studies, wastes, environmental burdens, and environmental deterioration were commonly

used as the requirements posed by the environment. However, it seems that these value concepts were too broadly defined. It should be noted that the definition of the value guided the preferred decisions and behavior from practitioners. If the value concept was defined too broadly, the implications for the customers (contractors, clients and the environment) would be very minimal. As can be seen in previous chapters, even in the concept of global climate change, a single aspect in environmental sustainability, there were many issues that should be considered in the value chain of precast concrete production and erection. The premise of this study was that when considering the environment as values, more systematic approach including a narrowed-down definition of value was required.

2. The evaluation of environmental values tended to use technical, computational approaches to count value (Norton and Steinemann 2001). Many of the models that have been used in valuing climate change were based on conventional economic cost-benefit analysis (Nordhaus 1993, 1994, 1997; Peck and Teisberg 1992; Manne et al. 1995). However, even the advocates of such technical, computational approaches admitted that such approaches were unable to capture large-scale, ecological values (Freeman 1993, p. 485). Conventional economic analysis gave less importance to flows that would take place in the future (Broome 1992; Price 1993, 1996). According to Padilla (2004), the application of conventional economic analysis of environmental impact removed the analysis of future impact because of its negligible present value. However, it should be noted that if future generations have certain rights that should be respected, these rights should be included in the analysis (Padilla 2002). Similarly, the objects being valued (e.g. companies, production process, construction process and products) with continuous improvement plans should achieve better results than objects without such plans. The purpose of this study, however, was not to assess the success and failure of such economic analysis of environmental values, but rather to take tentative steps towards a new approach to assess environmental values, with the contribution of the lean concept.
3. The LCA methodology used to measure the environmental impacts had several drawbacks that could not be overlooked. These drawbacks included credibility of the environmental information, the single sign used to measure the environmental information and lack of ratings on the continuous improvement plan, which have all been discussed in Sect. 11.2.2. This evaluation method could be improved by re-defining the concept of value and improving the assessment methodology with the contribution of the lean concept.
4. The interpretation of the environment as customer was limited. Most researchers agreed that the environment should be a potential customer to the facility (e.g. Ofori 1992; Porter and van der Linde 1995; Huovila and Koskela 1998). There was a trade-off from the environmental point of view, if the owner and the contractors had different economic and social priorities, which both differentiated from the environmental priorities (Huovila and Koskela 1998). Porter and van der Linde (1995) stated that successful environmentalists and companies would reject the old trade-offs and build on the underlying economic logic that linked the environment, resource productivity, innovation and competitiveness.

Based on the lean concept, the trade-off relationship between different parties and different sustainability factors (economic, social and environmental) did not always stand up. Firstly, there were some areas that the product, process and facility could be improved while achieving both economic and environmental sustainability (e.g. the management of idling trucks that could be improved by applying the JIT concept; the production management that could be improved by focusing on a smooth work flow). As can be seen in Chaps. 7 and 9, these areas represented a large proportion of the effective carbon that could be achieved by applying the lean concept. Both economic and environmental benefits could be achieved by eliminating these non-value adding activities. In addition, one concept that could not be overlooked here was environmental tolerance – the environment had the ability to endure unfavorable environmental impacts. Economic factors could be prioritized in production and construction activities that were conducted under the tolerance limits. Although the lean concept did not contribute directly to estimate the tolerance limits of the environment, it proposed a similar conception – to use lean to reflect how much the products/process should affect the environment. If the products/process outperformed the threshold established by the lean concept, it may be suggested that economic priorities could be focused upon.

11.5 Implication I: Precasters

As implied by the ranking of the non-value adding activities in different precast concrete factories, there were no universally applicable rules to follow for the precasters to achieve low-carbon production. Different precasters might face different problems that caused the increase of the carbon emissions level. The non-value adding activities that might cause an increase in the carbon emissions level were identified in Chap. 7 and a case study was described in Chap. 8.

This section highlights the implications that the precasters may adopt in future production to achieve low-carbon production, based on the observations from the 17 precast concrete factories.

Besides the many non-value adding activities identified in the precast concrete factories, there remained a few fundamental problems that must be addressed in order to achieve low-carbon production. These fundamental problems can be seen from the formula used in Sect. 8.5:

The amount of carbon = Effective carbon + Waste (raw materials, finished products and inappropriate production arrangements)

The embodied carbon of raw materials (from cradle to gate) was the largest source of emissions in the precast concrete columns, as can be seen in Table 11.4. Both steel and concrete contributed significantly to the overall embodied carbon of the precast concrete columns. It was therefore important to design the precast concrete columns to be both economically and environmentally efficient to reduce

Table 11.4 The sources of carbon emissions of the precast concrete column

Category	Carbon emissions ($kg\ CO_2$/column)	Percentage (%)
Manufacture of raw materials	464.4079	76.10
Transport of raw materials (International)	105.4563	17.30
Concrete plant operation	25.5200	4.20
Transport of concrete and steel (Local)	10.8038	1.80
Precast concrete production	3.4015	0.60
Total	609.5895	100.00

the usage of concrete and steel. Precasters might not conduct any research relating to alternative designs that could reduce the use of raw materials due to the costs associated with the research. The responsibility then fell onto the government authorities (e.g. The Prefabrication Technology Centre of HDB) to develop such designs.

In addition, the use of green building materials (e.g. the green concrete) should be advocated by the precasters. Incentive schemes should be provided by the government to encourage the precasters to use such green building materials. As could be seen from the case study in Chap. 10, the G35 green concrete was only S$5 more expensive than the normal concrete. This type of concrete could be largely used in low-rise and medium-rise public housing projects without compromising the strength of the structure.

As can be seen from Table 11.4, transportation of raw materials had large impact on the level of carbon emissions of the precast concrete products. 17.30 % of the total embodied carbon emissions was generated by international transport of raw materials. International transport of raw materials should be specially noted by Singapore and other countries without local supplies of raw materials. This amount of carbon emissions could be readily reduced for countries with local supplies of raw materials (e.g. China, U.S., UK).

11.6 Implication II: Contractors

As implied by the comprehensive list of the non-value adding activities in the construction sites, there were no universally applicable rules to follow for the contractors to achieve low-carbon erection of precast concrete products. Different contractors might face different problems that caused the increase of the carbon emissions level. The non-value adding activities that might cause an increase in the level of carbon emissions were identified in Chap. 9 and a case study was described in Chap. 10.

This section provides the implications that the contractors may adopt in future production to achieve low-carbon production, based on the observations from the 30 construction sites.

There seems to be two possible strategies for the contractors to adopt to reduce carbon emissions. First, deviations from the erection cycle should be eliminated. The effective carbon for one erection cycle was significantly lower than the effective carbon for one production cycle (8.53 kg CO_2 compared to 609.59 kg CO_2). However, this does not mean that the erection cycle was not important. In fact, the contractors were less effective when managing the carbon emissions. Carbon emissions generated from non-value adding activities in the construction sites were 2.45 times bigger than the effective carbon of the erection cycle. Non-value adding activities in the erection cycle represented great efficiency loss that should be brought to the full attention of the contractors.

Training and development for human resources relating to the knowledge of the lean concept should be improved. It should be noted that this was not relevant to the competency of the employees. In the construction sites with or without competent employees, as rated by the interviewees, there were non-value adding activities. It is therefore proposed that two types of team should be appropriately trained by the contractors. The first type of team is the Kaizen team, who will conduct visual observations in the construction sites. By conducting the value stream mapping process and using basic tracking tools, non-value adding activities can be identified by the Kaizen team, based on which quick fixes can be executed. The second type of team is the Lean team, who will conduct a higher level of value stream mapping with value added focus. While the Kaizen team usually conducts short-term visit in the construction sites, the Lean team should base the value stream mapping process on a long-term basis. Continuous improvement plans provided by the Lean team can be used in future projects to achieve better performance.

11.7 Implication III: Regulatory Authorities

This study identified sources of carbon emissions that were not normally recorded in carbon labelling programmes. The implications of this study to regulatory authorities could be seen from three levels:

First of all, regulatory bodies might refer to this study to improve the code of practice for precast concrete products. For example, the Singapore Productivity and Standards Board proposed a code of practice for precast concrete slab and wall panels in 1999 (Singapore Productivity and Standards Board 1999). The Hong Kong government recommended the use of a code of practice for precast concrete construction (HKSAR 2003). However, such codes of practice were mainly related to technical issues. With the rising recognition of sustainable development, many emerging issues should be added to these codes of practice. It is therefore proposed that the lean concept be briefly introduced in these codes of practice for the precasters to focus on the efficiency of their production activities. While this study addressed the contribution of lean to low-carbon production, the use of lean in these codes of practices might not be restricted to this area. As indicated by this study, some of the sustainability factors were used interchangeably. The lean concept could also be used to improve the performance of other sustainability factors. For example, sustainability factors such as operational energy, transport energy, waste and production efficiency could also be improved by applying the lean concept.

Green construction material is not a simple concept to indicate the construction materials with low embodied carbon. Green building is also not a simple fusion of green design, techniques and materials. Both concepts involve a holistic approach to achieve sustainable development either in the product or the project life cycle. Although the concept of life cycle was popularly used by a majority of the professionals, most concentration has been focused on design and technical-related areas that could be applied in the design, production and construction stage. However, this study indicated that there was a trend that the project management process, i.e. the planning sessions and regular progress meetings to ensure a smoother flow or handover of the green baton, could offer a low-cost low-barrier avenue and should be appreciated by precasters, contractors and regulatory authorities.

Many regulatory authorities might start to take the lean concept into consideration when evaluating the environmental impacts of products. For example, the Singapore Green Labelling Scheme (SGLS) initiated by the Singapore Environment Council used the following descriptive text and associated promotional collaterals for precast concrete products (SEC 2010):

Eco Friendly Building Material
SGLS User Agreement Number
Recycled Content: xx %
Carbon Emission Value: xx CO_2 kg/m^3

With the contribution of the lean concept, the descriptive text and associated promotional collaterals could be revised to:

Eco Friendly Building Material
SGLS User Agreement Number
Recycled Content: xx %
Carbon Emission Value: xx CO_2 kg/m^3
Lean Value/Score: xx CO_2 kg/m^3 or xx % of the Carbon Emissions Value

This would help the customer to truly identify the environment-friendly products in the market. The lean value/score could also be used by the precasters to identify their potential improvements.

11.8 Validation of Results

As presented in Chap. 6, the validation of the results followed three processes, which were the review of the calculation process, comparison with previous research and external review.

The calculation of the lean implementation process involved two steps. First, the value stream mapping process was conducted for the production cycle and erection cycle of the precast concrete products. This was to ensure that the inputs of carbon emissions due to raw materials and energy were systematically considered without any missing inputs. The value streams of the production cycle and erection cycle were obtained from real-life case studies in one precaster and one contractor in the Singapore precast concrete industry. The inputs of raw materials and energy were confirmed with the precaster and contractor to ensure that there were no missing inputs.

The calculation process followed the IPCC guidelines for national greenhouse gas inventories (IPCC 2008). The Tier 1 approach using world-wide average emission factors was adopted for most construction materials. However, it should be noted that these emission factors were on a "cradle-to-gate" basis. This strategy was adopted due to the similarities of the production processes of concrete in different countries. Reasonable assumptions were made to simulate the transportation process. By doing so, this study best simulated the production and erection process of the precast concrete products in the Singapore precast concrete industry. Based on the production and erection process, carbon emission values of both the production cycle and the erection cycle were calculated.

The emission factors used in this study complied with those presented in a series of LCA studies. The information sources have been provided in Sects. 8.3.2 and 10.4. One important thing that should be noted was the emission factors for Portland cement. Traditional Portland cement manufacturing would produce almost 1 tonne CO_2 per tonne of cement produced. However, as technology improved (e.g. carbon sequestration and the use of cement kiln dust), the emission factor could be significantly reduced. This study therefore took steps to obtain more up-to-date information to fully represent current practices. If the researchers would like to use traditional data, the emission factor of Portland cement should be revised using the respective figure, e.g. 819 kg/t as proposed by the British Precast Concrete Federation (2009).

In addition, external reviews of the results were conducted by the executive engineer of the precaster, the senior production manager of the contractor and two professional groups (three LCA professionals in each group). The results were approved by these professionals. However, as many world-wide average emission factors were used in this study, the accuracy of the quantifications should be focused upon. For example, as it was extremely difficult to pin-point exactly where the raw materials originated from, domestic transport of raw materials in the country of origin/production was not considered in this study. Making reasonable assumption were the common strategies taken in LCA studies. More importantly, the reader should not lose sight of the focus of this study, i.e. to apply the lean concept to identify and eliminate the non-value adding activities in precast concrete production and erection to achieve low-carbon production and low-carbon erection. The calculation process was less significant to this study than the identification process. Using world-wide emission factors and making reasonable calculation assumptions did not change the focus and the significance of this study.

References

American Institute of Architects (2007) Architects and climate change. [Online] Available at: http://www.aia.org. Accessed 20 Feb 2008

Bae JW, Kim YW (2007) Sustainable value on construction project and application of lean construction methods. In: Pasquire C, Tzortzopoulos P (eds) Proceedings of IGLC-15, Michigan, USA

Balcomb JD, Curtner A (2000) Designing sustainable buildings: MCDM023, a tool for making right decisions. In: Proceedings of solar 2000, solar powers life, share the energy. Solar Energy Society, Madison

Ball J (2002) Can ISO 14000 and eco-labelling turn the construction industry green? Build Environ 37(4):421–428

BRE (2010) Environmental profiles methodology. [Online] Available at: http://www.bre.co.uk/greenguide/page.jsp?id=2106. Accessed 15 Feb 2010

British Precast Concrete Federation (2009) The concrete industry sustainability performance report, 1st report. The Concrete Centre, Surrey

Broome J (1992) Counting the costs of global warming. White Horse Press, Cambridge

Callicott JB (1984) Non-anthropocentric value theory and environmental ethics. Am Philos Q 21:299–309

Callicott JB (1986) On the intrinsic value of nonhuman species. In: Norton BG (ed) The preservation of species. Princeton University, Princeton, pp 138–172

Callicott JB (1995) Environmental ethics: overview. In: Reich WT (ed) The preservation of species. Princeton University, Princeton, pp 138–172

Callicott JB (2002) The pragmatic power and promise of theoretical environmental ethics: forging a new discourse. Environ Value 11(1):3–25

Construction Industry Research and Information Association (1995) Environmental impact of materials, volume A: summary. CIRIA Special Publication, London

Dickson B (2000) The ethicist conception of environmental problems. Environ Value 9(2):127–152

EPA (2003) Lean manufacturing and environment. [Online] Available at: http://www.epa.gov/lean/performance/index.htm. Accessed 14 Nov 2010

Ferng J, Price ADF (2005) An exploration of the synergies between Six Sigma, total quality management, lean construction and sustainable construction. Int J Six Sigma Compet Advant 1(2):167–187

Freeman R (1993) The measurement of environmental and resource values. Resources for the Future, Washington, DC

Gluch P, Baumann H (2004) The life cycle costing (LCC) approach: a conceptual discussion of its usefulness for environmental decision-making. Build Environ 39(5):571–580

Grant T, Macdonald F (2009) Life cycle assessment as decision support: a systemic critique. In: Horne R, Verghese K, Grant T (eds) Life cycle assessment: principles, practice and prospects. CSIRO Publishing, Collighwood

Groover MP (2010) Principles of modern manufacturing, 4th edn. Wiley, New York

Hemmelskamp J, Brockmann KL (1997) Environmental labels – the German 'Blue Angel'. Futures 29(1):67–76

Hertwich EG, Pease WS, Koshland CP (1997) Evaluating the environmental impact of products and production processes: a comparison of six methods. Sci Total Environ 196(1):13–29

HKSAR (2003) Code of practice for precast concrete construction 2003. Buildings Department, Hong Kong

Huovila P, Curwell S (2007) Sustainability assessment of building design, construction and use. In: Curwell S, Deakin M, Symes M (eds) Sustainable urban development. Routledge, New York

Huovila P, Koskela L (1998) Contribution of the principles of lean construction to meet the challenges of sustainable development. In: Formoso CT (ed) Proceedings of IGLC-6, Guaruja, Brazil

IPCC (2008) 2006 IPCC guidelines for national greenhouse gas inventories, volume 2: energy. [Online] Available at: http://www.ipcc-nggip.iges.or.jp/public/2006gl/vol2.html. Accessed 15 Jan 2009

Kaplan R, Norton D (1996) The balanced scorecard. Harvard Business School Press, Boston

Karl H, Orwat C (1999) Economic aspects of environmental labelling. In: Folmer H, Tietenberg T (eds) The international yearbook of environmental and resource economics 1999/2000: a survey of current issues. Elgar, Cheltenham, pp 107–133

Lapinski AR, Horman MJ, Riley DR (2006) Lean process for sustainable project delivery. J Constr Eng Manage 132(10):1083–1091

Luo Y, Riley D, Horman MJ (2005). Lean principles for prefabrication in Green Design-Build (GDB) projects. In: Russell K (ed) Proceedings of IGLC-13, Sydney

Manne A, Mendelsohn R, Richels R (1995) MERGE: a model for evaluating regional and global effects of GHG reduction policies. Energy Policy 23(1):17–34

Nahmens I (2009) From lean to green construction: a natural extension. In: Ariaratnam ST, Rojas EM (eds) Proceedings of the 2009 construction research congress, Seattle, pp 1058–1067

Nash RF (1989) The rights of nature. University of Wisconsin, Madison

Nordhaus WD (1993) Rolling the dice: an optimal transition path for controlling greenhouse gases. Resour Energy Econ 15(1):27–50

Nordhaus WD (1994) Managing the global commons: the economics of climate change. MIT Press, Cambridge, MA

Nordhaus WD (1997) Discounting in economics and climate change. Climate Change 37(2):315–328

Norton B (1991) Thoreau's insect analogies: or, why environmentalists hate mainstream economists. Environ Ethics 13(3):235–251

Norton BG, Steinemann AC (2001) Environmental values and adaptive management. Environ Value 10(4):473–506

Ofori G (1992) The environment: the fourth construction project objective? Constr Manage Econ 10(5):369–395

O'neill J, Spash CL (2000) Appendix: policy research brief: conceptions of value in environmental decision-making. Environ Value 9(4):521–536

Padilla E (2002) Intergenerational equity and sustainability. Ecol Econ 41(1):69–83

Padilla E (2004) Climate change, economic analysis and sustainable development. Environ Value 13(4):523–544

Peck SC, Teisberg TJ (1992) CETA: a model for carbon emissions trajectory assessment. Energy J 13(1):55–77

Porter ME (1985) Competitive advantage: creating and sustaining superior performance. Free Press, New York

Porter ME, van der Linde C (1995) Green and competitive: ending the stalemate. Harv Bus Rev 75(5):120–134

Price C (1993) Time, discounting and value. Blackwell, Oxford

Price C (1996) Discounting and project appraisal: from the bizarre to the ridiculous. In: Kirkpatrick C, Weiss J (eds) Cost-benefit analysis and project appraisal in developing countries. Edward Elgar, Cheltenham, pp 90–101

Riley D et al (2005) Lean and green: the role of design-build mechanical competencies in the design and construction of green buildings. In: Tommelein T (ed) Proceedings of the 2005 construction research congress, San Diego, pp 116–120

Rolston H (1988) Environmental ethics. Temple University, Philadelphia

Satterfield T (2001) In search of value literacy: suggestions for the elicitation of environmental values. Environ Ethics 10(3):331–359

SEC (2010) Overview of the green label scheme. [online] Available at: http://www.sec.org.sg/awards/greenlabel/overview. Accessed 6 Mar 2010

Singapore Productivity and Standards Board (1999) CP 81: 1999 – code of practice for precast concrete slab and wall panels. Singapore Productivity and Standards Board, Singapore

Smith TM, Fischlein M, Suh S, Huelman P (2006) Green building rating systems: a comparison of the LEED and Green Globes systems in the US. [Online] Available at: http://www.usgbc.org/docs/LEEDdocs/LEED_RS_v2-1.pdf. Accessed 12 Aug 2008

Trusty W (2001) Life cycle assessment. An oral presentation at NRC – VTT seminar. Athena Institute, Ottawa

Vijayan A, Kumar A (2005) A review of tools to assess the sustainability in building construction. Environ Prog 24(2):125–132

Winch GM (2002) Managing construction projects: an information processing approach. Blackwell Science, Oxford

Chapter 12
Conclusions and Recommendations

This chapter presents the summary, conclusions and recommendations of this study. Limitations of this study and suggestions for future research are also presented in this chapter.

12.1 Summary

In this section, each part of the study is summarized. The structure of the study is shown in Fig. 1.3.

12.1.1 Part I: Literature Review

Part I covered the research background of this study, which included sustainable development, lean production philosophy and the precast concrete industry. Chap. 2 reviewed the concept of sustainable development in three components: economic, social and environmental sustainability. The review showed that it is possible to reduce carbon emissions from the perspective of management. Chap. 3 reviewed the concept, principles and development of the lean production philosophy. It is found that these pieces of information were used to develop the comprehensive list of non-value adding activities for both precasters and contractors. In addition, the applicability of the lean production philosophy to solve "green" problems was examined in Chap. 3. Chapter 4 reviewed the production and erection processes of the precast concrete products in order to identify the previous applications of the lean production philosophy in the precast concrete factories and examine the applicability of the lean production philosophy to reduce carbon emissions in the value chain (cradle-to-gate) of the precast concrete products. Two pilot studies were also conducted to examine the applicability of the lean production philosophy.

P. Wu and S.P. Low, *Lean and Cleaner Production: Applications in Prefabrication to Reduce Carbon Emissions*, DOI 10.1007/978-3-642-42062-7_12,

It was found that a few lean principles have been applied in the precast concrete factories as isolated applications. As there were many areas that could be improved by the lean production philosophy, it seems that the lean concept might be applied in the precast concrete production process to reduce carbon emissions.

12.1.2 Part II: Theoretical Background

Part II covered the theoretical background of this study, which included sustainable science, model of manufacturing, economic explanation of production and the environment, and more importantly, the core of environmental management. Part II aimed to bridge the lean production philosophy with sustainable development on two platforms, which were the precast concrete industry and global climate change. The TFV framework proposed by Koskela (2000) was adopted in this study. The design of this study followed the core of environmental management, which included the science of ecology, the dynamic equilibrium and the systems concept (i.e. global, social and human systems). Chapter 6 covered the methodology of this study which was divided into two steps: identification and assessment. A questionnaire research approach was adopted to identify the non-value adding activities. Case study research approach was also used to quantify the lean improvements in precast concrete factories and construction sites. The core of this chapter was to create a method to identify the non-value adding activities in the value chain of the precast concrete products, which would be further tested and examined, both quantitatively and qualitatively.

12.1.3 Part III: Lean Applications by Precasters

Part III examined the lean applications by precasters from four perspectives: site layout management, supply chain management, production management and stock management. In Chap. 7, the non-value adding activities from these four perspectives were identified and rated. The importance of the non-value adding activities (which was represented by the degree of severity in this study) was identified from semi-structured interviews with seventeen precasters in Singapore. Non-parametric analysis of these non-value adding activities was also provided in Chap. 7. Mitigation strategies and actions were provided for the precasters to improve. Chapter 8 quantified the lean improvements on carbon emissions level by applying the lean production philosophy in one precaster in Singapore. The contribution of this study to the carbon inventory of the Singapore construction industry was also provided in Chap. 8.

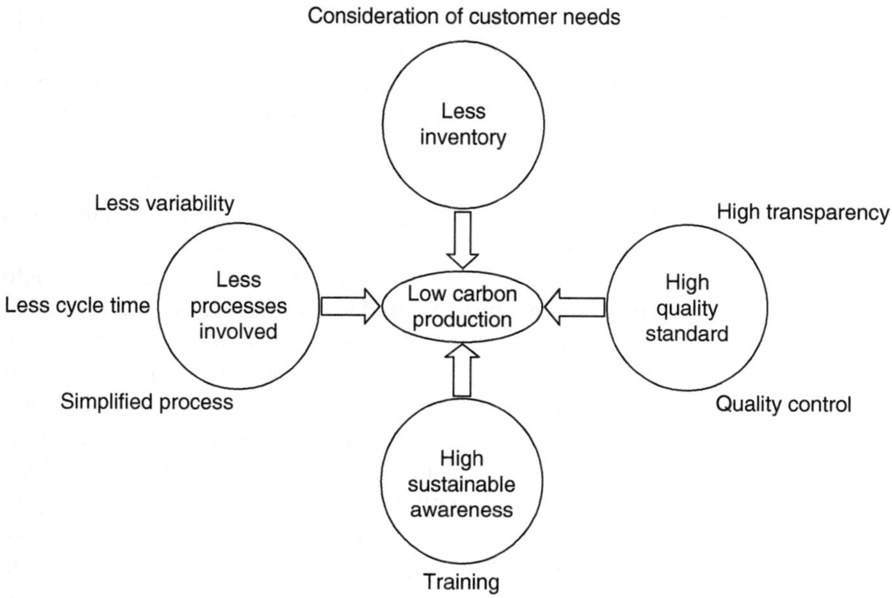

Fig. 12.1 How lean production principles help to reduce carbon emissions

12.1.4 Part IV: Lean Applications by Contractors

Part IV examined the lean application by contractors from four perspectives: site layout management, transportation management, stock management and erection management. The importance of the non-value adding activities was identified from semi-structured interviews with thirty contractors in Singapore in Chap. 9. Parametric analysis of these non-value adding activities was also provided in Chap. 9. Mitigation strategies and actions were provided for the contractors to improve. Chapter 10 quantified the lean implementation by examining how much carbon emissions could be reduced by applying the lean production philosophy in one contractor in Singapore.

12.1.5 Implications and Conclusions

The lean production philosophy has proved to be effective in productivity improvement (Voss and Robinson 1987; Wantuck 1989; Lim and Low 1992). In this study, lean production principles can be used to reduce carbon emissions by eliminating waste of raw materials, waste of finished products and inappropriate production/installation arrangements (see research objective 1). Carbon emissions caused by these activities can be significantly reduced if appropriately managed (see research objective 2). This is illustrated in Fig. 12.1.

Compared with other carbon reduction technologies, such as research into environment-friendly materials, passive designs, etc., it is found that lean production principles can provide a viable, affordable and beneficial solution to both the precasters and the contractors to achieve low carbon emissions (see Sects. 8.5 and 10.6).

Lean production principles can be used as stand-alone applications, which means that the entrance barrier is relatively low. Companies and organizations can choose one or several principles which can be best applied in the organization to achieve less energy consumption and low carbon emissions. However, it should be noted that while some lean principles can produce instantaneous results, some of the improvement processes are based on a "continuous improvement" basis that may not produce instantaneous results. Without instantaneous results, companies and organizations may be reluctant to apply lean production principles for their daily activities. In addition, the organizational structure and decision-making process may have to change due to these additional considerations, which may affect short-term performance. Thus, before applying the lean production philosophy to their daily activities to achieve energy efficiency and reduce carbon emissions, companies and organizations should be fully prepared for the impending changes and challenges (which can be achieved by using the VSM tools) and keep in mind that on a "continuous improvement" basis, a lean environment can ultimately be created. This study provided a list of the non-value adding activities in the production and installation processes of a specific precast concrete product. It can be used by both precasters and contractors to improve their environmental performance, especially in the aspect of carbon emissions (see research objective 3).

Lean application from the precasters' point of view is investigated because precast concrete components are widely adopted around the world and the market share is growing annually. There are similarities between the production of precast concrete components and the manufacturing processes (where lean originated) for lean to be applied with minimum modifications. For both the precasters and the contractors, lean application can be used to provide possible solutions so that works can be done right the first time without the involvement of non-value adding activities. Precasters can adopt the lean production principles to reduce carbon emissions from waste of raw materials, waste of finished products and inappropriate operations. The amount of carbon emissions that can be reduced by applying the lean production philosophy in one specific precast concrete product was assessed based on a detailed LCA study (see research objective 4).

As explained earlier, when applying lean production principles, companies and organizations should be fully prepared for potential changes and impacts to their organizational structure and daily activities. This study examined contribution of the lean concept in developing a relative measurement of the environmental impacts (see research objective 5), which might affect the performance of precasters and contractors. Based on a "continuous improvement" basis, a lean environment can be achieved in the end. It should be noted that although applying a single lean production principle may not be totally new in the construction industry, it is a real challenge to implement the lean thinking to solve problems.

12.2 Main Findings

This study aims to apply the lean production principles to reduce carbon emissions in the precast concrete industry and provide a management tool-kit for both precasters and contractors to reduce carbon emissions at lower investment costs. Chapters 7 and 9 examined the applications of lean by both precasters and contractors and provided possible areas that could be improved to achieve either low-carbon production or low carbon construction. Since these improvements are mostly from the perspective of management, the investment costs are very low, compared to technological improvements that are used to achieve low-carbon production or low-carbon construction. Chapters 8 and 10 studied the applications of lean by one precaster and one contractor respectively. Management actions that could be taken to reduce carbon emissions were also provided in Chaps. 8 and 10. Therefore, the research aim was properly addressed.

In particular, the research objectives that were set out in Sect. 1.3 were addressed as follows:

1. Objective 1 was to identify the non-value adding activities which are closely related to unnecessary carbon emissions in the precast concrete industry. The non-value adding activities in both precast concrete production and construction were identified in Chaps. 7 and 9. The relative importance of these non-value adding activities was also provided for both precasters (see Sect. 7.7) and contractors (see Sect. 9.7).
2. Objective 2 was to quantify the improvements by examining how much carbon emissions can be reduced. Case studies relating to the use of the lean production principles to reduce carbon emissions, including one precaster and one contractor, were provided in Chaps. 8 and 10 respectively. The amount of carbon emissions that could be reduced by applying the lean production principles were assessed, both quantitatively and qualitatively (see Sects. 8.5 and 10.6).
3. Objective 3 was to provide a management tool-kit for both precasters and contractors. In Chaps. 7 and 9, the non-value adding activities were ranked based on their degree of severity. Non-parametric tests (for precasters) and parametric tests (for contractors) were conducted for these non-value adding activities. This ranking can be used as a check list for the precasters and contractors to focus on the most important activities, which can help improve efficiency while achieving low-carbon production or low-carbon construction.
4. Adding to the Singapore-specific carbon inventory by identifying the embodied carbon of precast concrete components. The embodied carbon of a specific type of precast concrete column used in Singapore was estimated in Chap. 8. The estimation method and procedure were provided in Sect. 8.3. Future studies that aim at completing the Singapore-specific carbon inventory for construction materials can refer to this study for the method and procedure used.
5. Developing a relative measurement of the environmental impacts, especially in carbon-related evaluation. The relative measurement of the environmental impacts was investigated in Chap. 11. Such relative measurement was

completed with the introduction of the lean concept in many environmental labeling programmes and was referred to as the lean score or lean benchmarking in this study (see Sect. 11.2). The applicability of this relative measurement to construction materials other than precast concrete products was also examined in Sect. 11.3.

12.3 Contributions to Theory and Knowledge

The following theoretical conclusions can be drawn from this study, as be seen in Table 12.1:

1. This study re-examined the concept of value in the lean production philosophy, especially in the Transformation-Flow-Value (TFV) framework proposed by Koskela (2000). The concept of value has been broadened to include the environment in previous studies. However, it was found that such extensions were mostly too broadly defined. As the extensions will be used to guide preferred decisions and behaviours of the practitioners, these should be discussed in details. This study shows that the value concept in lean can be extended to include environmental issues, i.e. carbon emissions in this study. Such extension adds value to many stakeholders, including suppliers, contractors, regulatory authorities and more importantly, the environment. Previous researchers have challenged the value concept in lean to be unitary. However, as suggested by this study, it may be stated that the value concept might be unitary at a point in time, but it is not impossible to extend the value concept by carefully considering the link between the concept and new areas of research (e.g. carbon emissions in this study).

Table 12.1 Contributions to theory and knowledge

Theory and knowledge	New thinking proposed in this study
Concept of value (Koskela 1992, 2000; Koskela et al. 1997)	The value concept might be unitary at a point in time, but it is not impossible to extend the value concept by carefully considering the link between the concept and new areas of research
Environmental management (Enrlich and Enrlich 1972; Meadows et al. 1972; Garlauskas 1975)	If the products/processes outperformed the threshold posed by the lean concept, it may be suggested that economic priorities could be focused upon
The evaluation of environmental values (e.g. Drèze and Stern 1987)	Lean benchmark can be integrated in the evaluation to improve public discourse
The lean concept	The lean concept does not only provide an envelope to encompass various manufacturing and production theories, but also provides a platform on which many relevant theories and principles can be further tested and developed

2. The evaluation of environmental values tended to use technical, computational approaches to count value (Norton and Steinemann 2001). Conventional economic analysis gave less importance to flows that take place in the future (Broome 1992; Price 1993, 1996). This study shows that the lean production philosophy can be used to evaluate environmental values to overcome such problems. This is a relative measurement by building up a benchmark that will take place in the future. It is useful and helpful to introduce the lean concept to guide the evaluation of environmental changes to improve upon public discourse. Economic evaluation of the environment can therefore be improved with the contribution of the lean concept. However, it should be noted that this study does not shift completely from technical, computational approaches but rather serve as an evaluative heuristic to discuss other possible approaches to solve environmental problems.
3. The old trade-off relationship between different parties is challenged by the lean concept. This point has been explained in Sect. 11.4.
4. Derived from the lean production philosophy, Value Stream Mapping (VSM) can be used to identify the types of waste in the value stream. In fact, by combing both the concept of VSM and sustainability science, this study provided an integrated framework to analyze the possible actions that can be taken to meet the challenge of sustainable development. The science of ecology, the dynamic equilibrium and the systems concept in environmental management were tested on the platform of lean to address the impact of global climate change on the precast concrete production and installation cycle. Therefore, it may be argued that the lean concept does not only provide an envelope to encompass various manufacturing and production theories, but also provides a platform on which many relevant theories and principles can be further tested and developed.
5. Various factors that are related to sustainability in the precast concrete industry were identified in this study. These factors can contribute to the studies of the precast concrete industry.

12.4 Contributions to Practice

The implications for precasters, contractors and regulatory authorities have been discussed in Sects. 11.5, 11.6 and 11.7. Practical recommendations for precasters and contractors have been made in Sects. 11.5, 11.6, and 11.7. However, it should be noted that there are a few practical conclusions and recommendations that can be drawn from this study:

1. The various non-value adding activities in both precast concrete factories and construction sites need to be well managed by the precasters and contractors. The most important non-value adding activities that can increase the level of carbon emissions were provided in Sects. 7.7 and 9.7. Each non-value adding activity determines one of the aspects of precasters' production or contractors' erection practices.

2. The consideration for applying the lean production philosophy to reduce carbon emissions in precast concrete factories and construction sites needs to conduct a VSM process at the very start. Although it does not seem possible that the precasters and contractors face all the types of non-value adding activities, the VSM process should be conducted to fully identify the impact brought about by reducing the non-value adding activities. However, this study suggests the strategies to rationally choose the non-value adding activities with high priorities as proposed in Chaps. 7 and 9. This study therefore proposes that precasters and contractors can manage the non-value adding activities both synergistically and individually.
3. Precasters and contractors who are preparing to apply the lean concept to reduce carbon emissions should pay attention to the dynamic nature of both the concept of lean and carbon emissions. This is related to the principle of "continuous improvement" in the lean production philosophy. In practice, precasters and contractors should continuously monitor and respond appropriately to the application.
4. Good understanding of the lean concept may provide opportunities for the precasters and contractors to meet other challenges posed by sustainable development. The various non-value adding activities, the methods and the procedures to quantify the carbon emissions can be used as a platform for the precasters and contractors to deal with other sustainable challenges, such as solid wastes, noise and efficiency.
5. This study will be especially useful when the precasters or contractors in Singapore decide to follow the Singapore Green Labelling Scheme, the BCA Green Mark Scheme as provided for in the Building Control (Buildability) Regulations 2011.

    - Singapore Green Labelling Scheme. This study provided sources of carbon emissions that can be reduced when producing precast concrete columns (see Sects. 7.7 and 9.7). The applicability of the lean production philosophy to identify and eliminate the non-value adding activities for other precast concrete products and construction materials has also been identified (see Sect. 11.3). By applying the lean production philosophy to the production process, the precasters can achieve a lower carbon emissions value which is indicated in the green label (see Sect. 11.7).
    - BCA Green Mark Scheme. As suggested by the discussions in Sect. 2.3, this study will be useful for contractors to improve their performance on a few assessment criteria listed in the BCA Green Mark Scheme (e.g. implement environmental friendly programmes, environmental friendly products and building quality assessment). It should be noted that the BCA Green Mark Scheme can also benefit from this study by including benchmarking and continuous improvement, two important principles of lean, in the assessment criteria. This strategy has already been taken by the Green Globes in its latest revised version (see Sect. 2.3). The new BCA Green Mark 4.0 has been released in August, 2010 (BCA 2012). The new version introduced a few

assessment criteria which have been discussed in this study, including the use of green concrete, sustainable products and sustainable practices during construction. It is believed that the contractors will improve their performance based on the new BCA Green Mark 4.0.

- The legislation on Buildability was introduced by the Building and Construction Authority (BCA) in 2001 under the Building Control Act to promote buildable design through greater adoption of prefabricated, modular and standardized building components (BCA 2011). Following an amendment to the legislation, there are minimum requirements in Buildable Design Score and Constructability Score that need to be met by the designers and the contractors respectively. This study will be useful for the designers, developers and contractors to fulfill these minimum requirements, especially in the section on "Good Industry Practices". For example, by following the lean benchmarking process (see Fig. 11.1), the contractors can obtain the allocated points from assessment criteria (d), which is "to conduct monthly work study sessions, to scrutinize and improve the work process on site, as well as minimizing wastage and improve productivity" (BCA 2011).

12.5 Limitations of the Research

This study focuses on construction activities which are related to the precast concrete industry. This is because the precast concrete industry and the manufacturing industry have many similarities and the lean philosophy, which has been adopted in the manufacturing industry for many years, can also be applied in the construction industry. However, it should be noted that there remains many differences between the construction industry and the manufacturing industry which may hinder the application of the lean production principles. The cultural differences between the two industries cannot be neglected when identifying the applicability of lean production principles in the construction industry to reduce carbon emissions. It is therefore proposed that further studies take these differences into consideration when identifying the applicability of lean production principles in the construction industry.

Although this study aims to develop a management tool-kit for both the precasters and contractors to reduce carbon emissions in their daily production and construction activities, this study might face some limitations from the data collection and data analyzing process. The data was collected from Singapore-based precasters and contractors, which might not present the best lean applications. The population size of the precasters was 20 and in order to achieve accuracy in this study, a sample size of 20 was required. However, as two precasters refused to join the study, the sample size was reduced to 18, which might adversely affect the accuracy of this study. Similarly, the sample size of contractors was small (=30) and the small sample size therefore rendered it harder to identify the significant relationships from the data. In order to overcome the problems that were brought about by the small sample size, several strategies were adopted. Firstly, precasters and contractors from higher financial grade were preferred, as they might present the Singapore precast concrete industry due to their large market share. Secondly, specific analysis and case studies were conducted for cross-referencing purposes. The results of this study were not solely based on the survey but rather a combination of the survey, specific analysis and case studies. In addition, it should be noted that the purpose of the survey and its analysis is not for hard scientific proof but rather for indications of trends in the data. Small sample size of precasters and contractors in higher financial grade might be used for such purpose.

Due to the sampling method adopted, the samples were selected from the population in a nonrandom way. The results might be biased because the interviewees might only present their best practices. Case study research was therefore added to overcome this problem. In the data analyzing process, the applications were tested in the production process for precast concrete columns. Due to the minor differences between the production processes of different precast concrete products, the applications may need to be revised based on the specific features of the production process of other precast concrete products.

In addition, Singapore has not fully developed its entire carbon inventory. The world-wide average data relating to embodied carbon was adopted in this study, where Singapore-specific data was not available. However, it should be noted that Singapore is a country which relies heavily on imports of raw materials and products. Hence, the embodied carbon may be quite different from the world-wide averages, when taking transportation into consideration. For example, as it is extremely difficult to pin-point exactly where the raw material originates from, domestic transport of raw materials in the country of origin/production is currently not considered in this study. More importantly, this study aims to calculate the CO_2 emissions using the emission factors. However, it should be noted that global climate change can be caused by both CO_2 and non-CO_2 gases. Emissions from non-CO_2 gases are strongly dependent on the technology used, which is referred to as the Tier 3 approach in this study. In order to assess the impact of GHG emissions to the global climate, both CO_2 and non-CO_2 should be taken into consideration.

The quantification of carbon emissions seemed to be another limitation of this study. It should be noted that although certain applications of the lean production principle can be verified by quantifying its contribution in reducing carbon emissions, there are some other principles whose contributions are difficult to identify. For example, by training operators using eco-operation manuals, carbon emissions may be reduced in precast concrete factories and in construction sites. However, how much carbon emissions can be reduced depends heavily on the programs that are provided and whether the employees can really learn from these programs. As presented in this study, many non-value adding activities are caused by inexperienced employees. The reasons that lead to inexperienced employees should be investigated in future studies. In addition, the detailed quantification process for the lean principles with a "non-technical" background should be improved.

12.6 Suggestions for Future Research

Following this study, some future research may be suggested for both theoretical and practical aspects.

In this study, the TFV framework was analyzed and used to identify the non-value adding activities in both the precast concrete factories and the construction sites. The value concept of the TFV framework was refined to include the environment as one important consideration. However, it should be noted that the development of the theory for lean production and lean construction would never end. Many theories that contribute to the concept of lean, including Scientific Management Theory and Systems Theory, should be examined. A suggestion can be made for future studies to use other frameworks to examine the contributions of the lean production philosophy to meet the challenges of sustainable development.

According to Ballard and Howell (1998), the applicability of lean principles to construction might seem to require that construction's differentiating characteristics be softened or explained away. This was the strategy adopted in this study to make construction more like manufacturing where lean originated. Construction projects using precast concrete products require prefabrication and assembly. By choosing construction projects using precast concrete products, the construction process is simplified because much work is completed in the prefabrication process in precast concrete factories. However, it should be noted that not all construction projects have so many similarities with the manufacturing process. Ballard and Howell (1998) proposed two steps in order to fully implement lean in the construction industry, which are: (1) claiming from construction what actually belongs to contemporary product manufacturing and minimizing construction's peculiarities in order to take advantage of lean techniques developed in manufacturing, and (2) developing lean techniques adequate to dynamic construction, the remainder that resists the first approach. Therefore, a suggestion can be made to future studies to go beyond the precast concrete industry. Construction projects other than the precast concrete projects in the construction industry should be analyzed to identify the contributions of lean to sustainable development in the construction industry.

Moreover, this study chose carbon emissions as the sustainability factor. However, as stated in Sect. 5.2, in the context of sustainability science, there are three pillars to consider; which are the global system, social system and human system. In addition, sustainable development has three pillars, which are economic, social and environmental sustainability. In the context of environmental management, a systems management of environment, disruptions, effects, human ecosystem, technological and engineering controls, and legal controls should be advocated. Future studies are suggested to analyze the contribution of lean to sustainable development from other dynamic aspects. This proposed work should be based on an integrated literature of the targeted dynamic aspect from the three strands, which are: (1) sustainability science, (2) sustainable development and (3) systems management in environmental management. This will be of significance to plugging a gap in literature relating to the contributions of lean to sustainable development, and will help to provide lessons and experiences for construction companies to meet the challenges posed by sustainable development.

This study chose precast concrete columns as the research target. Although the production processes of precast concrete products are very similar to each other, there remains differences, e.g. in the precasting method. For example, dry mix that has a relatively low moisture content can be used in the manufacture of concrete pipe in rotary packerhead machines while wet mix is often adopted for other precast concrete products. In addition, the production process of construction materials other than precast concrete products (e.g. timber, steel and plastic) is different. The contributions of lean to the sustainable development of these construction materials need to be investigated. Future studies are suggested to: (1) use other precast concrete products, and (2) use construction materials other than precast concrete products to identify whether the differences in the production process may affect the applicability of lean to reduce carbon emissions.

The last suggestion for future work is that, since both the concept of lean and sustainable development in construction are evolving with the changing situation in the construction industry, future research may incorporate new emerging issues to identify the relationship between lean and sustainable development.

References

Ballard G, Howell G (1998) Shielding production: an essential step in production control. ASCE J Constr Eng Manage 124(1):18–24

BCA (2011) Annex B: Development of the precast concrete component industry in Singapore. [online] Available at: http://www.bca.gov.sg/publications/BuildabilitySeries/others/rscp_anxb.PDF. Accessed 20 July 2011

BCA (2012) BCA green mark certification standards for new buildings, GM version 4.0, Aug 2010 issue. [online] Available at: http://bca.gov.sg/GreenMark/green_mark_criteria.html. Accessed 25 Jan 2012

Broome J (1992) Counting the costs of global warming. White Horse Press, Cambridge

Dreze J, Stern N (1987) The theory of cost-benefit analysis. In: Auerbach AJ, Feldstein M (eds) Handbook of public economics, vol II. Elsevier Science Publishers, Amsterdam

Enrlich PE, Enrlich A (1972) Populations, resources, environment. W.H. Freeman, San Francisco

Garlauskas AB (1975) Conceptual framework of environmental management. J Environ Manage 3(3):185–203

Koskela L (1992) Application of the new production philosophy to construction. CIFE technical report#72, Stanford University, Stanford

Koskela L (2000) An exploration towards a production theory and its application to construction. Technical Research Centre of Finland, Espoo

Koskela L, Laurikka P, Lautanala M (1997) Rapid construction as a change driver in construction companies. In: Alarcon L (ed) Lean construction. A.A. Balkema Publisher, Rotterdam

Lim LY, Low SP (1992) Just-in-time productivity for construction. SNP Publishers, Singapore

Meadows DH, Meadows DL, Randers R, Behren WW III (1972) The limits of growth. Universe Books, New York

Norton BG, Steinemann AC (2001) Environmental values and adaptive management. Environ Value 10(4):473–506

Price C (1993) Time, discounting and value. Blackwell, Oxford

Price C (1996) Discounting and project appraisal: from the bizarre to the ridiculous. In: Kirkpatrick C, Weiss J (eds) Cost-benefit analysis and project appraisal in developing countries. Edward Elgar, Cheltenham, pp 90–101

Voss CA, Robinson SJ (1987) Application of just-in-time manufacturing techniques in the United Kingdom. Int J Oper Prod Manage 7(4):46–52

Wantuck KA (1989) Just-in-time for America: a common sense production strategy. Forum Ltd., Milwaukee

Appendices

Appendix 1: Questionnaire for Precasters in the Singapore Precast Concrete Industry (Pilot Studies)

Study of Precast Concrete Production, Lean Production Principles and Carbon Emissions

Introduction

- Our research is about applying just-in-time/lean principles in the construction industry to reduce carbon emissions. This is driven by the increasing pressure from the industry worldwide to address the problem of global climate change. The topic has mutual benefit, for both the research fraternity and the industry.
- Basically, the concept is about reducing or eliminating non-value adding activities in the production life cycle. There are several tools or techniques to achieve this objective, including the pull system (manufacture when needed, or ordered by the customer), quality control, elimination of waste, employee involvement (training, education), uninterrupted work flow (to smooth the work flow, nothing break down to damage the process), etc.
- The philosophy is that by eliminating fuel/energy consumption in these non-value adding activities, carbon emissions in the production stage can be reduced.

This Survey

The Aim of This Survey

- Obtain information on production processes of in your Organization to see whether some of the principles mentioned above have already been applied on site. If not, what are the further improvements that can be recommended?

P. Wu and S.P. Low, *Lean and Cleaner Production: Applications in Prefabrication to Reduce Carbon Emissions*, DOI 10.1007/978-3-642-42062-7,

- Provide information on what current research is focusing on. While traditional carbon reducing techniques focus on applying new environment-friendly materials and techniques, these usually entail a long duration and possibly high implementation costs. This study focuses on improving management processes and/or systems to achieve low carbon emissions. Statistics have shown that carbon emissions can be reduced by up to 15 % when appropriate management processes and/or systems are adopted.

Potential Benefits

- Pre-empt the potential regulations for sustainable construction and buildings. Although these might not be taking place right now, it is always strategic to take precautionary measures before these regulations are introduced or even before it is too late to take the necessary measures.
- Lower implementation costs.
- Most importantly, and as stated previously, this study aims to reduce carbon emissions by cutting down energy/fuel consumption in non-value adding activities. It would be beneficial to the company in the economics sense.

Production Type – The Pull System/Push System

The type/cause of inventory (The pull system – manufactures exactly the amount when needed or ordered by customers. The push system – manufactures, stores and delivers to the customer).

- Handling materials from suppliers.
  1. When handling materials from suppliers, are they used immediately? Or,
  2. They are delivered to the storage area, picked out from the storage area and delivered to site.
  3. Can you provide the site layout plan of the precast yard/factory, or we can arrange for a visit to the site.
  4. Zero inventory is not always achievable. Are there any measures that you have taken to reduce the inventory level?

  Notes:

- Handling the finished products
  1. Is the prefabrication based on order (the pull system) or based on the push system, which means that you produce and keeps them until they are sold.
  2. If the push system is used, how large is the stockpile? (size, measurements, site plan?)
  3. How much time would you need if you have to identify and select the specific precast concrete components that are required by the customer?
  4. During the sorting out process, is there any damage?

  Notes:

Quality Control During Production

1. How long have you been applying quality control procedures (assuming that you have)?
2. Compare with previous years, how many percentages do you actually save on costs arising from quality control?
3. If you are implementing the quality control concept from the start, how many stages are there in the evolving process for quality control?
4. Can you offer the information statistically on how much would be saved stage by stage?
5. What are the measures in the quality control procedures? For example, Quality assurance (plan-do-check-act), Failure testing, statistical control (Six Sigma level of quality).

Notes:

Delivery System

1. How many types of materials do you need right now to carry out the production process?
2. How do you arrange for that?
3. What kind of ordering system are you using right now?
4. Can you provide the supplier information, including locations, numbers for us to calculate and to find out a better way to arrange for their deliveries to your precast yard?

Notes:

Unnecessary Movements

View the site layout plan to see if there are any unnecessary movements during production.

Or if possible, please allow us to visit your precast yard to understand the production processes.

Notes:

Training System

Are you providing any training programs for your employees to learn more about sustainable construction, carbon emissions, etc?

Or are you expecting them to learn all these on their own?

Or are you expecting them to learn all these through on-the-job (OJT) training after they have been employed (such as regular checking of the vehicles, eco-driving, etc)?

Notes:

Recycle or Reuse

If products are damaged in the precast yard/factory or on site, what would you do with the damaged components with respect to recycling and reuse?

Notes:

Appendix 2: Questionnaires for Precasters in the Singapore Precast Concrete Industry (Empirical Study)

	Probability					Impact				
1. Site layout management	1	2	3	4	5	1	2	3	4	5
The probability is rated using five scales: (1) Never; (2) Rarely; (3) Occasionally; (4) Often; (5) Always.										
The level of impact on energy consumption and carbon emissions is rated from 1 (insignificant effect); 2 (Minor detrimental effect); 3 (moderate detrimental effect); 4 (significant detrimental effect) to 5 (catastrophic effect).										
1.1 Building materials										
1.1.1 Improper specifications of building materials										
1.1.2 Inaccurate estimation of quantities require										
1.1.3 Does not think of alternative designs that minimize the use of materials										
1.1.4 Does not think of green building materials										
1.2 No overall consideration of building site facilities										
1.2.1 No overall consideration of equipment used in future precast concrete production processes										
1.2.2 No overall consideration of infrastructure used in future precast concrete production processes										
1.3 Does not comply with mandatory statutory requirements										
1.4 Does not pay full attention to contractors' and subcontractors' requirements										
1.4.1 Duration										
1.4.2 Office space										
1.4.3 Maximum number of men on site										
1.4.4 Services required										
1.5 Inappropriate design of temporary works and services										
1.5.1 Space for access										
1.5.2 Tower crane's fully blocked area										

(continued)

(continued)

1. Site layout management	Probability					Impact				
	1	2	3	4	5	1	2	3	4	5
1.5.3 Clearance of the blocked area										
1.5.4 Siting of static plant										
1.5.5 Parking of mobile plant										
1.6 Over provide material storage										
1.6.1 Secure store										
1.6.2 Weatherproof store										
1.6.3 Open store										
1.7 Inappropriate management of the site layout										
1.7.1 Site layout plan is not tested for economic and efficient production										
1.7.2 Site layout plan is not sent to contractors, subcontractors and general foreman										
1.7.3 Site layout plan is not placed on the notice board for information										
1.7.4 changes to the site layout plan are not notified immediately										

1.8 General questions relating to site layout management

1.8.1 Are supervisors appropriately trained relating to the use of materials, plants and equipment?
1.8.2 Is environmental performance considered when designing the site layout?
1.8.3 Is there any statutory regulation relating to carbon emissions when designing the site layout?
1.8.4 Is there any statutory regulation relating to low energy consumption when designing the site layout?
1.8.5 Has reconstruction happened before because of failing to meet statutory requirements?
1.8.6 Is the site layout plan sent to contractors and subcontractors and placed on the notice board for information?
1.8.7 Is the site layout designed in such a manner that would allow the materials to be used in "first in first out"?
1.8.8 Has the site changed since it was first built?
1.8.9 What is the percentage of the storage area to the total area?

Notes:

2. Supply Chain management	Probability					Impact				
	1	2	3	4	5	1	2	3	4	5
2.1 Selecting suppliers										
2.1.1 Large quantity supply base which leads to inflexibility										
2.1.2 No long-term contract to achieve loyalty between suppliers and precasters										
2.1.3 Transportation is not taken into consideration										
2.1.4 No quality audits of the supplier										
2.2 JIT management process										
2.2.1 Demand fluctuations										
2.2.2 Not fully prepared for the arrival of raw materials										
2.2.3 Lack of both advance order and confirmation order										
2.2.4 Unsatisfied data exchange with suppliers										

(continued)

(continued)

2. Supply Chain management	Probability					Impact				
	1	2	3	4	5	1	2	3	4	5
2.3 General questions related to supply chain management										
2.3.1 Is single sourcing adopted in the supply chain?										
2.3.2 Is small lot size adopted in the supply chain?										
2.3.3 Is long-term contract awarded to achieve loyalty?										
2.3.4 Is the Singapore precast concrete industry vulnerable to supply disruptions?										
2.3.5 Is the company operating under a stable production schedule?										
2.3.6 Between the expected costs of small lot delivery and the savings of reduced inventory, has the company conducted any trade-off investigation?										
2.3.7 What is the current delivery method adopted in the company?										
2.3.8 How does the company evaluate suppliers?										
Notes:										

3. Production management	Probability					Impact				
	1	2	3	4	5	1	2	3	4	5
3.1 Materials										
3.1.1 Waste of raw materials in the production process										
3.1.2 Raw materials do not meet specifications										
3.1.3 Materials damaged during handling										
3.1.4 Unnecessary materials handling										
3.1.5 Too much inventory in factory										
3.1.6 Loss of materials in factory										
3.2 Cycle time										
3.2.1 Wait time for inspection										
3.2.2 Wait time for the delivery of materials										
3.2.3 Wait time for labor										
3.2.4 Wait time for equipment										
3.3 Finished products										
3.3.1 Repair due to damaged products during inventory										
3.3.2 Repair due to damaged products when handling										
3.3.3 Double-handling or delivery due to unsatisfied quality or specifications										
3.4 Operations										
3.4.1 Inappropriate site layout										
3.4.2 Equipment breakdown										
3.4.3 Inappropriate selection of equipment										
3.5 Human resources										
3.5.1 Inadequate work crews										
3.5.2 Inexperienced employees										
3.5.3 Lack of supervision										
3.6 General questions relating to production management										
3.6.1 Is the company aware of the regulations related to environmental performance?										
3.6.2 Is the company currently taking any action to reduce the non-value adding activities listed?										
3.6.3 Is the company currently taking any action to reduce energy consumption and carbon emissions in the factory?										
3.6.4 Are the customer requirements taken into consideration when designing and manufacturing precast concrete products?										

(continued)

(continued)

3. Production management	Probability 1	2	3	4	5	Impact 1	2	3	4	5
3.6.5 Are the manufacturing equipment carefully maintained during the production life cycle?										
3.6.6 When selecting the operator for a specific kind of equipment, is he trained on sustainable operations?										
3.6.7 Are there any training programmes for the operators who work in the manufacturing process?										
3.6.8 Is there an employee performance rating system that takes environment-friendly operations as one consideration?										
3.6.9 Is the company conducting any research on how to improve future production?										
3.6.10 Are there any internal periodic meetings to discuss improvement?										
3.6.11 Do the internal periodic meetings involve external professional of specific improvement issues?										
3.6.12 What is the production type of this factory?										
Notes:										

4. Stock management	Probability 1	2	3	4	5	Impact 1	2	3	4	5
4.1 Layout and equipment										
4.1.1 Lack of well order stockyard										
4.1.2 Insecure stockyard that leads to the loss of products										
4.1.3 Inappropriate selection of equipment										
4.2 Staffing arrangement										
4.2.1 Inappropriate staffing arrangement (a crane driver, a banksman, a charge-hand)										
4.3 Stacking arrangement										
4.3.1 Unclear identification marks										
4.3.2 Inappropriate battens										
4.3.3 Unclear working instructions										
4.4 Stock records										
4.4.1 Lack of periodic stock checks										
4.4.2 Lack of computer stock control										
4.5 Loading, unloading and delivery										
4.5.1 Unclear delivery notes which lead to wrong delivery										
4.5.2 Lack of sufficient care which lead to damage										
4.5.3 Lack of routine inspection prior to release of products										
4.6 General questions relating to stock management										
4.6.1 Is identification mark clearly provided to avoid wrong delivery?										
4.6.2 Has wrong delivery happened due to defective identification marks?										
4.6.3 Is the delivery note orally made or provided by written documents?										
4.6.4 Are routine inspections conducted before the release of products?										
4.6.5 Have products been damaged due to inappropriate stacking?										
4.6.6 Are working instructions provided to employees?										
4.6.7 Are the working instructions orally provided or provided by written documents?										
4.6.8 Are periodic stock checks conducted?										
4.6.9 Is computer stock control system adopted for stock management?										
4.6.10 Is the driver appropriately trained to provide sufficient care to the precast concrete products?										
Notes:										

Appendix 3: Questionnaires for Contractors in the Singapore Precast Concrete Industry (Empirical Study)

	Probability					Impact				
1. Site layout management	1	2	3	4	5	1	2	3	4	5
The probability is rated using five scales: (1) Never; (2) Rarely; (3) Occasionally; (4) Often; (5) Always.										
The level of impact on energy consumption and carbon emissions is rated from 1 (insignificant effect); 2 (Minor detrimental effect); 3 (moderate detrimental effect); 4 (significant detrimental effect) to 5 (catastrophic effect).										
1.1 Construction materials										
1.1.1 Improper specifications of the precast concrete products										
1.1.2 Inaccurate estimation of quantities required										
1.1.3 Does not think of green building materials										
1.2 No overall consideration of building site facilities										
1.3 Does not comply with mandatory statutory requirements										
1.4 Does not pay full attention to the construction requirements										
1.4.1 Duration										
1.4.2 Office space										
1.4.3 Maximum number of men on site										
1.4.4 The use of other equipment and plants										
1.4.5 Services required										
1.5 Inappropriate design of temporary works and services										
1.5.1 Space for access										
1.5.2 Tower crane's fully blocked area										
1.5.3 Clearance of the blocked area										
1.5.4 Siting of static plant										
1.5.5 Parking of mobile plant										
1.6 Over provide storage area										
1.6.1 Secure store										
1.6.2 Weatherproof store										
1.6.3 Open store										
1.7 Inappropriately planning the site layout										
1.7.1 The site layout is not tested for economic and efficient construction										
1.7.2 The layout is not sent to subcontractors and general foreman.										
1.7.3 The layout is not placed on the notice board for information										
1.7.4 Changes to the site layout are not notified immediately										
1.8 General questions relating to site layout management										
1.8.1 Are the supervisors/project managers appropriately trained relating to the use of materials, plant and equipment?										
1.8.2 In the current site layout design, is environmental performance taken into consideration?										
1.8.3 Is there any statutory regulation relating to low energy consumption/low carbon emissions when designing and building the construction site?										
1.8.4 Are the site offices fully occupied at the moment?										

(continued)

(continued)

1. Site layout management	Probability 1 2 3 4 5	Impact 1 2 3 4 5
1.8.5 Are the precast concrete columns used in a "first in first out manner"?		
1.8.6 Is the site layout placed on the notice board for information?		
Notes:		

2. Transportation management	Probability 1 2 3 4 5	Impact 1 2 3 4 5
2.1 Damages during transportation		
2.1.1 No skilled attention to the detail of supports and frames		
2.1.2 The driver is not aware of a few typical damages during transportation, such as:		
Damages due to the wrong placement of batten		
Damages caused by flexing the vehicle bed and torsion induced by road surfaces		
Cracking caused by vibration of the elements		
Damages to slender sections		
Broken nibs and bottom corners of panels		
Damages caused by the use of slings and chains		
Damages caused by the omission of soft packing pieces between units		
Damages caused by contact with low bridges en route from works to site		
2.1.3 Inappropriate packings and supports		
2.1.4 No standing instructions, such as:		
Statutory requirements		
The balance of the vehicle		
Provision of supports		
Insertion of packing pieces		
Bearer location		
2.2 Selecting supplier		
2.2.1 Large quantity supply base		
2.2.2 No single sourcing supply with long-term contract		
2.2.3 Transportation is not taken into consideration		
2.2.4 No quality audits of the supplier prior to the award of the contract		
2.3 JIT management process		
2.3.1 Demand fluctuations		
2.3.2 Not fully prepared for the arrival of raw materials		
2.3.3 No advance order and confirmation order		
2.3.4 Insufficient data exchange with suppliers		
2.4 Delivery		
2.4.1 No accurate delivery notes		
2.4.2 Insufficient care		
2.4.3 Lack of routine inspection		
2.5 General questions relating to transportation management		
2.5.1 Is the transportation works subcontracted out?		

(continued)

(continued)

2. Transportation management	Probability 1 2 3 4 5	Impact 1 2 3 4 5
2.5.2 Does the contractor (or the subcontractors if the transportation job is subcontracted out) provide standing instructions to the drivers?		
2.5.3 Are the instructions made orally or through written documents?		
2.5.4 Is the operator aware of the typical damages to the precast concrete products during transportation?		
2.5.5 Is single sourcing adopted in the supply chain?		
2.5.6 Is a long-term contract awarded to achieve loyalty and reduce the risk of disruption?		
2.5.7 Is the contractor operating with a stable erection schedule?		
2.5.8 What is the current delivery method adopted by the contractor?		
2.5.9 Has the company conducted any trade-off investigation relating to the small lot deliveries and reduced inventory?		
2.5.10 Do you think that damages to the precast concrete products can be eliminated if sufficient care is provided?		
2.5.11 Did re-delivery happen due to unclear delivery note?		
2.5.12 Is the delivery note made orally or by written documents?		
2.5.13 Is the vehicle driver appropriately trained to provide sufficient care to the precast concrete products?		
2.5.14 Are routine inspections conducted before the release of the precast concrete products		
Notes:		

3. Stock management	Probability 1 2 3 4 5	Impact 1 2 3 4 5
3.1 Layout and equipment		
3.1.1 Lack of well order stockyard		
3.1.2 Lack of secured stockyard		
3.1.3 Inappropriate selection of equipment		
3.2 Staffing arrangement		
3.2.1 Inappropriate staffing arrangement (a crane driver, a banksman, a charge-hand)		
3.3 Stacking arrangements		
3.3.1 Unclear identification marks		
3.3.2 Inappropriate battens		
3.3.3 Unclear working instructions		
3.4 Stock records		
3.4.1 Lack of periodic stock checks		
3.4.2 Lack of computer stock control		
3.5 General questions relating to stock management		
3.5.1 Have the precast concrete been damaged due to inappropriate stacking?		
3.5.2 Are working instructions provided to employees who are in charge of loading and unloading activities?		
3.5.3 Are the working instructions provided by on-the-job trainings or through written documents?		
3.5.4 Are periodic stock checks conducted?		
3.5.5. Is a computer stock control system adopted for stock management?		
Notes:		

(continued)

(continued)

4. Erection management	Probability 1	2	3	4	5	Impact 1	2	3	4	5
4.1 Erection method statement										
4.1.1 No erection programme showing some or all aspects in the following:										
The sequence of erection of the components										
The associated activities of temporary works										
The provision of supports										
Available crane time										
Artificial lighting										
Others										
4.1.2 The programme is not recorded and modified with the design process										
4.1.3 The programme is not disseminated to site										
4.1.4 No illustration with sketches and diagram										
4.1.5 Lack of feedback system about the erection method										
4.2 Pre-erection preparation										
4.2.1 No pre-erection surveys										
4.2.2 Lack of preparatory work at construction site, such as:										
Provision of holding bolts										
Erection specialists										
Emphasis on erection accuracy										
4.2.3 Cease of erection by laws and regulations										
4.2.4 Inappropriate crane selection										
4.3 Erection operations										
4.3.1 Inappropriate siting of tower crane										
It can handle materials directly from the delivery point into the building structure										
The overlap of cranes to provide a large coverage area										
A consulting engineer hired to conduct the design of crane bases and track supports										
4.3.2 Inappropriate crane operation										
One suitably qualified and experienced driver -One suitably qualified and experienced banksman										
Frequent inspections of site conditions										
Immediate notification when site condition is changed										
Extreme care is paid to establish good communications										
IT facilities adopted to improve communications										
4.3.3 No special attention paid to joints and connections										
4.3.4 Week concrete worker at construction site										
4.3.5 No process to learn from experienced engineers and vehicle drivers										
4.3.6 Inadequate supervision during erection processes										
4.4 Maintenance of plant and equipment										
4.4.1 Lack of regular planned maintenance of equipment										
4.4.2 The maintenance schedule does not match the rate of use										
4.4.3 Lack of inspections and tests conducted at the frequency required										

(continued)

(continued)

4. Erection management	Probability 1	2	3	4	5	Impact 1	2	3	4	5
4.5 Legal aspects										
Cease of erection due to a breach of law and regulations during installation, such as:										
General provisions										
Health and Welfare										
The use of lifting appliances										
Construction maintenance and inspection										
Others										
4.6 General questions relating to erection management										
4.6.1 Is a detailed erection method statement provided by the contractor?										
4.6.2 Is the erection method statement published as written documents?										
4.6.3 Is the erection method designed by professionals who are familiar with erection activities?										
4.6.4 Does the contractor emphasize on erection accuracy?										
4.6.5 During the pre-erection stage, are there laws and regulations that the contractor must abide?										
4.6.6 Has it happened before that due to a breach of law and regulations, the erection activities were ceased?										
4.6.7 Are there standard crane operating manuals which the operator can refer to?										
4.6.8 Does the maintenance schedule of the equipment match the rate of use?										
4.6.9 How did the contractor use the precast concrete products (deliver-store-use or deliver-use)										

MIX
Papier aus verantwortungsvollen Quellen
Paper from responsible sources
FSC® C105338

If you have any concerns about our products,
you can contact us on
ProductSafety@springernature.com

In case Publisher is established outside the EU,
the EU authorized representative is:
Springer Nature Customer Service Center GmbH
Europaplatz 3, 69115 Heidelberg, Germany

Printed by Libri Plureos GmbH
in Hamburg, Germany